FOOD SCIENCE AND TECHNOLOGY

NUTRACEUTICALS

FOOD APPLICATIONS AND HEALTH BENEFITS

Food Science and Technology

Additional books and e-books in this series can be found on Nova's website under the Series tab.

FOOD SCIENCE AND TECHNOLOGY

NUTRACEUTICALS

FOOD APPLICATIONS AND HEALTH BENEFITS

ANITA KUMARI

AND

GULAB SINGH

EDITORS

nova

Medicine & Health

New York

DOI: https://doi.org/10.52305/ZDNY4862

NOTICE TO THE READER

Additional color graphics may be available in the e-book version of this book.

Library of Congress Cataloging-in-Publication Data

ISBN: 978-1-68507-488-3

Published by Nova Science Publishers, Inc. † New York

Affectionately dedicated

to

My Husband Sushil Kumar Daksh
and Children Aayat & Aayan

(Dr. Anita Kumari)

Contents

PREFACE

Nutraceutical and functional foods are in trend nowadays. Consumers are showing great interest in adapting nutraceutical/functional foods in their regular diet because they not only provide good nutrition but also have therapeutic properties. Various nutraceuticals have different mechanisms of action and are hence used as alternative pharmaceuticals for mitigating health problems. Research in the field of health-promoting foods has been escalating since the last decade. In recent years, a growing interest related to nutraceuticals has arisen among different vulnerable groups of the society because they act as an alternative to modern medicine.

This book will provide the latest scientific information on different aspects of nutraceuticals and functional foods. From production to processing, and marketing of nutraceuticals, one must have knowledge of national regulations on nutraceuticals. The contents of this book have been formulated by considering the syllabus of UG and PG level in various central, state and other deemed and private universities and the book is a valuable resource for students, researchers, academicians, food technologists, food scientists, nutritionists, health professionals and for those who are involved in the agriculture, pharmaceutical and food processing sectors.

Students and leading researchers can find information related to nutraceuticals, functional foods, strategies for formulation of functional foods, the role of nanotechnology in the food industry, extraction of nutraceuticals, role of probiotics, prebiotics, synbiotics for health benefits, bioactive proteins and peptides, role of nutraceuticals in disease management, role of omega 3 fatty acid and herbs in health and diseases as well as role of nutraceuticals in food packaging in a single book. The book will prove to be a valuable reference book for graduate as well as post-graduate students in the fields of Food Science and Nutrition, Clinical Nutrition and Dietetics, Applied Nutrition, Nutrition Biology, Food Science and Nutrition Food Technology/Food Engineering, Food Biotechnology, Food Microbiology, Post-Harvest Technology, Biochemical Engineering and Life Sciences.

Tables and schematic figures have also been given in the book chapters for better understanding and to simplify the complex matter. References including textbooks, journals, and relevant websites are given. Teachers, students and researchers can have direct access to the references used. This book is a complete package for students, researchers and industry personnel working in this field.

Anita Kumari
Gulab Singh

LIST OF ABBREVIATIONS

8-OHDG	8-hydroxy-2' -deoxyguanosine
AA	Arachidonic acid
ADHD	Attention deficit hyperactivity disorder
AITC	Allyl isothiocyanate
ALA	Alpha-linoleic acid
ALT	Alanine aminotransferase
AMD	Age-related Macular Degeneration
ASU	Avocado/Soy unsaponifiable
ATP	Adenosine triphosphate
Bcl- xL	B-cell Lymphoma-extra large
Bcl-2	B-cell Lymphoma 2
BCRP	Breast Cancer Resistance Protein
BHA	Butyrated Hydroxy Anisole
BHT	Butyrated Hydroxy Toluene
BIS	Bureau of Indian Standards
BNDF	Brain derived neurotrophic factor
BP	Blood Pressure
BP	British Pharmacopoeia
CAGR	Compound Annual Growth Rate
CC	Counter Current
CCC	Counter Current Chromatography

CCE	Counter Current Extraction
CCK	Cholecystokinin
cGMP	Current Good Manufacturing Practice
CHD	Coronary heart disease
CMC	Carboxy Methyl Cellulose
CNS	Central Nervous System
COX	Cyclooxygenases
COX	Cyclooxygenase
CRP	C-reactive protein
CVD	Cardio vascular diseases
DA	Dopamine
DHA	Docosahexaenoic acid
DM	ry mass
DNA	Deoxyribonucleic acid
DSHEA	Dietary Supplement Health and Education Act
EAE	Enzyme Associated Extraction
ECCG	Epigallocatechin-3 Gallate
EFSA-European Food Safety Authority	
ENS	Enteric Nervous system
EPA	Ecosapantenoic acid
FAO	Food & Agricultural Organization
FDA	Food and Drug Administration
FLIP	FLICE-like Inhibitory Protein
FMCG	Fast-moving Consumer Goods
FNFC	Foods with Nutrient Function Claims
FOS	Fructooligosaccharides
FOSHU	Food for Specified Health Uses
FPH	Fish Protein Hydrolysate
FRAP	Ferric reducing antioxidant power
FSSA	Food Safety and Standards Authority
FSSAI	Food Safety and Standards Authority of India
FWGE	Fermental Wheat Germ Extract
GAGs	Glucosamine Sulfate
GI	Gastrointestinal tract

GLP	Glucagon-like peptide
GLP-1	Glucagon-like peptide
GLVs	Green leafy vegetable
GMO	Genetically modified organism
GOS	Galacto-oligosaccharides
GRAS	Generally recognized as safe
GTP	Guanosine 5'-triphosphate
HCAs	Heterocyclic amines
HDL	High-density lipoprotein,
HDL-C	High Density Lipoprotein
HOMA-IR	Homeostasis model assessment- Insulin resistance
HPCCC	High Performance Counter Current Chromatography
HPLC	High Performance Liquid Chromatography
HPLC-MS	High-performance liquid chromatography-mass spectrometry
HPLC-UV	High-performance liquid chromatography-UV detector
HSCCC	Hydrostatic Counter Current Chromatography
IAP	Inhibitor of Apoptosis
IBD	Inflammatory Bowel Disease
IBS	Irritable bowel syndrome
IBW	Inflammatory bowel disease
ICMR	Indian Council of Medical research
IL	Interleukins
iNOS	Inducible Nitric Oxide Synthase
IP	Indian Pharmacopoeia
ITCs	Isothiocyanates
LAB	lactic acid bacteria
LC/MS	Liquid Chromatography mass spectrometry
LC-PUFAs	Long chain- polyunsaturated fatty acid
LDL	Low density lipoprotein
LDL-C	Low Density Lipoprotein- cholesterol
LPS	Lipopolysaccharides
MAE	Microwave Associated extraction
MF	Membrane Filtration

MIC	Minimum inhibitory concentration
MRP	Multidrug Resistance Associated protein
MUFA	Monounsaturated fatty acid
NAFLD	Nonalcoholic fatty liver disease
NF-κB	Nuclear Factor kappa-light-chain-enhancer of activated B cells
NMDA	N-methyl-D-aspartate
NO	Nitric Oxide
NSBH	Non-specific bronchial hyper-responsiveness
O/W	Oil/Water
OEO	Oregano Essential Oil
OFS	Oligofructose
ORAC	Oxygen radical absorbance capacity
PC	Partition Chromatography
PDDAT	Primary Degenerative types of Dementia of Alzheimer's type
PEF	Pulse Electric Field
PG	Prostaglandins
PGE2	Prostaglandins E2
PLE	Pressurized liquid extraction
PUFA	Poly Unsaturated Fatty Acid
PVA	PolyVinyl Alcohol
RCT	Randomized Controlled Trial
RDA	Recommended dietary allowance
RecHaN	Resource Centre on Health Supplements and Nutraceuticals
RNA	Ribonucleic acid
ROS	Reactive oxygen species
SCFA	Short chain fatty acid
S-CO2	Super Critical Carbon dioxide
SDA	Stearidonic acid
SDAT	Senile Dementia of Alzheimer's Type
SFA	Saturated fatty acid
SFE	Supercritical Fluid Extraction

SNL	Solid lipid nanoparticle
SOD	Superoxide dismutase
TAG	Triacylglycerol
TC	Total cholesterol
TCM	Traditional Chinese medicine
TEAC	Trolox equivalent antioxidant capacity
TLC	Thin Layer Chromatography
TNF	Tumor Necrosis Factor
TRAP	Total radical-trapping antioxidant parameter
TX	Thromboxane
UAE	Ultrasonic Associated Extraction
UCP1	Uncoupling Protein 1
USFA	Unsaturated fatty acid
USP	United State Pharmacopoeia
UV	Ultraviolet
VCAM-1	Vascular cell adhesion protein 1
VLDL	Very low density lipoproteins
WHO	World Health Organization
WPC	Whey protein concentrate

In: Nutraceuticals
Editors: A. Kumari and G. Singh
ISBN: 978-1-68507-488-3

Chapter 1

INTRODUCTION TO NUTRACEUTICALS

***Vandana Sabharwal*[1], *Anita Kumari*[2,*],**
***Deepika Pawar*[2] *and Nidhi Bhardwaj*[3]**

[1]Department of Food & Nutrition and Food Technology, Institute of Home Economics, University of Delhi, Delhi, India

[2]Department of Nutrition Biology, School of Interdisciplinary and Applied Life Sciences, Central University of Haryana, Mahendergarh, Haryana, India

[3]Department of Human Genetics and Molecular Medicine, School of Health Science, Central University of Punjab, Bathinda, Punjab, India

ABSTRACT

With rapidly increasing knowledge in the field of nutrition and allied sciences, there has been a dramatic change in the concepts about food and its use in promotion of good health. Within these trends, nutraceuticals have captivated attention because of their significant contribution towards boosting health, improving immunity and protection from diseases which in turn results in reduced health care cost.

[*] Corresponding Author's E-mail: anikum@cuh.ac.in.

Compounds obtained from plant or animal foods are used to make nutraceuticals, which are concentrated to deliver therapeutic and nutritional benefits in appropriate pharmaceutical form. They can be classified based on their availability, chemical properties and mechanisms of action. Many nutraceutical compounds have proved to be preventive and curative for various chronic conditions because of their anti-oxidative, anti-inflammatory, anti-cancer and osteogenetic properties. These beneficial effects of nutraceuticals are attributed to the myriad of bioactive substances present in them, and varied mechanisms involved.

INTRODUCTION

Food has a significant impact on normal functioning of the body, maintaining good health, and risk reduction for many health problems. Since olden days, traditional foods and herbs have been a part of the comprehensive approach to wellbeing. Hippocrates, the famous Greek physician put forth the philosophy of food as medicine, and quoted "Let food be the medicine and medicine be the food" [1]. With rapidly increasing knowledge in the field of nutrition and allied sciences, there has been a dramatic change in the concepts about food and its use in promotion of good health. With these trends, many novel concepts are emerging such as phytonutrients, phytotherapy, nutritional therapy and nutraceuticals [2-4]. Of these the term "Nutraceutical" has captivated attention because of their significant contribution towards boosting health, improving immunity and protection from diseases which in turn results in reduced health care cost [5, 6].

THE CONCEPT OF NUTRACEUTICALS

The concept of nutraceutical emerged from studies conducted in the three countries namely United Kingdom, Germany, and France, which concluded that to attain good health, dietary factors were felt more important by consumers than hereditary factors or exercise.

In 1989, Stephen DeFelice, president of the Foundation of Innovation Medicine and a nutritionist by profession defined nutraceutical as "A food, or part of a food, that provides medical or health benefits, including prevention and/or treatment of a disease" [7]. Nutraceutical is the blend of two terms "Nutrition" and "Pharmaceutical." Compounds obtained from plant or animal foods are used to make nutraceuticals, which are concentrated to deliver therapeutic and nutritional benefits in appropriate pharmaceutical form [8]. The popularity of nutraceuticals is not only because they are a good source of energy and nutrients but also because of the range of medicinal benefits provided by them. Nutraceuticals help to detoxify our body and restore our healthy digestion [9]. When effectively used, they have the potential to prevent and even cure diseases provided they have high bioavailability and are used safely [10]. In the market, nutraceuticals are available in concentrated forms made from a single substance or as a combination. These may be marketed as powders, pills, capsules, and tinctures, etc.

In the literature, many terms such as functional or medical foods, nutritional supplements and dietary supplements are mistaken for nutraceuticals. These terms can be understood from different perspectives for example foods with strong or specific purpose are more often defined as functional food [11]. More specifically, functional foods are defined as "Food products to be taken as part of the usual diet in order to have beneficial effects that go beyond what are known as traditional nutritional effects" [12]. Examples of functional foods include milk and milk products; yogurt; fruits and vegetables especially oranges, kiwi; whole grains like oat, barley, buckwheat, etc. [13, 14]. Dietary supplements include amino acids, vitamins, minerals, herbs, etc., which have more specific health roles to play as they are expected to improve the intake of these nutrients [15].

While dietary supplements are not used in prevention and treatment of health issues [16], nutraceuticals are expected to have more intense results in prevention as well as cure of diseases. Nutraceuticals are described as "Products extracted, purified or produced from a plant, animal or marine source or produced from dried, powdered, or pressed plant material which renders physiological benefit, or to provide protection against chronic disease" [17].

Why Nutraceutical Seems Attractive?

The nutraceutical market is growing rapidly. This growth is due to various reasons. The consumers are getting more health conscious and exploring ways to achieve better health through their diet. However, because of busy lifestyles many fail to adhere to balanced diets. With these perspectives, nutraceuticals have several benefits such as:

- Consuming nutraceuticals can enhance the nutritional quality of the daily diets, which at times fail to fulfil nutrient requirements of an individual.
- My promoting good health, they can boost immunity and increase the life span for individuals
- Nutraceuticals are considered more natural and safer for consumption in comparison with medicines, which are apparently believed to have harmful effects.
- Nutraceuticals are preferred for consumption by people with specific deficiencies or special needs.
- They are easily accessible from grocery stores and can be bought without prescription.
- They offer prevention against chronic disorders, such as diabetes, cardio-vascular diseases and even cancers.

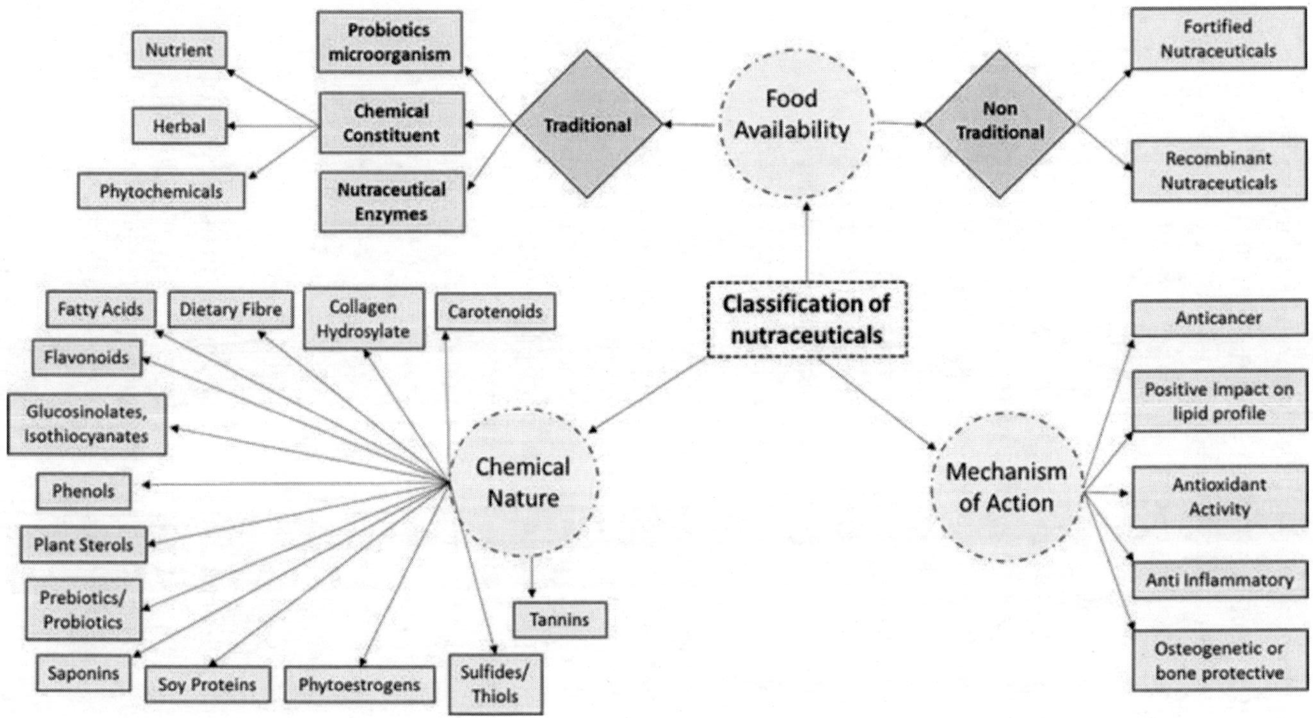

Figure 1. Classification of Nutraceuticals.

CLASSIFICATION OF NUTRACEUTICALS

Nutraceuticals can be classified on the basis of food availability, chemical nature and the mechanism of action of the bioactive substances. The classifications are not mutually exclusive but overlapping. The detailed classification of nutraceuticals is presented in Figure 1 [18].

Food Availability

- Traditional nutraceuticals: The food which is not subjected to any manual changes is known as traditional nutraceuticals. They are simply natural, whole foods with some potential health benefits like vegetables and fruits, whole grains, milk and milk products, meat and fish. They contain several natural components that benefit beyond the bounds of basic nutrition like saponins in soy, lycopene in tomatoes and omega-3 fatty acids in salmon. They can be further grouped as probiotic microorganisms; chemical constituents including nutrients, herbs and phytochemicals; and nutraceutical enzymes. Probiotic microorganisms are live bacteria or microorganisms which when consumed in adequate quantities, have the potential to benefit the health of the host [19]. Research studies have proven that the consumption of probiotics decreases the risk of allergy, cancer, asthma, and several other infections [20]. Probiotics also have the potential to treat lactose intolerance. This is because they can produce *ß*-galactosidase, an enzyme that can hydrolyse lactose into its component sugars [21]. Nutrients are the chemical compounds or substances like antioxidants, vitamins, minerals, fatty acids, and amino acids which are present in food and act as a nutraceutical. These nutrients are helpful in the management and prevention of specific diseases. For example, vitamin E is a lipid-soluble vitamin which is useful in protecting the cells through the damage caused by free radicals because of its antioxidant property. It is known to reduce the incidence of cancer, arthritis, aging, cataracts, etc. The use of botanical products or herbs like aloe vera, wheatgrass, ginger, garlic, turmeric in the form of concentrate or extract comes under the category of herbals. Traditionally herbals are used in the form of decoction to cure acute and chronic diseases. Numerous nutraceuticals are present in herbal plants, which have a therapeutic property. For example, the active compound present in turmeric is curcumin which has great health benefits. Phytochemicals are also referred to as the compounds phytonutrients that naturally occur in plants like anthocyanidins, terpenoids, phytoestrogens, carotenoids, phytosterols, glucosinolates, flavonoids, polyphenols and isoflavonoids. They are mostly present in whole grains, fruits, vegetables, legumes, herbs, spices, nuts, and seeds. It has tremendous therapeutic potential in treating various diseases or physiological disorders. Enzymes are also known as biocatalysts. They are proteins in nature and are produced by our cells. They are beneficial for accelerating metabolic processes and dealing with health problems such as constipation, diarrhoea and gastroesophageal reflux disease which are related to the digestive tract. Example lipase is an enzyme produced by Rhizopus species through controlled fermentation. Lipase aids in improved digestion by facilitating hydrolyses of long chain fatty acids.

- Non-traditional nutraceuticals: Non-traditional nutraceuticals are foods to which nutrients or ingredients have been added or they are obtained by agricultural breeding. Many scientists in the past have successfully invented techniques to make nutrient rich crops and furthermore research studies are being conducted to improve the nutritional quality of the crops [22, 23]. They are further categorized into fortified and recombinant nutraceuticals. Fortified nutraceuticals are obtained mainly from agricultural breeding by increasing the nutrient content of the crops. It involves the application of biotechnology and genetic engineering. Few examples are flour with increased levels of calcium, folic acid, iron and milk fortified with cholecalciferol [24, 25]. Another example is gold kiwifruit which is genetically modified for increasing its ascorbic acid content. Lutin which is obtained from corn, avocado, egg yolk, spinach having potential benefits as an anti-cancer agent falls under this category [26, 27].

Chemical Nature

The sources and uses of nutraceuticals based upon their chemical nature is given in Table 1.

Table 1. Sources and Uses of Nutraceuticals based upon Chemical Nature

Chemical Constituents	Sources	Uses
Carotenoids	Yellow and orange colour fruits and vegetables	They have antioxidant and anti-inflammatory properties. It also helps in improving immunity, vision, cognition, and functioning of the heart.
Collagen hydroxylates	Obtained from connective tissues of bovine such as cartilage, tendons, bone and skin.	They have antioxidant, anti-inflammatory, anti-aging and antitumor properties. They also help to prevent obesity.
Soluble dietary fibre	Beans, oats, barley, some vegetables and fruits like peas, apples, citrus fruits, carrots.	It slows down digestion and helps in weight management, lowering the risk of heart disease, regulates blood sugar, and helps reduce low density lipoprotein (LDL) cholesterol.
Insoluble dietary fibre	Whole-wheat, bran, whole pulses and beans, nuts, and vegetables like carrot, cauliflower	They hasten gastric emptying time which helps in easing constipation. They also have anticancer properties.
Fatty acids (Especially omega-3 fatty acids and MUFA	Fish oils, flaxseeds, walnuts, green leafy vegetables, soya	They are anti-inflammatory. They support functioning of brain and reduce cholesterol disposition

Chemical Constituents	Sources	Uses
Flavonoids	Tea, fruit, vegetables, grains, legumes, nuts, and wine	They have antioxidant, anti-inflammatory, immune-modulator, antiviral, antiallergic, and anticarcinogenic properties.
Glucosinolate isothiocyanates	Broccoli, cabbage, cauliflower, rapeseed, mustard	They are anti-inflammatory and anticarcinogenic.
Phenols	Black and green tea, citrus fruits, vegetables such as red cabbage, cherry tomato, and spices such as pepper and parsley	They have antitumor, antioxidant, anti-inflammatory, antimicrobial, antihypertensive and hepatoprotective properties.
Plant sterols	Vegetable oils, nuts, and seeds	They lower LDL cholesterol levels.
Prebiotics	Fructo-oligosaccharides and inulin	They help to improve immunity and promote gastrointestinal health
Probiotics	*Lactobacillus, bifidobacterial* present in yogurt, other dairy and non-dairy food products	They enhance immunity and helps in digestion.
Saponins	Beans like soyabeans, chickpeas	Effective against colon cancer. They help in reducing cholesterol level.
Soya Proteins	Soyabean and its products like tofu	It lowers LDL cholesterol levels. They also act as antioxidants, and protect from various cancers like prostate, breast and bowel cancer.
Phytoestrogens	Soyabean, fruits like plum, pear, apple; vegetables like beans, sprouts, cabbage, spinach, wine and tea	They are effective in maintaining bone mineral density and prevent bone loss. They also help reduce menopausal symptoms.
Sulphides/Thiols	Cruciferous vegetables	They help to maintain healthy immunity.
Tannins	Tea, grapes, barley, fruits of the forest and the sorghum	They act as antioxidants.

Mechanism of Actions

Various research studies have indicated that nutraceuticals help in the prevention of chronic diseases with many other advantages like delaying aging and increasing life expectancy. Research studies have demonstrated the efficacy of many nutraceutical supplements for protection against obesity, cardiovascular disease, cancer, diabetes, and osteoporosis. This section will discuss various classification of nutraceuticals based on biological activities in nutraceuticals.

- Anticancer Activity: Cancer development is a long-term process that leads to uncontrolled or abnormal growth of cells all over the body, called metastasis, and is associated with many complex factors. Nutraceuticals have the ability to prevent cancer by various mechanisms [28]. Carotenoids and lycopene possess anti-oxidative properties which are effective against cancer. They act as oxygen quenchers and decrease oxidative stress. Another mechanism by which nutraceutical helps preventing cancer is by controlling DNA damaging factors in cells and prevention of DNA transcription in tumours. Cruciferous vegetables lower the chances of colorectal and lung cancer by blocking enzymes that promote tumour growth [29]. Flavonoids present in citrus fruit protect against cancer because of their ability to act as an antioxidant. Similarly, cancer preventive properties of isoflavones found in soya products, polyphenolic phytochemicals and epigallocatechin gallate found in tea and curcumin from curry have been well demonstrated. Soybean is a unique food in this respect which renders protection against lung, breast, colon, rectal uterine, and prostate cancers. Beta-carotene found in yellow, orange, and green leafy vegetables and fruits such as carrots, spinach, coriander, lettuce, mango, oranges, also possess anticancer properties [30].
- Positive impact on lipid profile: Nutraceuticals are considered useful in the management of excessive low-density lipoproteins (LDL) and total cholesterol (TC) in blood, a condition often termed as hypercholesterolemia [31-33]. This is achieved through various mechanisms like inhibition of cholesterol synthesis, absorption, or enhanced excretion of cholesterol. Plant sterols are effective in lowering lipid profiles by decreasing the intestinal absorption of cholesterol. Soluble dietary fibres help in regulating lipid profile because they are fermented by the microorganisms

present in the large intestine [34]. Berberine, a plant alkaloid, has also been found to have cholesterol-lowering activity [35]. Curcumin helps in increased cholesterol excretion thereby reducing TC, LDL and triglycerides levels in the blood [36].

- Anti-oxidant properties: Various chronic diseases such as hypertension, cancer, cardiovascular disease, atherosclerosis and diabetes are caused due to cumulation of free radicals in the body which results in creating oxidative stress in the body [37]. Foods such as fruits and vegetables are rich sources of vitamins and phytochemicals and therefore considered a great source of antioxidants [38]. Vitamin C, vitamin E, and phenolic compounds have the unique ability to remove free radicals from our body [39]. Vitamin E and C also protect cells from lipid peroxidation [40]. Beetroot contains betalain and phenolic compounds which protect low-density lipoproteins from oxidation [41]. They safeguard the liver from damage and also help in maintaining blood pressure levels [42, 43]. Dried fruits and nuts such as pistachios, walnuts contain polyphenolic compounds which act as antioxidants and help to regulate blood glucose levels as well as lowering risk factors causing heart problems [44-46]. Isoflavones and lignans present in dates can act as antioxidants and also help control diabetes because of their capability to module insulin secretion [47]. Another example is ginger extract and quercetin [48].
- Anti-inflammatory: Inflammation is the body's response to injury or irritation which is distinguished by pain, redness and swelling. Acute inflammation can result in chronic diseases. Nutraceuticals have demonstrated to possess anti-inflammatory properties which can be helpful in prevention and management of chronic diseases especially the one caused due to inflammation. For example, curcumin can act as an anti-inflammatory agent because of its ability to inhibit important enzymes that mediate inflammatory processes like cyclooxygenase-2, lipoxygenase, and inducible nitric oxide synthase [49]. Omega-3 PUFA in fish oil is another anti-inflammatory nutraceutical which suppresses the atherogenic

activation of vascular endothelial cells and hence having beneficial effects in rheumatoid arthritis, inflammatory bowel disease, and among some asthmatics [50]. Probiotics are also known to be anti-inflammatory; their effect being modulated by modulation of the NF-κB signalling pathway, inflammatory cytokines, and the regulatory T cell response [51]. Ginger, cinnamon, peppermint, and lycopene are also well researched to reduce inflammation [52].

- Osteogenetic or bone protective: Bone loss or osteoporosis, is a slowly progressive disease that results from poorly regulated production of cytokines which promote inflammation. The nutraceuticals identified from diet sources such as butein, cardamon, curcumin, embelin, diosgenin, gambogic acid, genistein, plumbagin, quercetin, etc., can modulate pathways for cell signaling and decelerate or reverse osteoporosis [53].

CONCLUSION

The nutrition transition led by industrialization and urbanization has a tremendous impact on the health and wellbeing of individuals. One of the possible solutions for promoting good health among the masses is encouraging use of nutraceuticals which has the potential for preventing and treating many health problems. Numerous food sources like fruits, vegetables, herbal plants, seeds, nuts, whole grains, and spices of different plants are used in the production of nutraceuticals. Nutraceuticals are made up of a myriad of bioactive substances which contribute to the various beneficial effects through various mechanisms which individually or collectively help improve the health of individuals. On the whole, nutraceuticals have given rise to a new era of research in nutrition and allied disciplines which offers huge potential for the food industry to grow as well.

REFERENCES

[1] Helal, N. A., Eassa, H. A., Amer, A. M., Eltokhy, M. A., Edafiogho, I., Nounou, M. I. (2019). Nutraceuticals' Novel Formulations: The Good, the Bad, the Unknown and Patents Involved. *Recent Pat. Drug Deliv. Formul.*, 13: 105-156.

[2] Bland, J. S. (1996). Phytonutrition, phytotherapy and phytopharmacology. *Altern. Ther. Health Med.*, 2: 73-76.

[3] Berger, M. M., Shenkin, A. (2006). Vitamins and trace elements: Practical aspects of supplementation. *Nutrition*, 22: 952-955.

[4] Bagchi, D. (2006). Nutraceuticals and functional foods regulations in the United States and around the world. *Toxicol.*, 221: 1-3.

[5] Ramaa, C. S., Shirode, A. R., Mundada, A. S., Kadam, V. J. (2006) Nutraceuticals-an emerging era in the treatment and prevention of cardiovascular diseases. *Curr. Pharm. Biotech.,* 7: 15-23.

[6] Biesalski, H. K. (2001). Nutraceuticals: the link between nutrition and medicine. In: Kramer, K., Hoppe, P. P., Packer, L., editors. *Nutraceuticals in health and disease prevention*. New York: Marcel Dekker Inc, 1-26.

[7] DeFelice, S. L. (1995). The nutraceutical revolution: Its impact on food industry R&D. *Trends in Food Science and Technology*, 6:59-61.

[8] Maddi, V. S., Aragade, P. D., Digge, V. G., Nitalikar, M. N. (2007). Importance of nutraceuticals in health management. *Pharmacognosy Reviews.* 1(2):377.

[9] Chauhan, B., Kumar, G., Kalam, N., Ansari, S. H. (2013). Current concepts and prospects of herbal nutraceutical: A review, *Journal of Advanced Pharmaceutical Technology and Research*, 4: 428.

[10] Ali, A., Ahmad, U., Akhtar, J., Badruddeen, Khan M. M. (2019). Engineered nano scale formulation strategies to augment efficiency of nutraceuticals. *Journal of Functional Foods*, 62:103554.

[11] Zeisel, S. H. (1999). Regulation of "Nutraceuticals." *Science*, 285: 185-186.

[12] Whitman, M. (2001). Understanding the perceived need for complementary and alternative nutraceuticals: lifestyle issues. *Clin. J. Oncol. Nurs.*, 5: 190-194.

[13] Hathcock, J. (2001). Dietary supplements: How they are used and regulated. *J. Nut.*, 131: 1114-1117.

[14] Dureja, H., Kaushik, D., Kumar, V. (2003). Developments in nutraceuticals. *Ind. J. Pharmacol.,* 35: 363-372.

[15] Roberfroid, M. B. (2002). Global view on functional foods. *European perspectives*, *Br. J. Nut*. 88: 133-138.

[16] Gibson, R. A., Makrides, M. (2000). Polyunsaturated fatty acid requirements of term infants. *Am. J. Clin. Nut.*, 71: 251-255.

[17] Functional Foods in Japan, *Medical Food News*, No. 6. (2000). Available at https://www.foodnavigator.com/Article/2000/11/06/Functional-foods-in-Japan. Accessed on 20 September, 2021.

[18] Chávarri, M. (2020). *Introductory Chapter: Nutraceuticals as an Alternative to Maintain a Healthy Lifestyle*. IntechOpen, Available at https://www.intechopen.com/chapters/69721. Accessed on 19 September 2021.

[19] Michail, S., Sylvester, F., Fuchs, G. and Issenma, R. (2006). Clinical efficacy of probiotics: Review of the evidence with focus on children, clinical practice guideline. *J. Pediatr. Gastroenterol. Nutr.*, 43(4): 550-557.

[20] Lenoir-Wijnkoop, I., Sanders, M. E., Cabana, M. D., Caglar, E., Corthier, G., Rayes, N., Sherman, P. M., Timmerman, H. M. (2007). Probiotic and prebiotic influence beyond the intestinal tract. *Nutr. Rev.*, 65:469-489.

[21] Pineiro, M. and Stanton, C. (2007). Probiotic Bacteria: Legislative framework requirements to evidence basis. *J. Nutr.*, 137: 850S-853S.

[22] Singh, J., Sinha, S. (2012). Classification, regulatory acts and applications of nutraceuticals for health. *Int. J. Pharma Bio Sci.,* 2: 177-187.

[23] Sapkale, A. P., Thorat, M. S., Vir, P. R., Singh, M. C. (2012). Nutraceuticals - Global status and applications: A Review. *Int. J. Chem. Pharm.*, 1: 1166-1181.

[24] Ottaway, P. B. (2008). *Food Fortification and Supplementation: Technological, Safety and Regulatory Aspects*; Woodhead Publishing: Cambridge, UK.

[25] Street, A. (2015). Food as Pharma: Marketing Nutraceuticals to India's Rural Poor. *Crit. Public Health*, 25: 361-372.

[26] Casey, C. F., Slawson, D. C., Neal, L. R. (2010). Vitamin D supplementation in infants, children, and adolescents, *Am. Fam. Physician*, 81: 745-748.

[27] Beck, K., Conlon, C. A., Kruger, R., Coad, J., Stonehouse, W. (2011). Gold kiwifruit consumed with an iron fortified breakfast cereal meal improves iron status in women with low iron stores: A 16-week randomized controlled trial. *Br. J. Nutr.*, 5: 101-9.

[28] Cencic, A., Chingwaru, W. (2010). Antimicrobial agents deriving from indigenous plants. *RPFNA*, 2: 83-92.

[29] Rajam, R. P., Murugesan, G., Vadivelkumar, K., Menachisundaram, V. (2019). Nutraceuticals - A review. *World Journal of Pharmaceutical Research*, 8 (11): 1354-1374.

[30] Thomasset, S. C., Berry, D. P., Garcea, G., Marczylo, T., Steward, W. P., Gescher, A. J. (2007). Dietary polyphenolic phytochemicals - Promising cancer chemopreventive agents in humans? A review of their clinical properties. *Int. J. Cancer,* 120: 451-458.

[31] Dav, G., Santilli, F., Patrono, C. (2010). Nutraceuticals in Diabetes and Metabolic Syndrome. *Cardiovasc. Ther.*, 28: 216-226.

[32] Bulló, M., Juanola-Falgarona, M., Hernández-Alonso, P., Salas-Salvadó, J. (2015). Nutrition attributes and health effects of pistachio nuts. *Br. J. Nutr.*, 113: S79-S93.

[33] Bolling, B. W., Chen, C. Y. O., McKay, D. L., Blumberg, J. B. (2011). Tree nut phytochemicals: Composition, antioxidant capacity, bioactivity, impact factors. A systematic review of almonds, Brazils, cashews, hazelnuts, macadamias, pecans, pine nuts, pistachios and walnuts. *Nutr. Res. Rev.,* 24, 244-275.

[34] Biagi, M., Minoretti, P., Bertona, M., Emanuele, E. (2018). Effects of a nutraceutical combination of fermented red rice, liposomal berberine, and curcumin on lipid and inflammatory parameters in

patients with mild-to-moderate hypercholesterolemia: An 8-week, open-label, single-arm pilot study. *Arch. Med. Sci. Atheroscler. Dis.*, 3: e137-e141.

[35] Xiong, Z., Cao, X., Wen, Q., Chen, Z., Cheng, Z., Huang, X., Zhang, Y., Long, C., Zhang, Y., Huang, Z. (2019). An overview of the bioactivity of monacolin K/lovastatin. *Food Chem. Toxicol.*, 131, 110585.

[36] Surampudi, P., Enkhmaa, B., Anuurad, E., Berglund, L. (2016). Lipid Lowering with Soluble Dietary Fiber. *Curr. Atheroscler. Rep.*, 18: 75.

[37] Fonsêca, D., Salgado, P., de Aragão Neto, H., Golzio, A., Caldas Filho, M., Melo, C., Leite, F., Piuvezam, M., Pordeus, L., Barbosa Filho, J., et al. (2016). Ortho-eugenol exhibits anti-nociceptive and anti-inflammatory activities. *Int. Immunopharmacol.* 38, 402-408.

[38] Srinivasan, K. (2014). Antioxidant Potential of Spices and Their Active Constituents. *Crit. Rev. Food Sci. Nutr.*54, 352-372.

[39] Lushchak, V. I. (2014). Free radicals, reactive oxygen species, oxidative stress and its classification. *Chem. Biol. Interact.*, 24, 164-175.

[40] Liguori, I., Russo, G., Curcio, F., Bulli, G., Aran, L., Della-Morte, D., Gargiulo, G., Testa, G., Cacciatore, F., Bonaduce, D., et al. (2018). Oxidative stress, aging, and diseases. *Clin. Interv. Aging*, 13: 757-772.

[41] Danwilai, K., Konmun, J., Sripanidkulchai, B., Subongkot, S. (2017). Antioxidant activity of ginger extract as a daily supplement in cancer patients receiving adjuvant chemotherapy: A pilot study. *Cancer Manag. Res.*, 9: 11-18.

[42] Cao, C., Pathak, S., Patil, K. (2018). *Antioxidant Nutraceuticals: Preventive and Healthcare Applications*, 1st ed.; Taylor & Francis Group: Milton, UK. 428.

[43] Tesoriere, L., Allegra, M., Butera, D., Livrea, M. A. (2004). Absorption, excretion, and distribution of dietary antioxidant betalains in LDLs: Potential health effects of betalains in humans. *Am. J. Clin. Nutr.*, 80: 941-945.

[44] Krajka-Ku'zniak, V., Szaefer, H., Ignatowicz, E., Adamska, T., Baer-Dubowska, W. (2012). Beetroot juice protects against Nnitrosodiethylamine-induced liver injury in rats. *Food Chem. Toxicol.,* 50: 2027-2033.

[45] Jeszka-Skowron, M., Zgoła-Grze´skowiak, A., Stanisz, E., Wa´skiewicz, A. (2017). Potential health benefits and quality of dried fruits: Goji fruits, cranberries and raisins. *Food Chem.*, 221: 228-236.

[46] Hernández-Alonso, P., Bulló, M., Salas-Salvadó, J. (2016). Pistachios for Health. *Nutr. Today*, 51: 133-138.

[47] Coles, L. T., Clifton, P. M. (2012). Effect of beetroot juice on lowering blood pressure in free-living, disease-free adults: A randomized, placebo-controlled trial. *Nutr. J.,* 11, 106.

[48] Naveen, J., Baskaran, V. (2018). Antidiabetic plant-derived nutraceuticals: A critical review. *Eur. J. Nutr.,* 57: 1275-1299.

[49] Menon, V. P., Sudheer, A. R. (2007). Antioxidant and anti-inflammatory properties of curcumin. *Adv. Exp. Med. Biol.,* 595:105-125.

[50] Frank, M., Sacks, Hannia Campos. (2006). Polyunsaturated Fatty Acids, Inflammation, and Cardiovascular Disease: Time to Widen Our View of the Mechanisms, *The Journal of Clinical Endocrinology & Metabolism*, 91 (2): 398-400.

[51] Ojha, S., Goyal, S., Sharma, C., Arora, S., Kumari, S., Arya, D. S. (2013). Cardioprotective effect of lycopene against isoproterenol-induced myocardial infarction in rats. *Hum. Exp. Toxicol.,* 32, 492-503.

[52] Liu, X., Wu, Y., Li, F., Zhang, D. (2015). Dietary fiber intake reduces risk of inflammatory bowel disease: Result from a meta-analysis. *Nutr. Res.*, 35, 753-758.

[53] Pandey, M. K., Subash, C., Gupta, Karelia, D., Gilhooley, P. K., Shakibaei, M., Aggarwal, B. B. (2018). Dietary nutraceuticals as backbone for bone health, *Biotechnology Advances*, 36 (6): 1633-1648.

In: Nutraceuticals
Editors: A. Kumari and G. Singh
ISBN: 978-1-68507-488-3

Chapter 2

NUTRACEUTICAL ALTERNATIVE TO PHARMACEUTICAL

Neha Singh[1], Rajni Chopra[2] and Meenakshi Garg[1*]
[1]Department of Food Technology, Bhaskaracharya College of Applied Sciences, University of Delhi, Delhi, India
[2]Department of Food Science and Technology, National Institute of Food Technology Entrepreneurship and Management, Sonipat, Haryana, India

ABSTRACT

Nutraceuticals claim physiological benefits, i.e., why they are called alternatives to pharmaceuticals. They are formulated from active components isolated either from plant sources (phytocomplexes) like herbs or from animal origin like fish, which, when concentrated and dispensed in an appropriate pharmaceutical form could prevent or treat certain pathological conditions. Further, nutraceutical may cover a wide spectrum of commodities such as isolated nutrients, botanicals, dietary supplements, functional foods, and health supplements. The nutraceutical market is going to upsurge in the coming years due to the

[1*] Corresponding Author's E-mail: meenakshi.garg@bcas.du.ac.in.

requirement of the public, which transformed from a "want" for preventive health into a "need" in the current scenario. Thus, it integrates into the economic growth strategy of any country.

Additionally, end-users, peculiarly those who belong to higher socio-economic and upper-middle-class sections, perceive nutraceuticals as a proactive medicine to repress the pre-clinical health conditions. Therefore, they could be a potential toolkit to be used 'beyond the diet but before the drugs' to impede or cure the emerging infectious and non-infectious diseases. There is enormous demand for nutraceutical-type products by the general public in the current market due to their natural origin, minimal adverse effects, and their wide availability over the counter in supermarkets, pharmacies, or health product retailers. Henceforth, safe and biologically available nutraceuticals may work as pharmaceutical products for the treatment of many diseases and also improve the physiological functions of the body.

1. Introduction

The integration of two phrases quoted by Hippocrates, "Natural forces within us are the true healers of disease" and "let food be thy medicine and medicine be thy food," define unified power of body's defense mechanism with food as a therapeutic agent in healing and rewiring the body disorders or disease. Health is an active state which is protected by an array of the body's defense mechanisms and our daily diet. Therefore, food and drug from natural origin have greater importance in the public healthcare center. The body's five defense systems, namely, angiogenesis, regeneration, microbiome, DNA protection, and immunity, are common denominators of health. These systems are taking charge of maintaining our health and resisting the threat we all face as part of our daily routine. Each of the defense systems is persuaded by food on our plate [1].

Medical science has long recognized the significance of diet in harmonizing our health along with pharmaceuticals. In this way, the "Nutraceutical" perspective received an applicable and reliable outlook to attain good health and mitigate the risk of pathological conditions. The word nutraceutical confides to a nutrient and pharmaceutical approach where bioactive substances from natural origin with clinically proven health benefits could prevent, treat or cure disease, improve overall health, delay the aging process, and increase life expectancy. Therefore, clinically safe and bioavailable nutraceuticals could be used as an alternative to pharmaceuticals in managing the disease before they set in.

Stephen Defelice, a nutritionist, and pharmacist, president of the Foundation of Innovation Medicine (1989), elucidated the word nutraceutical as a food, or part of a food, that proffer medical or health benefits, including prevention and treatment of a disease. A nutraceutical may cover a broad spectrum of commodities that vary from isolated nutrients, dietary supplements, and specific diets to genetically engineered designer foods, herbal products, and processed foods such as cereals, soups, and beverages [2]. They are formulated from active components isolated either from plants (phytocomplexes) or from animal origin. When these functional components are concentrated and dispensed in an appropriate pharmaceutical form, they could avert or treat certain clinical conditions [3].

Pharmaceuticals are drugs primarily used for treatment after the pathological condition has occurred and sometimes may have side effects. In contrast, Nutraceuticals act as a proactive medicine to repress and reduce the risk related to lifestyle disorders. Therefore, they could be incorporated in the daily diet regime as a potential toolkit in a region ranging from 'beyond the diet but before the drugs' because they proffer both nutritional and functional properties of food extracts along with healing effects of natural active constituents [4].

In the last decades, the demand for nutraceutical products has been resurged by the general public because of their natural origin, minimal adverse effects, and their availability over the counter in the supermarket, pharmacies, or health product retailers. Apart from this, they have been available in the market in various different forms (Figure 1).

So, the nutraceutical market will be an upsurge in coming years owing to the requirement of the public, which transformed from "want" for preventive health into a "need" in the current scenario.' Thus, it integrates into the economic growth strategy of any country.

However, regulations related to nutraceutical or dietary supplements vary from country to country. Therefore, harmonization in regulation is critically essential for consumers' safety from misleading claims of nutraceuticals. Nevertheless, the nutraceutical industry faces many challenges: drug-food interactions, self-medication, bioavailability, safety, clinical efficacy, and so forth. Henceforth, safe and biologically available nutraceuticals may reduce the dependence on pharmaceutical products and direct towards a more sustainable and integrated approach to improve the body's physiological functions.

This chapter outlined various topics: sources of plant-based nutraceuticals, animal-based nutraceuticals, nutraceutical market in India and other countries, regulations and safety, challenges, clinical efficacy, and case studies related to the nutraceuticals.

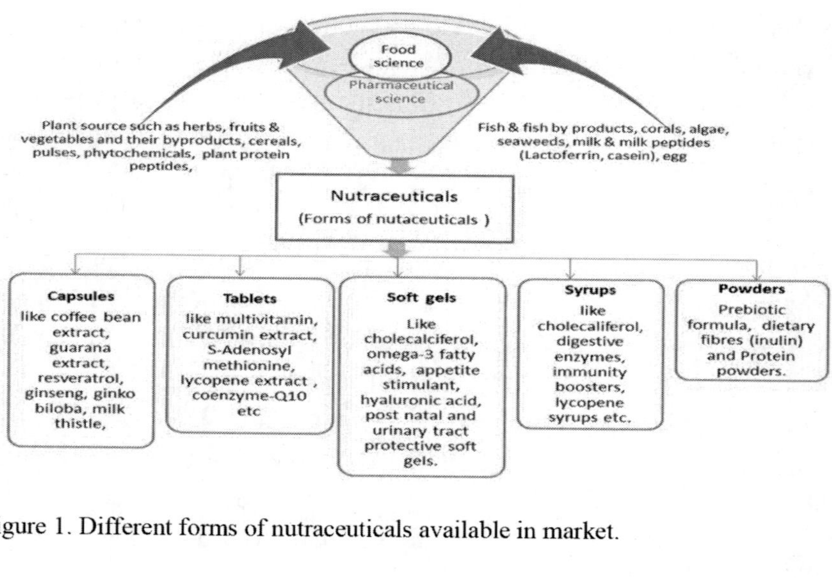

Figure 1. Different forms of nutraceuticals available in market.

1.1. Plant Based Nutraceuticals

Nutraceuticals from plant origin have emerged as new medicinal therapy (phytotherapy) to reverse the damage caused by a sedentary lifestyle and faulty eating habits. Plant nutraceuticals are recognized as safe because people believe that "nature is the true healer and safer" than pharmaceutical drugs. In addition, bioactive constituents from plants such as curcumin from turmeric, gingerol from ginger, anthocyanins, caffeine, melatonin, etc., have therapeutic value, prevent the incidence of chronic diseases, and have action potential against various microbial infections. Sometimes, they may possess curative properties. The few examples of widely available nutraceuticals that people frequently use are ginseng, melatonin, fenugreek, curcumin, resveratrol, garcinia, and quercetin.

Further, some nutraceuticals have well-established scientific data on their efficacy, toxicity, mode of action, and bioavailability, while others are still anecdotal. Thus extensive research is required in this arena to assess their clinical efficacy, safety, and toxicity. Nevertheless, Table 1 and Table 2 outlined various plants and plants' proteins/peptides that have nutraceutical potential to attenuate the symptoms of chronic diseases.

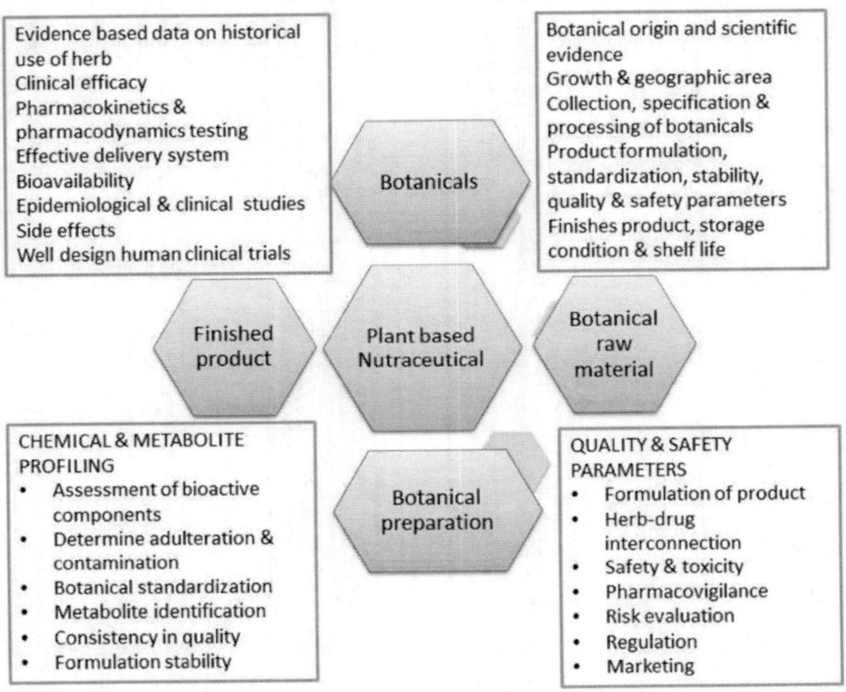

Figure 2. Parameters consider in formulation of nutraceuticals.

Table 1. Nutraceuticals from plant origin

Scientific Name	Common Name	Plant Parts	Nutrients/ Bioactive compound	Clinical efficiency	References
Digitaria exilis and Digitaria iburua	Acha and iburu native to Africa	Whole grain cereal	High in sulphur containing amino acids (methionine and cysteine), dietary fibre and resistant starch	Anti- diabetic due to Low glycemic index, cholesterol lowering effect, alleviates constipation.	[5, 6]
Avena sativa	Oats	Oat bran	Beta glucan	Lowers glucose level, reduce total and LDL cholesterol which reduces risk of coronary heart disease.	[7, 8, 9]
		Oat bran and oat gum	Soluble dietary fibre	Delays gastric emptying time, decrease nutrient absorption, alleviates constipation, increases satiety after meal.	
Malva sylvestris	Common mallow	Flowers and leaves	Phenols, flavonoids, carotenoids and ascorbic acid alpha- tocopherol	antiseptic, chemopreventive, chemotherapeutic agent, reduces the risk of cardiovascular diseases, enhance immune system and modulate degenerative conditions related to aging	[10, 11]
		Leaves	malvidin 3-glucoside	Inhibit ear oedema and leukocytes migration, future treatment for skin disorders	[12]

Scientific Name	Common Name	Plant Parts	Nutrients/ Bioactive compound	Clinical efficiency	References
Glycine max *Trifolium pretense* *Medicago sativa*	Soy red clover alfalfa	Seeds	Isoflavones (genistin and daidzin)	Reduces the risk of breast cancer in pre- and post-menopausal women, manage menopausal syndrome, alleviates constipation, can be taken by celiac individual, possible treatment for osteoporosis by inhibiting tyrosine kinase, decrease the risk of coronary heart disease by reducing LDL & triglyceride.	[13, 14]
Pseudocereal grains (Chenopodium quinoa Willd; *Amaranthus* sp.; *Fagopyrum* Sp.)	Quinoa, amaranth and buck Wheat	Seeds	High in protein content having albumins, globulins and glutelins as major fractions while, prolamine is the minor fraction. Rich in Lysine, Cystein and methionine. Good source of fibre, resistant starch, vitamins, minerals and phytochemicals (saponins, polyphenols, phytosterols, phytosteroids, and betalains)	Reduces obesity, glucose and cholesterol lowering effects, alleviates constipation, reduce malnutrition, reduce oxidative stress, can be consumed by celiac patients.	[15]

Table 1. (Continued)

Scientific Name	Common Name	Plant Parts	Nutrients/ Bioactive compound	Clinical efficiency	References
Arachis hypogaea	Peanuts		Containing Co-enzyme Q10 in copious quantity, they are good source of resveratrol, phenolic acids, flavonoids and phytosterols (beta sitosterol, campesterols and stigmasterol).	Block cholesterol absorption from diet and reduces the incidence of cardiovascular disease.	[16, 17]
Morus spp. *Vitis spp.* *Arachis hypogaea*	Mulberries, grapes, peanuts and red wine		Consist of all 20 amino acids having arginine in high amount. Resveratrol	Decrease or avert lipid peroxidation in pharmaceutical products, extends the lifespan, minimizes the risk of cancer, metabolic syndrome and neural degeneration.	[18, 19]
Linum usitatissimum	Flaxseed	Seeds	Secoisolariciresin-ol diglucoside (SDG), the chief lignin	Lower the serum cholesterol level, reduces the incidence of breast, prostate and colon cancer, slow down the onset of type 2 diabetes and impede oxidative stress as it bear antioxidant property	[20]
Nigella sativa & *Cinnamomum cassia*	Black cumin and cinnamon	Seeds and bark	Flavonoids and phenolic compounds	Effective in counteract hyperglycemia therefore, could be used as a adjuvant therapy along with antidiabetic treatment. Ability to normalize the lipid profile and kidney function. Provide protection to	[21]

Scientific Name	Common Name	Plant Parts	Nutrients/ Bioactive compound	Clinical efficiency	References
				the islets of pancreas by opposing immune function.	
Allium sativum	Garlic	Bulb and leaves	S-allyl-cysteine sulfoxide (alliin) Allicin, diallyl sulfide, diallyl disulfide, diallyl trisulfide, ajoene, and S-allyl-cysteine	Possess cardioprotective effect via activation of PPARγ targeting autophagy and suppression of hypoxia-induces necroptosis. Thus, could be promising therapeutic agent for the treatment of myocardial infraction and other cardiovascular disease.	[22]
				Effective in reducing inflammatory markers (IL-6, CRP and ESR) in peritoneal dialysis patient.	[23]
				Ameliorate complications of sickle-cell anemia.	[24]
				Exhibit immunomodulatory, anticancer, hepatoprotective, neuroprotective, antimicrobial and antioxidant activities, minimizes the glucose level (anti-diabetic), decrease obesity, reduce the risk of cardiovascular disease.	[25]

Table 1. (Continued)

Scientific Name	Common Name	Plant Parts	Nutrients/ Bioactive compound	Clinical efficiency	References
Nigella sativa & *Cinnamomum cassia*	Black cumin and cinnamon	Seeds and bark	Flavonoids and phenolic compounds	Effective in counteract hyperglycemia therefore, could be used as an adjuvant therapy along with antidiabetic treatment. Ability to normalize the lipid profile and kidney function. Provide protection to the islets of pancreas by opposing immune function.	[21]
Zingiber officinale	Ginger	Rhizome (roots)	Gingerol and shogaols (impart pungent taste to dried ginger)	Gingerol alleviates nausea, arthritis and pain. Both have anti-inflammatory, antioxidant, anti-allergic, antitumor and antimicrobial properties. Could be promising nutraceutical counteracting lifestyle metabolic disorders like diabetes, hypertension, obesity etc.	[26]
Terminalia catappa & *Colocasia esculenta*	tropical almond & Colocasia	Leaves	Flavonoids and phenols	Exhibits antioxidant and antimicrobial properties. Thus, synergistic effect of both herbs could be a used as antibiotic-resistance-modifying compound.	[27]
Terminalia catappa		Leaves	hydrolysable tannins (punicalin, punicalagin),	Bacteriostatic property, against gram-negative bacteria, traditionally used to treat hepatitis in India, antifungal against	[28, 29]

Scientific Name	Common Name	Plant Parts	Nutrients/ Bioactive compound	Clinical efficiency	References
			gallic acid and flavonoid C-glycosides	*candida* spp. Therefore could be used to treat fungal disease.	
Hibiscus sabdariffa	Karkaday	Flower	anthocyanines and flavonoids	Combat atherosclerosis by inhibiting serum stimulated proliferation of smooth muscle cells (SMC) which trigger apoptosis of SMC.	[30]
		Leaf	Polyphenolic extract rich in flavonoids	Delay atherosclerosis as it removes cholesterol from macrophages. Thus, could be used as anti-atherosclerotic agent	[31]
				Possesses hypoglycemic, hypolipidemic, and antioxidant effects, and induce tumor cell apoptosis, could inhibit the migration and invasion of human prostate cancer cells	[32]
				Antihyperlipidemic and antihypertensive effects as it may decreases serum total cholesterol, triglyceride and LDL cholesterol, increase HDL cholesterol, reduce glucose level.	[33, 34]
		Calyces		Possesses antidiabetic and antioxidant properties due to the inhibition of α-amylase and α-glucosidase activity.	[35]

Table 2. Plant Protein/peptides with nutraceutical properties

Food sources	Protein/peptide	Name/sequence	Biological activity	References
Cereal, lupin seed protein	conglutin gamma	Conglutin	Glucose lowering effect i.e., hypoglycemic	[36]
Common bean	non-digestible fractions (NDF) peptides	GLTSK, LSGNK, GEGSGA, MPACGSS and MTEEY	Possess anticancer property as it shows antiproliferative effect on human colon cancer cells by altering molecules entailed in cell cycle arrest or apoptosis.	[37]
Soy	soy hydrolysates	beta-conglycinin, basic gliobulin, glycinin	Decrease lipid accretion in adipocyte Hypocholesterolemic, reduces the serum cholesterol up to the level >4%	[38, 39]
Broccoli	Tripeptide, Tyr-Pro- Lys	YPK	ACE inhibitory property	[40]
Rice Rice bran	Albumin Albumin, globulin and glutelin	Oryzatensin (GYPMYPLR)	Immunomodulatory Anti- diabetic due to inhibition of α-glucosidase activity; anti-hypertensive due to ACE-inhibitionary enzyme.	[41, 42]

1.2. Animal Based Nutraceuticals

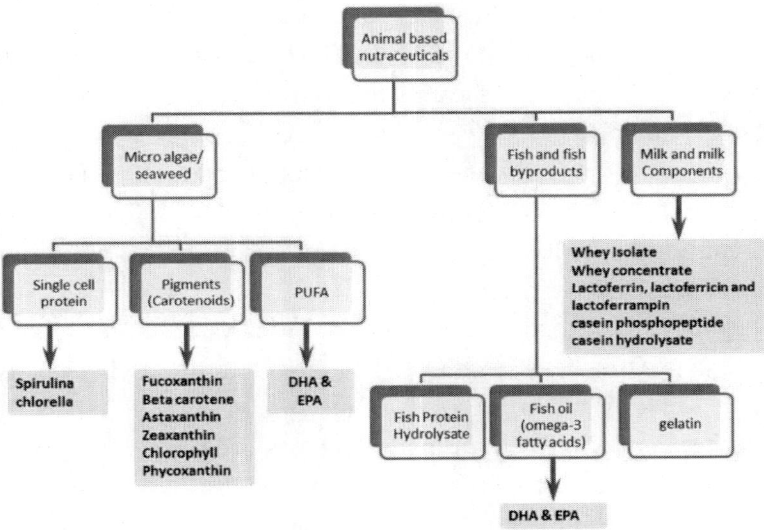

Figure 3. Nutraceutical from animal origin.

1.2.1. Nutraceuticals from Fish and Fish by-Products Sources

Fish Protein Hydrolysate (FPH) is a potential source of bioactive peptides for the nutraceutical and pharmaceutical segment. They have been described as an appropriate and potent source of proteins in the human diet due to their balanced amino acid composition and beneficial effect on gastrointestinal absorption. FPH extracted from blue whiting (*Micromesistius poutassou*), and brown shrimp (*Penaeus aztecus*) could be seen as potential appetite suppressive peptides as they were capable of stimulating the cholecystokinin releasing activity from STC-1 intestinal endocrine cells [43]. A randomized clinical study [44] evaluated the effect of FPH, i.e., Slimpro, acquired from blue whiting (*Micromesistius poutassou*) muscles on 120 slightly overweight subjects. FPH intake at a level of 1.4 & 2.8g for 90 days was effective in stimulating the secretion of cholecystokinin (CCK) and glucagon-like peptide (GLP-1) compared to the placebo arm. Which subsequently, produces a slimming effect due to decrease in body weight by reducing fat mass, not by draining impact, i.e., reduction in extracellular water content.

Inhibitory peptides (Val-Trp and Leu-Lys-Tyr) extracted from Antarctic krill peptide powder (AKPP) formulated by enzymatic hydrolysis of tail meat significantly reduced the systolic blood pressure by inhibiting angiotensin -1- converting (ACE) enzyme activity [45]. Furthermore, the oral intake of ACE inhibitory peptide (LPH-IV) obtained from loach (*Misgurnus anguillicaudatus*) could significantly alleviate hypertension by reducing systolic blood pressure thus may possess antihypertensive activity [46].

Additionally, the porcine skin gelatin hydrolysate has the potential to manage type-2 diabetes mellitus as it inhibits dipeptidyl peptidase IV (DPP-IV) activity and improves glycemic control [47].

Omega-3 fatty acids from marine sources offer many health benefits like reducing the risk of arthrosclerosis and cardiovascular diseases, improving the lipid profile, controlling cholesterol, lowering the triglycerides level, managing diabetes, and many more in a row. In recent Vitamin D and Omega-3 Trial (VITAL), oral intake of 840 mg/d of eicosapentaenoic acid (EPA) and docosahexaenoic acid (DHA) attenuates the incidence of a heart attack about 28%, fatal heart attack about 50%, and total coronary heart disease about 17%. In the ASCEND study (A Study of Cardiovascular Events in Diabetes), there was a 17% decrease in death due to cardiovascular disease with an oral intake of 840mg/d of EPA & DHA. In the REDUCE-IT study (the Reduction of Cardiovascular Events with Icosapent Ethyl–Intervention Trial), intake of 4g/d of EPA reported a 25% decrease in the primary endpoint of significant cardiovascular instances in subjects with high triglycerides levels (135-499 mg /dL) and were also taking a statin drug. Therefore, Compiled data from the above three studies, i.e., VITAL, ASCEND and REDUCE-IT, revealed that Omega-3 fatty acids (EPA & DHA) have the potential to decrease the risk of cardiovascular diseases [48]. Furthermore, the pooled evidence from systematic review and meta-analysis (involving 39 trials with 2875 subjects) outlined that omega-3 fatty acid was effective in elevating serum total antioxidant capacity, glutathione peroxidase activity, and reduced malondialdehyde in cases versus placebo arm. So, overall omega-3 fatty acids have a protective effect against oxidative stress, i.e., reactive oxygen species [49].

1.2.2. Microalgae and Seaweeds

Microalgae such as cholera and spirulina are frequently used as valuable nutraceutical raw materials isolated from marine origin. Blue-green algae spirulina (*Arthrospira*) has attained significant interest after NASA used it as a dietary supplement for astronauts. It has anti-inflammatory activity as it impedes the release of histamines from mast cells and could modulate the immune system [50]. Systematic review and meta-analysis of RCTs on 278 subjects indicated that spirulina supplementation significantly reduces weight, mainly in obese patients. Also, decrease cholesterol absorption, insulin resistance, oxidative stress, and inflammatory markers [51].

Marine algae contain sulfated polysaccharides (SPs) in ample amounts, like carrageenans in red algae, ulvans in green algae, and fucoidans in brown algae. These SPs from marine origin proffer various physiological benefits, mainly, immunological activity, antioxidant, anti-allergic, anticoagulant, and anticancer properties. The underlined beneficial effects make marine-derived SPs a potent source for formulating nutraceutical products [52].

Seaweed has become a boon for the nutraceutical industry as it possesses various health-beneficial bioactive agents. Polysaccharides from seaweed have been widely used because of their several expedient properties such as anti-inflammatory, antiviral, anti-cancerous, and anticoagulant activities [53]. For example, fucoxanthin, a carotenoid obtained from brown seaweed, has an anti-obesity and anti-diabetic effect. It ameliorates insulin resistance and lowers blood glucose levels by modulating cytokine secretions from white adipose tissues [54]. Fucoxanthin (Fx) is metabolized into fucoxanthinol and amarouciaxanthin after absorption, and the metabolized components are accumulated in visceral white adipose tissue (WAT). The neuroprotective, hepatoprotective, and photoprotective efficacy of Fx is due to its antioxidant potential. The anti-obesity property of Fx is based on the upregulation of thermogenesis by expression of uncoupling protein 1 (UCP1) and the acceleration of metabolism instigated by mitochondrial activation [55].

Seaweed is also a good source of protein containing all amino acids, particularly alanine, proline, glutamic acid, glycine, arginine, and aspartic acid. The protein content ranges from 5 to 47% on a dry basis. Further, the essential amino acids in algae constitute almost half of the amino acids, and their protein profile is similar to the profile of egg protein. Among all seaweeds (green, brown, and red seaweed), red seaweed contains protein in ample amounts, i.e., approximately 47%. Therefore, algae can be used as an alternative and cheap source of ingredients with high nutritional value to combat the challenges of protein malnutrition [56].

1.2.3. Milk and Milk Components

Advancement in separation techniques and enzyme technology in the dairy industry provides an opportunity to extract, concentrate or modify the cows' milk components so that these beneficial components could be used as nutraceuticals to improve the body's biological functions. Specific peptides within the dietary protein may exert specific physiological functions like peptides from casein (casein phosphopeptide) may increase the bioavailability of minerals (Ca & Zn) by improving their solubility. Also, decrease the activity of an angiotensin-converting enzyme (ACE) thus, reducing the vasoconstriction and alleviating the high blood pressure [57].

Systematic review and meta-analysis [58] revealed that casein phosphopeptide amorphous calcium phosphate (CPP-ACP) exhibited remineralization efficacy, decreased enamel surface roughness, and effective reconditioning of form, aesthetics, and function to treat white spot lesions. So, CPP-ACP seems to be a potential remedy to treat white spot lesions.

The Single-blind, placebo-controlled study was conducted by [59] to assess the effect of casein hydrolysate formulated by a protease from *Aspergillus oryzae*, having major ACE enzyme inhibitory peptides Val-

Pro-Pro (VPP) and IIe-Pro-Pro (IPP). The study comprised 131 subjects with high normal blood pressure and mild hypertensive. These were randomly divided into four groups, and two tablets of four different dosages of VPP and IPP (i.e., 0, 1.8, 2.5 & 3.5) were given to each subject daily for six weeks. A considerable reduction in systolic blood pressure was noticed at 6 weeks in the group taking 1.8 mg of dosage. At the same time, systolic blood pressure (SBP) in the active groups was lowered at both 3 weeks and 6 weeks after administration of either 2.5mg or 3.6mg of dosages in contrast to SBP prior to treatment. Further, the antihypertensive effect was significantly higher in mild hypertensive subjects compared to other subjects. Another novel ACE inhibitory peptide, YQKFPQYLQY (YQK), was obtained from hydrolysis of bovine casein using pepsin and trypsin. The oral ingestion of YQK was found to lower systolic blood pressure in hypertensive rats significantly. Therefore, it may have the potential to be used as an antihypertensive agent in nutraceuticals or functional foods, or pharmaceuticals [60]. Yak milk casein also contains an ACE inhibitory peptide (KYIPIQ), which elevates NO synthesis and expression of endothelial nitric oxide synthase in human vascular endothelial cells via activation of the protein kinase B (Akt) pathway. So, it can be a potential future therapeutic drug or nutraceutical for hypertension treatment [61].

Lactoferrin is a non-haem iron-binding glycoprotein that has been studied extensively. It is also a red fraction of minor bioactive milk protein. It has a wide range of physiological activities such as antimicrobial, immunological, anticancer, anti-allergic, antioxidant, and anti-inflammatory properties. Because of its therapeutic value, lactoferrin is considered a nutraceutical ingredient for the emerging food industry [62]. It also protects neonates against infection caused by a wide spectrum of pathogens. That's why exclusive breastfeeding is vital for infants during the first six months to provide immunity against microbial infection [63].

Lactoferrin-derived peptides, lactoferricin, and lactoferrampin exert more potent antimicrobial properties compared to intact protein. Especially, lactoferricin has shown potential antibacterial, antiparasitic and antifungal activity against various human diseases ranging from ocular infection to osteoarticular, dermatological and gastrointestinal diseases [64].

1.3. Nutraceutical Market in India and Other Countries

The evolution of the health-centered ecosystem has upswing the nutraceutical market in India and other countries. The market is cover under the umbrella of the wellness category (e.g., vitamins and minerals). Wellness stands on two pillars: a "preventive approach" and a "rejuvenating approach" to compensate for the stress of daily life. At present, the nutraceutical segment is urban-centric and is on the edge of breaking out and becoming a mighty pillar of the health, fitness, and wellness revolution. Functional food shares the most significant chunk in the nutraceutical market in India, followed by a dietary supplement.

The U.S. has been the largest growing nutraceutical market with its full potential. It grew to $65 billion from 50 billion with a compounded growth of 10% annually between 2010 and 2015. In Europe, the market has grown from $ 35 Bn in 2010 to $ 40 Bn in 2016 and is expected to grow by $51 Bn by 2021, which indicates an annual growth of 5%. Within Europe, Germany, France, and Switzerland covered 70% of the European market [65].

The graph of the nutraceutical market is ever-expanding from 1999 to 2019 and is expected to grow in the coming five to ten years. Figure 4. indicated that the nutraceutical business in the initial years from 1999 to 2002 increased from 39bn to 47 bn and from 2010 to 2014 market grew from 140 bn to 172 bn which turns to 241 bn industry in 2019. This statistic showed that the nutraceutical industry grew at 7% per annum from 1999 to 2002, while the growth was doubled at 14% per annum in the next few years up to 2010. In a recent scenario, $12-$15 bn is being added up per annum [65].

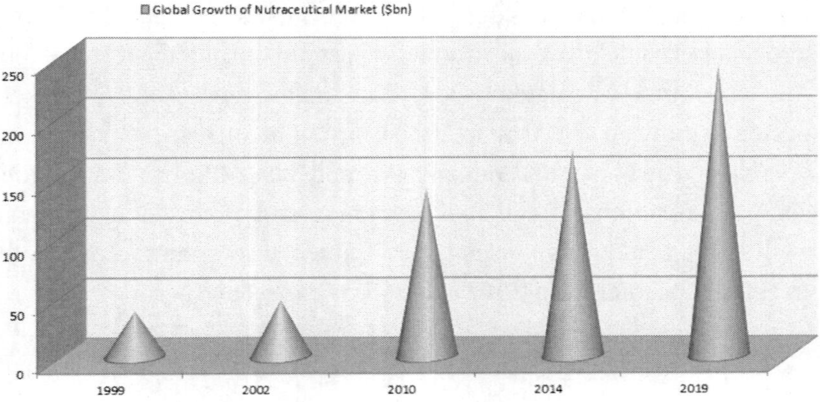

Figure 4. Global growth of nutraceutical market.

The Asian Pacific Nutraceutical Market size is expected to reach a CAGR of about 7.4% during the period 2021 to 2026. The booming consciousness among the consumers about the health benefits of nutritional diet, functional food, or dietary supplement leads to the expansion of the APAC nutraceutical market. India is expected to flourish in the Asia-Pacific health food sector as people's preferences switch from pharmaceutical to nutraceutical products. Further, within dietary supplements, the demand for protein and plant supplements is getting more consideration among Indian consumers while vitamins among Australian consumers. Beyond that, pediatric health supplements are a flourishing sector in China with the heavy sale. Major key players in the APAC nutraceutical market are ADM company, Cargill, Nestle, Groupe Danone, E.I. du Pont de Nemours and Company, Aland (Jiangsu) Nutraceutical Co., Ltd, General Mills.

The paradigm shift in lifestyle, increasing knowledge and awareness about health products, leads to the growth of domestic and international nutraceutical companies like HealthKart, Patanjali, Cargill, Nestle, and Amway in the Indian market. In the race of marketing, Patanjali is the forthcoming billion-dollar company in India. Besides this, other international ventures, namely, UK wellness giant, Holland, and Barrett arriving in the Indian market, witnessed the huge potential of the Indian nutraceuticals market. However, the surge in the incidence of lifestyle disorders or chronic disease is deemed to extend the nutraceutical products in the upcoming calculated years. The Indian nutraceutical market is expected to grow up to $10 bn in 2022 from $ 4 bn in 2015, which accounts for massive growth of 21% annually. Around 35% of Indian nutraceuticals encompass functional food and beverages and represent considerable growth in the coming five years [65]. Figure 5 shows the evolution of the nutraceutical market from 2017 and will be expected to grow.

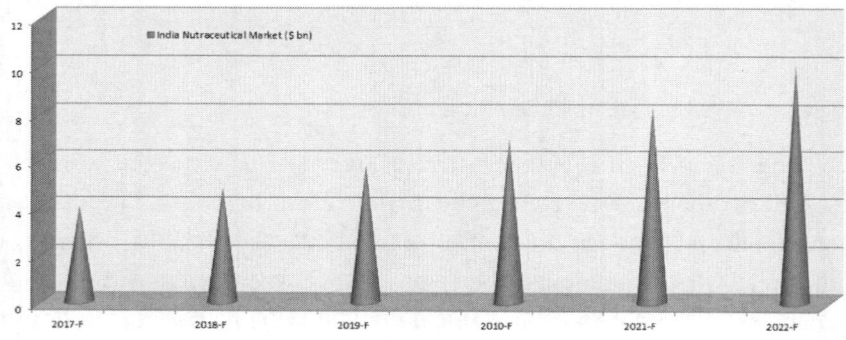

Figure 5. Indian nutraceutical market.

In the present scenario, where COVID-19 pandemic is subsisting all over the globe, the dependence on nutraceutical ingredients like herbal supplements or immune-boosting plants/supplements are augmenting to enhance the physiological function and to boost the immune system. During this epidemic, the nutraceutical market is expected to skyrocket due to consumers' demand for immune-boosting supplements. The coronavirus has been affecting all the industries, primarily, food and beverage sector. The long-term aftermath is unforeseeable; however, the key players in the nutraceutical segment have seen immense growth during the outbreak.

1.4. Clinical Efficacy

The concept of nutraceutical has emerged from historical usage of flora and fauna for their survival on earth, which certainly with coming period becoming a future of the food industry and gives a kick for innovation of salubrious and novel food product. As bioactive constituents found in the ecosphere have been reported to offer various health benefits such as reducing oxidative stress, managing diabetes, delaying aging, alleviating cholesterol levels, minimizing inflammation related to several diseases, and many more in a row. Due to the aforementioned benefits, nutraceuticals are seen as the miracle therapy that holds a bright future for public health. However, the clinical efficacy of these bioactive molecules and their application in the food and pharmaceutical industry are restricted because they are heat labile, have poor bioavailability, poor storage stability, and low solubility in fluids. Moreover, in the course of storage, indispensable bioactive compounds may be lost rapidly due to physicochemical reactions, chemical deterioration, and volatilization.

Few studies based on RCTs revealed the clinical efficacy of certain herbs in alleviating the symptoms of pathological conditions. For example, [66], carried out a double-blind, parallel-group, placebo-controlled, randomized clinical trial to evaluate the efficacy of nutraceutical mixture (composed of *Lagerstroemia speciosa, Berberis aristata, Curcuma longa,* Alpha-lipoic acid, Chrome picolinate, and Folic acid) on short term glucose and lipid-lowering effect in patients with altered fasting glucose. The results indicated that the tested combination of nutraceutical was found to improve several parameters like triglycerides (-34.7%), HDL-C (-13.7%), fasting plasma glucose (-13.4), and HOMA-Index (-25%) compared to a baseline value. Therefore, nutraceutical enhanced mildly but considerably many parameters linked with insulin resistance in patients afflicted by altered fasting glucose compared to both baseline and control groups with no side effects.

Although ample evidence on the functionalities of nutraceuticals was based on in-vitro cell line studies or application techniques that are different from oral administration, there is a scarcity of clinical evidence on the efficacy of nutraceutical products based on in-vivo studies. Despite the product, marketing, and physiological function it proffers, the functional claim of any nutraceutical product must be supported by a well-established and properly designed human clinical trial. The data are accessible from various sources though there is a lack of comprehensive scientific guidelines to design, conduct, and report human clinical trials to substantiate the health benefits of dietary supplements or foods. The basic methodology to design and conduct human clinical trials to evaluate the health benefits of the nutraceutical is similar to that of pharmaceutical study. Anyhow, there are a few aspects that require special attention while investigating nutraceuticals [67]. Therefore, generality concerning the efficacy of herbal medicine cannot be produced, and decisions must be expressed on a case-by-case basis [68].

In pharmacology, the solid and evidence-based data acquired from well-designed randomized controlled trials (RCTs) serve as the gold standard to evaluate the clinical efficacy of nutraceutical products. But, it is hard to employ in nutrition-based settings as it requires suitable placebo, appropriate masking, sufficient robust biomarkers, healthy population v/s patients' application, and clinical outcomes v/s maintenance of health or others [69]. Besides, systematic reviews and meta-analyses provide satisfactory clinical proofs for several herbs like flaxseed for hypertension, ginger for nausea induced by pregnancy, etc., yet the solid conclusion indicating efficacy is more challenging. In contrast, contradictory results on efficacy have also been delineated for various common herbs mainly, Aloe Vera for treatment of psoriasis, ginkgo for tinnitus, and Pomegranate to treat or prevent cardiovascular diseases. Even though several herbs, in particular, St Johns' wort, has been related to cause severe adverse effects. Therefore, it is not advisable to generalize the efficacy and safety of herbal therapy because several clinical trials have been executed with inadequate rigor and recorded details [70]. Dietary supplements are classified into six categories in accordance with its clinical efficacy [68]:

- The first category includes herbal dietary supplements whose documented clinical effectiveness has been manifested by systematic reviews and meta-analysis published by the Pubmed database/Cochrane library or by other authorized medical journals (where 'weight of evidence 'is satisfactory and' direction of evidence' is towards positive effects). Like ginger for mild nausea occurring during pregnancy [71] and horse chestnut for chronic venous insufficiency [72].

- The second group comprises products that are substantially used in specialized conditions; however, their use does not support by strong clinical evidence (i.e., 'weight of evidence' is satisfactory while 'direction of evidence indicates negative effects). Examples mainly include Echinacea for the common cold [73] or valerian for insomnia [74].
- The third category cover remedies that have exhibited a small effect, but their clinical efficacy has not been critically assessed, such as garcinia and green tea, advertised for weight loss [75, 76].
- The fourth category consists of dietary supplements whose clinical evidence revealed contradictory results, such as aloe vera for psoriasis [77].
- The fifth category consists of numerous dietary supplements which have shown promising clinical data. Still, the overall outcome is not engrossing and requires more appropriate and accurate studies to support their use, such as garlic as hypotensive, agnus castus for premenstrual symptoms, etc., [70].
- The last category involves a broad range of dietary supplements whose RCTs have not been assessed, like Yohimbe for erectile dysfunction [78].

Absorption and biological availability of dietary supplements hold the uppermost ladder in the clinical efficacy of a product. As they become a part of the daily regime but the bio-accessibility of nutraceuticals shall retain a significant degree of therapeutic value on oral administration. Many people use neutraceuticals daily; hence, administering supplements via the oral route is preferred over any other method because it is easy to take, non-invasive and cost-effective. However, the effectiveness of various nutraceuticals is often restricted due to inadequate dose efficiency when ingested orally. So, to counteract this problem and improve the efficacy of the nutraceutical through the oral route, advancement in the delivery system has been developed to increase the bioavailability of nutraceuticals through various techniques. Some of them are phospholipid-based delivery systems (Phytosomes, liposomes), emulsion-based delivery systems, chemical modification, and another emerging delivery system [79].

1.5. Regulations and Safety Aspects

Nutraceuticals exist in the gray area between pharmaceuticals and food because of the scarcity of clear information regarding clinical data claimed health benefits, safety profile, and shared regulatory system. Due to the lack of common nutraceuticals definition and inconsistency in the regulatory framework, it would be judicious to have an appropriate and univocal definition of nutraceuticals along with a shared regulatory system as it provides:

- An opportunity to assess the safety.
- Mechanism of action.
- Clinical efficacy of nutraceutical products.

The principal objective of nutraceutical regulation is to ensure that products are safe and prominently labeled without any misleading claims.

1.5.1. Global Regulatory Framework for Nutraceuticals

FDA regulates both finished dietary supplement products and dietary ingredients. It regulates dietary supplements under different regulations other than those enclosing "conventional" foods and drug products. The Dietary Supplement Health and Education Act (DSHEA) was enacted by U.S. Congress in October 1994 and signed into law on October 25, 1994, by President Bill Clinton to establish a regulatory framework to assure the safety of dietary supplements [81]. DSHEA given the U.S. Food and Drug Administration (FDA) authority to lay down a new framework for regulations covering manufacturing, health claims, and labeling of dietary supplements and forming governmental bodies like the Office of Dietary supplements to boost research on dietary supplements and Commission on Dietary Supplement Labels to look after the label claims of dietary supplements. Under this law, the word "dietary Supplements" means a product i.e., designed to supplement the diet containing one or more of the following dietary ingredients: a vitamin; a mineral; an herb or other botanical; an amino acid; a dietary substance used by humans to supplement the diet by accelerating the total dietary intake; or a concentrate, metabolite, constituent, extract, or combination of any ingredients mentioned above. According to DSHEA (1994), Adulterated or misbranded products are banned from sale by the manufacturers and distributors of dietary supplements and dietary components. Therefore, companies carrying out the business of dietary supplements are responsible for assessing the safety and labeling of their products prior to marketing to ensure that they meet all the criteria of DSHEA and FDA regulations. FDA is responsible for taking action against any hazardous or life-threatening dietary supplement product after it comes out on the market. In 2007, the Current Good Manufacturing Practice (cGMP) regulation (21CFR, part 111) was enforced by the FDA and the department of health and human services (HHS). The regulations notified that all the domestic and foreign firms that manufacture, package, label, or hold dietary supplements, including those engaged with testing, quality control, and distribution of dietary supplements in the U.S., must abide with the cGMP for quality control and safety. Despite this, the manufacturer, packer, or distributor whose name is mentioned on the label of a dietary supplement product marketed in the U.S. must submit reports related to any serious unfavorable events to the FDA. Withal dietary supplements that cover foods are supervised by the FDA's Centre for Food Safety and Applied Nutrition. Likewise, the FDA and Federal Trade Commission are responsible for enforcing regulations involving marketing, manufacturing, labeling, and advertising dietary supplements.

The Green Paper on Food Law (1997) paved the way for the foundation of European food law, followed by the 2000 White Paper on Food Safety, which declared some 80 proposals for new and improved legislation in this area. Basically, it envisaged the establishment of General food Law Regulation, laying down principles of food law, and the formation of the European Food Safety Authority (EFSA), which provide scientific guidance on issues based on scientific risk assessment with segregated responsibilities related to risk assessment, risk management, and risk communication. There is no specific regulatory framework for nutraceuticals that exist in EU food Law. As the General Food Law Regulation applies to all food, its general principle also encloses foods with added functional properties like 'nutraceuticals,' 'functional foods,' 'dietetic foods, and food supplements. The EU legislation on food supplements is directive 2002/46/EC which defines food supplements as "foodstuffs the purpose of which is to supplement the normal diet and are concentrated source of nutrients or other substances with nutritional or physiological effect, alone or in combination, marketed in dose form like capsules, pastilles, tablets, pills and other similar forms, sachets of powder, ampoules of liquid, drop dispensing bottle, and other similar forms of liquids, powders designed to be taken in measured small unit quantities." The main aim of EU legislation on food products covering functional food and nutraceuticals is 'safety' [80]. So, the judgment on the safety basis of legislation depends on risk analysis where scientific risk assessment is carried out by EFSA and risk management is executed by European Commission, member states, and European Parliament. In the risk management stage, both precautionary principle and other reasonable factors may be taken into consideration to select the most appropriate way to deal with an issue [82]. There is a marked inconsistency in the EU, which is associated with product claims by different countries and Recommended Daily Allowance (RDAs) of several nutraceuticals. Therefore, nutraceutical companies in Europe have established a coalition called Food Supplement Europe (FSE) to harmonize rules. The main objective of the harmonized regulations on those products listed in Directive 2002/46/EC is to protect consumers from potential health hazards caused by those products and ensure that they are not marketed with misleading claims or information [80].

Foods that provide health benefits are generally recognized as health foods in Japan, a term currently not defined by the law. The health claim system in Japan has evolved with the advancement in scientific knowledge, international trends in the treatment of health claims, and the incidence of health hazards related to healthy foods. Significant modifications in the treatment of health claims occurred with the establishment of FoSHU in 1991 and FNFC in 2001. There are only two classifications in Japan, one is drug including quasi drug, and another is food. In 2001, Foods with Health Claims (FHC) was established, a new regulatory framework that includes FoSHU and foods with nutrient function claims (FNFC). So FoSHU was incorporated into this system. In 2003, a Food safety committee was established which carry our safety assessment instead of the Ministry of Health, labor, and welfare (MHLW). Then, the FoSHU category was revised in 2005. Currently, FoSHU consists of four categories, namely, ordinary, qualified, standardized, and disease reduction claims. Finally, in 2009, the FODSU category was revised, which resulted in the introduction of five categories of FODSU:

- Medical foods for the ill
- Foods for a person having difficulty in swallowing
- Infant formula
- The formula for pregnant or lactating women and FoSHU

Primarily, the regulatory framework makes a rigid and consistent difference between drugs and foods. For example, the Pharmaceutical Affairs Act revised in 2014 banned terms like disease, prevention, treatment, cure, and diagnosis for food products, even in FOSHU and FNFC. Therefore, the establishment of the Food Safety Committee and Consumer Affair Agency takes charge to control the overall evaluation and approval of health claims on behalf of MHLW [83].

1.5.2. Nutraceuticals Regulation in India

In India, the Food Safety and Standard Authority of India (FSSAI) is responsible for enforcing nutraceuticals regulations under the Food Safety Standard Act (FSS Act, 2006). FSS Act, 2006 was enacted as a consolidated Act where all the eight Acts and Orders were repealed to establish a single regulatory body at the national level. This results in a paradigm shift from multi-level, multi-departmental control of food safety and nutrition to a single line of command with an outreach policy framework on food safety and nutrition. According to its preamble, the cornerstone of the Act is to ensure the availability of safe and wholesome food for human consumption. Ministry of Health and Family Welfare was nominated as the Administrative Ministry for FSS act, 2006. FSSAI was established as the Food Authority under the aegis of MoHFW in 2008 to enforce a provision of the new law. So, Food Authority is a single reference point to handle all the matters linked to food safety and standards.

Food Safety and standard (Health Supplements, Nutraceuticals, Food for Special Dietary Use, Food for Special Medical Purpose, Functional Food and Novel Food) Regulations, 2016 shall come into force by 1st January, 2018. As per regulations "Food business operator may extract, isolate and purify nutraceuticals from food or non-food sources, i.e., preparing amino acids and their derivatives by bacterial fermentation under controlled conditions; may prepare and sell the nutraceuticals in the food format of granules, powder, tablet, capsules, liquid, jelly or gel, semi-solids and other similar formats and may be packed in sachet, ampoule, bottle, and in other format as measured unit amounts excluding those formats that are intended for parenteral administration." These regulations cover eight categories of food:

- Food for special dietary use
- Health supplements
- Nutraceuticals
- Food for special medical purposes
- Food with added probiotic ingredients
- Food with added probiotic ingredients
- Novel food and Specialty food containing plant or botanical ingredients with a safe history of usage

Nutraceutical must contain any of the ingredients listed in Schedule I or Schedule II or Schedule IV or Schedule Vi or Schedule VII or schedule

VIII. Further, the amount of nutrients incorporated shall not exceed the Recommended Daily Allowance (RDA) as mentioned by the Indian Council of Medical Research (ICMR), and if standards are not specified, then standards are declared by Codex Alimentarius Commission shall practice. A person shall manufacture or sell nutraceuticals in India, which is not listed in these regulations only after taking the prior approval from Food Authority, but its safety has been established in India or any country.

For approval purposes, the food business operator shall apply to the food Authority along with a documented history of usage of at least 15 years in India or 30 years in the country of origin. Also, the person shall not use any ingredients which are not specified in Schedule VI as a nutraceutical. Ingredients that are listed in Schedule VI shall be used as nutraceuticals with standardization to specified marker compounds and at daily usage levels indicated therein. In case ingredients for which standardization of marker compound has been established shall use manufacturer specification or quality requirements and purity criteria as stated in regulation 3. Ingredients whose daily minimum and maximum usage limits have not been specified; then one can adopt usage level according to relevant scientific evidence and retain that document. Alongside, documented scientific data shall submit to the Food Authority as and when called for.

Besides this, labeling, presentation, and advertisement of the product shall not claim the nutraceutical has the property to prevent, cure, or treat any human disorder or disease. However, food business operator is allowed to quote a statement related to the structure or function or the general well-being of the body, and generally accepted scientific data should support the statement. In addition, the following information must be mentioned on the label of the nutraceutical package: the term "NUTRACEUTICAL"; its common name; quantity of each nutraceutical ingredient added in a product having a nutritional or physiological effect; the amount of nutrient must be specified in terms of percentage of the applicable RDA as indicate by ICMR; an advisory warning for recommended use, 'Not for medical use', any danger related to excess intake, precautions to be taken while consuming, any side effects, contraindications or possible product-drug interaction; remark stated that the product is required to be stored out of reach of children. Nevertheless, the manufacturer shall only use additives specified in Schedule VA or Schedule VE or Schedule VF in the formulation of nutraceuticals.

The International Alliance of Dietary/Food Supplement Association (IADSA) and Confederation of Indian Industries (CII) have formed a coalition under Resource Centre on Health Supplements and Nutraceuticals (RecHaN) where IADSA unify the association from 6 continents in the arena of food supplements to coordinate dialogues in Codex Alimentarius and assist the government and the private sector on relevant and effective regulation and policy. Its primary objective is to develop a science-based interactive platform for making policy and harmonized approach to legislation.

While CII is a business association of industries that provide a reference point to Indian Industry and the international business sector, CII via the Food and Agriculture Centre of Excellence (CII-FACE) gives an integrated approach to action-oriented programs and capacity building that addresses problems from farm to fork. It works in collaboration with the government, FSSAI, industry, and other stakeholders.

After the enforcement of nutraceutical regulation first time, the stakeholder may, in the beginning, face problems with the compliance requirements and need ongoing guidance. So, RecHaN, at its first initiative aid to strengthen the compliance efforts. This guidance is under FSS Act, 2006, and nutraceutical regulations. Above all, the document gives guidance on international regulation and best practices to ensure compliance in India.

1.6. Case Studies

Pineal gland secretes the essential hormone, melatonin, which exhibits endothelial safeguard activities like antioxidant property that scavenges free radicle by activating antioxidant enzymes defense system, improves the lipid and blood pressure, enhance bioavailability of nitric oxide (NO).

It also ameliorates functioning of vascular system in experimental hypertension, lessen intima infiltration and revitalize the production of nitric oxide. Thus, administration of melatonin may contribute to endothelium protective potential in many clinical conditions [84]. Meta-analysis of double-blind, placebo-controlled RCTs performed by [85] manifested the efficacy and safety of melatonin from external source in improving the nocturnal blood pressure. The meta-analysis included seven trials which comprise of 221 subjects who were taking melatonin 2-5 mg daily for 7-90 days. The result of study showed that controlled-release melatonin significantly reduce both night Systolic blood pressure [–6.1 (95% CI –10.7, –1.5) mmHg; $P = 0.009$] and night Diastolic blood pressure [–3.5 (95% CI –6.1, –0.9) mmHg; $P = 0.00$] compared to fast-release melatonin. Chronic intake of beta-blockers may be responsible for insomnia as it repress the endogenous nighttime melatonin secretion. However, melatonin supplementation may resolve this problem as explain by [86] in his study. The study design was Randomized, double-blind, placebo-controlled, parallel –group where 16 hypertensive patients treated with atenolol or metoprolol. The research outcome showed that hypertensive participants on 3 weeks melatonin supplementation during night time has significantly improves the quality of sleep by accelerating the total sleep time (+36 min; P=0.046), increasing sleeping efficacy (+7.6%; P=0.046) and declining sleep onset latency to stage 2 (-14 min; P= 0.001) in contrast to placebo.

Avemar™, fermental wheat germ extract (FWGE) nutraceutical has been approved as a "dietary food for special medical purpose for cancer patients" in Europe. The research directed by [87] reported the adjunct use of FWGE to treat high-risk skin melanoma patients. It is a randomized, pilot, phase II clinical trial where effectiveness of dacarbazine (DTIC) based supportive chemotherapy on survival parameters of stage III melanoma patients was compared with same treatment supplemented with intake of 8.5g FWGE for 1year. The FWGE supplemented group showed significant difference in both progression-free (PFS) and overall survival compared to control arm, determined after 7 year long term follow-up period. Therefore, Avemar supplementation as an adjuvant is recommended in stage III skin melanoma patients.

1.7. Challenges

Nutraceuticals have surpassed from conventional 'food guide pyramid' to recently termed 'Phytochemical intricate hierarchy' of bioactive components that are extracted or isolated from food matrix and further packed into various delivery material. These bioactive molecules, mainly, gingerol from ginger; capsaicin from capsicum; allicin from garlic, onion, shallots; lycopene from tomatoes and omega -3 from marine origin, have been reported to play a pivotal role in alleviating the symptoms of chronic disease and improve the physiological functioning of the body. However, the jeopardy of overdose of active compounds may be associated with ingestion. Therefore, it is always advisable to take supplement in limited amount or consult with your physician prior to its use.

Since the last decade, the nutraceutical industry has been branched out because public demand has been shifted from 'want' for preventive health into 'need' for preventive health. Regardless of the fact that nutraceutical industry is rapidly growing segment but also battling with various challenges notably [88]

- Variation in genesis of raw material: The frequently used nutraceutical herb, ginseng exemplifies how variability in origin of raw material affects the product quality and mislead the consumer. As ginseng has many varieties like wild ginseng, Brazilian ginseng, Southern ginseng, Indian ginseng, Malaysian ginseng, prickly ginseng and Pacific ginseng but neither of these belongs to genus Panax which is the actual ginseng. All of the named items are sold under the name of ginseng. However, only Panax ginseng encompass actual ginseng which includes South China ginseng (*P. notoginseng*), Korean ginseng (*P. ginseng*) and American ginseng (*P. quinquefolius*) [89].

- Paucity of information related to molecular interaction between active components of phytochemicals within the same plant.
- Variation in raw material processing
- Lack of authentication of active ingredients because of inaccessibility of marker compound or reference material. For example, it is found that certain varieties of star anise have high amount of anisatin, a neurotoxin, which may cause health hazard if they are mixed or misidentified with edible star anise [90].
- Incompetency and variability in quality control standards.
- Scarcity of well-established and proof-based clinical trials.
- Insufficient data reporting on safety and toxicity.
- Inadequacy in standardization of extraction procedures.
- Contamination or adulteration of nutraceuticals with other phytoconstituents like pyrrolizidine alkaloids, metals (As, Pb and Cd), mycotoxins, pharmaceutical drugs and pesticides. In certain cases, generally used dietary supplements such as weight loss and muscle-building supplements are more likely to be contaminated with banned substances in order to improve the clinical efficacy [91].

Besides above stated challenges, another one is the bioavailability of active compounds in order to exert their therapeutic effects in body. Upon ingestion, bioactive compounds in nutraceuticals come across gastrointestinal environment where they undergo various physiochemical changes and may lose their bioactivities completely. Therefore, effective delivery system is necessary to protect the activity of these bioactive agents from various physiochemical reactions occurring in gut. This improves the oral efficacy and biological function of nutraceuticals to maintain the overall health [79].

Therefore, to tackle the challenges of pioneering nutraceutical sector, it is important to follow well-establish clinical trial, reproducibility of result, meticulous quality control, toxicological and pharmacological testing, safety evaluation and unified regulations that can be applies to all nutraceuticals.

CONCLUSION

The portmanteau word "Nutraceutical" reside under the canopy of health and wellness category. The wellness category has two powerful pillars which maintain and modulate the overall health. The first pillar takes charge of "preventive action" that prevents people from various chronic & acute diseases before they set in. While the second pillar, focusses on "rejuvenation approach" to compensate the stress load of everyday life. Predominantly, nutraceutical products are popular among middle and upper-middle class population. The consumers inclined towards them due to their natural origin as they believe that nature is a true healer and safer. The bioactive compounds from plant and animal origin are the fundamental raw materials for the formulation of nutraceutical products such as garcinia, curcumin from turmeric, resveratrol from berries, allicin from garlic &onion, omega-3 fatty acids capsules from fish, Vitamin D from linolein (sheep wool) and so forth. The branches of nutraceutical products are proliferating in every direction of health and wellness category that ranges from products for improving physiological health to psychological health. Henceforth, "nutraceutical is an alternative to pharmaceutical." Nutraceutical industry is a merger of two robust industries i.e., Pharma industry and Food & beverages industry. The major key drivers for the maturation of ever emerging nutraceutical sector are the social media, advertisement, health practitioners, doctors and the consumer needs accompanied with increased knowledge and awareness about the beneficial effects of nutraceuticals products. However, nutraceutical industry is also facing various challenges like clinical efficacy, bioavailability, effective delivery system, regulation and safety. There are many nutraceutical ingredients (like melatonin, omega-3 fatty acids, lactoferrin, fish oil and fish protein isolate) which are intensively evaluated and found to be effective in improving the physiological health. In current scenario, there is requirement for unified definition of word "NUTRACEUTICAL" and single regulatory framework for import and export of nutraceutical products or ingredients throughout the world. It is also important that government should also support research in the nutraceutical arena by providing sufficient funds. Moreover, research centers should be well equipped with modern technological instrument to assess their safety and other aspects. Besides this, nutraceutical is a safer and nature-oriented approach to restore the deteriorated health and keep you away from synthetic drugs that's why it is known as alternative to pharmaceuticals.

REFERENCES

[1] Li, W. W. (2019). *Eat to Beat Disease: The New Science of how Your Body Can Heal Itself.* Hachette UK.

[2] DeFelice S. L. The nutraceutical revolution: its impact on food industry R&D. *Trends Food Sci Technol.* 1995;6:59–61.

[3] Maddi V. S., Aragade P. D., Digge V. G., Nitalikar M. N. Importance of nutraceuticals in health management. *Pharmacognosy Reviews.* 2007;1(2):377.

[4] Santini A., Tenore G. C., Novellino E. Nutraceuticals: a paradigm of proactive medicine. *Eur J Pharm Sci.* 2017;96:56–61.

[5] de Lumen, B. O., Thompson, S., & Odegard, W. J. (1993). Sulfur amino acid-rich proteins in acha (Digitaria exilis), a promising underutilized African cereal. *Journal of Agricultural and Food Chemistry,* 41(7), 1045-1047.

[6] Jideani, I. A., & Jideani, V. A. (2011). Developments on the cereal grains *Digitaria exilis* (acha) and *Digitaria iburua* (iburu). *Journal of food science and technology,* 48(3), 251-259.

[7] Shen, X. L., Zhao, T., Zhou, Y., Shi, X., Zou, Y., & Zhao, G. (2016). Effect of Oat β-Glucan Intake on Glycaemic Control and Insulin Sensitivity of Diabetic Patients: A Meta-Analysis of Randomized Controlled Trials. *Nutrients,* 8(1), 39.

[8] Whitehead, A., Beck, E. J., Tosh, S., & Wolever, T. M. (2014). Cholesterol-lowering effects of oat β-glucan: a meta-analysis of randomized controlled trials. *The American journal of clinical nutrition,* 100(6), 1413–1421.

[9] Mälkki, Y., & Virtanen, E. (2001). Gastrointestinal Effects of Oat Bran and Oat Gum: A Review. *LWT - Food Science and Technology,* 34(6), 337–347. doi:10.1006/fstl.2001.0795.

[10] Razavi, S. M., Zarrini, G., Molavi, G., & Ghasemi, G. (2011). Bioactivity of malva sylvestris L., a medicinal plant from Iran. *Iranian journal of basic medical sciences,* 14(6), 574–579.

[11] Barros, L., Carvalho, A. M., & Ferreira, I. C. (2010). Leaves, flowers, immature fruits and leafy flowered stems of Malva sylvestris: a comparative study of the nutraceutical potential and composition. *Food and Chemical Toxicology,* 48(6), 1466-1472.

[12] Prudente, A. S., Loddi, A. M., Duarte, M. R., Santos, A. R., Pochapski, M. T., Pizzolatti, M. G., ... & Cabrini, D. A. (2013). Pre-clinical anti-inflammatory aspects of a cuisine and medicinal millennial herb: *Malva sylvestris* L. *Food and chemical toxicology,* 58, 324-331.

[13] Chen, M., Rao, Y., Zheng, Y., Wei, S., Li, Y., Guo, T., & Yin, P. (2014). Association between soy isoflavone intake and breast cancer risk for pre-and post-menopausal women: a meta-analysis of epidemiological studies. *PloS one,* 9(2), e89288.

[14] Meghwal, M., & Sahu, C. K. (2015). Soy isoflavonoids as nutraceutical for human health: an update. *Journal of Cell Science & Therapy*, 6(1), 1.

[15] Martínez-Villaluenga, C., Peñas, E., & Hernández-Ledesma, B. (2020). Pseudocereal grains: Nutritional value, health benefits and current applications for the development of gluten-free foods. *Food and Chemical Toxicology,* 137, 111178.

[16] Arya, S. S., Salve, A. R., & Chauhan, S. (2016). Peanuts as functional food: a review. *Journal of food science and technology,* 53(1), 31-41.

[17] Lopes, R. M., Agostini-Costa, T. D. S., Gimenes, M. A., & Silveira, D. (2011). Chemical composition and biological activities of Arachis species. *Journal of agricultural and food chemistry,* 59(9), 4321-4330.

[18] Gülçin, İ. (2010). Antioxidant properties of resveratrol: a structure–activity insight. *Innovative Food Science & Emerging Technologies,* 11(1), 210-218.

[19] Hector, K. L., Lagisz, M., & Nakagawa, S. (2012). The effect of resveratrol on longevity across species: a meta-analysis. *Biology letters,* 8(5), 790–793.

[20] Moree, S. S., & Rajesha, J. (2011). Secoisolariciresinol diglucoside-a phytoestrogen nutraceutical of flaxseed: synthesis and evaluation of antioxidant potency. *Free Radicals and Antioxidants,* 1(4), 31-38.

[21] Kaur, G., Invally, M., Khan, M. K., & Jadhav, P. (2018). A nutraceutical combination of Cinnamomum cassia & Nigella sativa for Type 1 diabetes mellitus. *Journal of Ayurveda and integrative medicine,* 9(1), 27-37.

[22] Yue, L. J., Zhu, X. Y., Li, R. S., Chang, H. J., Gong, B., Tian, C. C., ... & Wang, D. J. (2019). S-allyl-cysteine sulfoxide (alliin) alleviates myocardial infarction by modulating cardiomyocyte necroptosis and autophagy. *International journal of molecular medicine,* 44(5), 1943-1951.

[23] Zare, E., Alirezaei, A., Bakhtiyari, M., & Mansouri, A. (2019). Evaluating the effect of garlic extract on serum inflammatory markers of peritoneal dialysis patients: a randomized double-blind clinical trial study. *BMC nephrology,* 20(1), 26.

[24] Takasu, J., Uykimpang, R., Sunga, M. A., Amagase, H., & Niihara, Y. (2006). Aged garlic extract is a potential therapy for sickle-cell anemia. *The Journal of nutrition,* 136(3), 803S-805S.

[25] Shang, A., Cao, S. Y., Xu, X. Y., Gan, R. Y., Tang, G. Y., Corke, H., Mavumengwana, V., & Li, H. B. (2019). Bioactive Compounds and Biological Functions of Garlic (*Allium sativum* L.). *Foods* (Basel, Switzerland), 8(7), 246.

[26] Semwal, R. B., Semwal, D. K., Combrinck, S., & Viljoen, A. M. (2015). Gingerols and shogaols: Important nutraceutical principles from ginger. *Phytochemistry,* 117, 554-568.

[27] Chanda, S., Rakholiya, K., Dholakia, K., & Baravalia, Y. (2013). Antimicrobial, antioxidant, and synergistic properties of two

nutraceutical plants: *Terminalia catappa* L. and *Colocasia esculenta* L. *Turkish Journal of Biology,* 37(1), 81-91.

[28] Garba, L., & Hafsat, G. L. (2013). Bacteriostatic Effect of *Terminalia catappa* Leaves Extract on Clinical Isolates of Gram Negative Bacteria. *Asian Journal of Applied Sciences,* 1(4).

[29] Terças, A. G., Monteiro, A. S., Moffa, E. B., Dos Santos, J., de Sousa, E. M., Pinto, A., Costa, P., Borges, A., Torres, L., Barros Filho, A., Fernandes, E. S., & Monteiro, C. A. (2017). Phytochemical Characterization of *Terminalia catappa* Linn. Extracts and Their antifungal Activities against Candida spp. *Frontiers in microbiology,* 8, 595.

[30] Lo, C. W., Huang, H. P., Chan, K. C., Wu, C. H., & Wang, C. J. (2010). *Hibiscus sabdariffa* extract induced apoptosis of proliferating smooth muscle cell. *Journal of food biochemistry,* 34(3), 549-563.

[31] Chen, J. H., Wang, C. J., Wang, C. P., Sheu, J. Y., Lin, C. L., & Lin, H. H. (2013). *Hibiscus sabdariffa* leaf polyphenolic extract inhibits LDL oxidation and foam cell formation involving up-regulation of LXRα/ABCA1 pathway. *Food chemistry,* 141(1), 397-406.

[32] Chiu, C. T., Chen, J. H., Chou, F. P., & Lin, H. H. (2015). *Hibiscus sabdariffa* Leaf Extract Inhibits Human Prostate Cancer Cell Invasion via Down-Regulation of Akt/NF-kB/MMP-9 Pathway. *Nutrients,* 7(7), 5065–5087.

[33] Sabzghabaee, A. M., Ataei, E., Kelishadi, R., Ghannadi, A., Soltani, R., Badri, S., & Shirani, S. (2013). Effect of *Hibiscus sabdariffa* Calices on Dyslipidemia in Obese Adolescents: A Triple-masked Randomized Controlled Trial. *Materia socio-medica,* 25(2), 76–79.

[34] Gurrola-Díaz, C. M., García-López, P. M., Sánchez-Enríquez, S., Troyo-Sanromán, R., Andrade-González, I., & Gómez-Leyva, J. F. (2010). Effects of Hibiscus sabdariffa extract powder and preventive treatment (diet) on the lipid profiles of patients with metabolic syndrome (MeSy). *Phytomedicine,* 17(7), 500-505.

[35] Ademiluyi, A. O., & Oboh, G. (2013). Aqueous extracts of Roselle (*Hibiscus sabdariffa* Linn.) varieties inhibit α-amylase and α-

glucosidase activities in vitro. *Journal of medicinal food,* 16(1), 88-93.

[36] Magni, C., Sessa, F., Accardo, E., Vanoni, M., Morazzoni, P., Scarafoni, A., & Duranti, M. (2004). Conglutin γ, a lupin seed protein, binds insulin in vitro and reduces plasma glucose levels of hyperglycemic rats. *The Journal of nutritional biochemistry,* 15(11), 646-650.

[37] Vital, D. A. L., De Mejía, E. G., Dia, V. P., & Loarca-Piña, G. (2014). Peptides in common bean fractions inhibit human colorectal cancer cells. *Food chemistry,* 157, 347-355.

[38] Martinez-Villaluenga, C., Bringe, N. A., Berhow, M. A., & Gonzalez de Mejia, E. (2008). β-Conglycinin embeds active peptides that inhibit lipid accumulation in 3T3-L1 adipocytes in vitro. *Journal of agricultural and food chemistry,* 56(22), 10533-10543.

[39] Jenkins, D. J., Mirrahimi, A., Srichaikul, K., Berryman, C. E., Wang, L., Carleton, A., ... & Kris-Etherton, P. M. (2010). Soy protein reduces serum cholesterol by both intrinsic and food displacement mechanisms. *The Journal of nutrition,* 140(12), 2302S-2311S.

[40] Lee, J. E., Bae, I. Y., Lee, H. G., & Yang, C. B. (2006). Tyr-Pro-Lys, an angiotensin I-converting enzyme inhibitory peptide derived from broccoli (*Brassica oleracea* Italica). *Food chemistry,* 99(1), 143-148.

[41] Takahashi, M., Moriguchi, S., Yoshikawa, M., & Sasaki, R. (1994). Isolation and characterization of oryzatensin: a novel bioactive peptide with ileum-contracting and immunomodulating activities derived from rice albumin. *Biochemistry and molecular biology international,* 33(6), 1151-1158.

[42] Uraipong, C., & Zhao, J. (2018). In vitro digestion of rice bran proteins produces peptides with potent inhibitory effects on α-glucosidase and angiotensin I converting enzyme. *Journal of the Science of Food and Agriculture,* 98(2), 758-766.

[43] Cudennec, B., Ravallec-Plé, R., Courois, E., & Fouchereau-Peron, M. (2008). Peptides from fish and crustacean by-products hydrolysates stimulate cholecystokinin release in STC-1 cells. *Food Chemistry,* 111(4), 970-975.

[44] Nobile, V., Duclos, E., Michelotti, A., Bizzaro, G., Negro, M., & Soisson, F. (2016). Supplementation with a fish protein hydrolysate (*Micromesistius poutassou*): effects on body weight, body composition, and CCK/GLP-1 secretion. *Food & nutrition research,* 60(1), 29857.

[45] Hatanaka, A., Miyahara, H., Suzuki, K. I., & Sato, S. (2009). Isolation and identification of antihypertensive peptides from antarctic krill tail meat hydrolysate. *Journal of food science,* 74(4), H116-H120.

[46] Li, Y., Zhou, J., Huang, K., Sun, Y., & Zeng, X. (2012). Purification of a novel angiotensin I-converting enzyme (ACE) inhibitory peptide with an antihypertensive effect from loach (*Misgurnus anguillicaudatus*). *Journal of agricultural and food chemistry,* 60(5), 1320-1325.

[47] Huang, S. L., Hung, C. C., Jao, C. L., Tung, Y. S., & Hsu, K. C. (2014). Porcine skin gelatin hydrolysate as a dipeptidyl peptidase IV inhibitor improves glycemic control in streptozotocin-induced diabetic rats. *Journal of Functional Foods,* 11, 235-242.

[48] Kris-Etherton, P. M., Richter, C. K., Bowen, K. J., Skulas-Ray, A. C., Jackson, K. H., Petersen, K. S., & Harris, W. S. (2019). Recent Clinical Trials Shed New Light on the Cardiovascular Benefits of Omega-3 Fatty Acids. *Methodist DeBakey cardiovascular journal,* 15(3), 171–178. https://doi.org/10.14797/mdcj-15-3-171.

[49] Heshmati, J., Morvaridzadeh, M., Maroufizadeh, S., Akbari, A., Yavari, M., Amirinejad, A., & Sepidarkish, M. (2019). Omega-3 fatty acids supplementation and oxidative stress parameters: A systematic review and meta-analysis of clinical trials. *Pharmacological Research,* 149, 104462.

[50] Karkos, P. D., Leong, S. C., Karkos, C. D., Sivaji, N., & Assimakopoulos, D. A. (2011). Spirulina in clinical practice: evidence-based human applications. *Evidence-based complementary and alternative medicine: eCAM,* 2011, 531053. https://doi.org/ 10.1093/ecam/nen058.

[51] Moradi, S., Ziaei, R., Foshati, S., Mohammadi, H., Nachvak, S. M., & Rouhani, M. H. (2019). Effects of spirulina supplementation on obesity: A systematic review and meta-analysis of randomized clinical trials. *Complementary therapies in medicine,* 47, 102211.

[52] Ngo, D. H., & Kim, S. K. (2013). Sulfated polysaccharides as bioactive agents from marine algae. *International Journal of Biological Macromolecules,* 62, 70-75.

[53] Tanna, B., & Mishra, A. (2019). Nutraceutical potential of seaweed polysaccharides: Structure, bioactivity, safety, and toxicity. *Comprehensive Reviews in Food Science and Food Safety,* 18(3), 817-831.

[54] Miyashita, K., Nishikawa, S., Beppu, F., Tsukui, T., Abe, M., & Hosokawa, M. (2011). The allenic carotenoid fucoxanthin, a novel marine nutraceutical from brown seaweeds. *Journal of the Science of Food and Agriculture,* 91(7), 1166-1174.

[55] Miyashita, K., Beppu, F., Hosokawa, M., Liu, X., & Wang, S. (2020). Nutraceutical characteristics of the brown seaweed carotenoid fucoxanthin. *Archives of Biochemistry and Biophysics,* 108364.

[56] Černá, M. (2011). Seaweed proteins and amino acids as nutraceuticals. In *Advances in food and nutrition research* (Vol. 64, pp. 297-312). Academic Press.

[57] Steijns, J. M. (2001). Milk ingredients as nutraceuticals. *International Journal of Dairy Technology,* 54(3), 81-88.

[58] Ma, X., Lin, X., Zhong, T., & Xie, F. (2019). Evaluation of the efficacy of casein phosphopeptide-amorphous calcium phosphate on remineralization of white spot lesions in vitro and clinical research: a systematic review and meta-analysis. *BMC oral health,* 19(1), 1-11.

[59] Mizuno, S., Matsuura, K., Gotou, T., Nishimura, S., Kajimoto, O., Yabune, M., ... & Yamamoto, N. (2005). Antihypertensive effect of casein hydrolysate in a placebo-controlled study in subjects with high-normal blood pressure and mild hypertension. *British Journal of Nutrition,* 94(1), 84-91.

[60] Xue, L., Wang, X., Hu, Z., Wu, Z., Wang, L., Wang, H., & Yang, M. (2018). Identification and characterization of an angiotensin-converting enzyme inhibitory peptide derived from bovine casein. *Peptides,* 99, 161-168.

[61] Lin, K., Ma, Z., Ramachandran, M., De Souza, C., Han, X., & Zhang, L. W. (2020). ACE inhibitory peptide KYIPIQ derived from yak milk casein induces nitric oxide production in HUVECs and diffuses via a transcellular mechanism in Caco-2 monolayers. *Process Biochemistry.*

[62] Iglesias-Figueroa, B. F., Espinoza-Sánchez, E. A., Siqueiros-Cendón, T. S., & Rascón-Cruz, Q. (2019). Lactoferrin as a nutraceutical protein from milk, an overview. *International Dairy Journal,* 89, 37-41.

[63] Czosnykowska-Łukacka, M., Orczyk-Pawiłowicz, M., Broers, B., & Królak-Olejnik, B. (2019). Lactoferrin in human milk of prolonged lactation. *Nutrients,* 11(10), 2350.

[64] Bruni, N., Capucchio, M. T., Biasibetti, E., Pessione, E., Cirrincione, S., Giraudo, L., ... & Dosio, F. (2016). Antimicrobial activity of lactoferrin-related peptides and applications in human and veterinary medicine. *Molecules,* 21(6), 752.

[65] Assocham 2017. Available at: https://mrssindia.com/uploads/reports/pdf/assocham-knowledge-reporton-nutracuticals-released-at-3rd-national-symposium-6.pdf (accessed June 12, 2019).

[66] Cicero, A., Fogacci, F., Morbini, M., Colletti, A., Bove, M., Veronesi, M., Giovannini, M., & Borghi, C. (2017). Nutraceutical Effects on Glucose and Lipid Metabolism in Patients with Impaired Fasting Glucose: A Pilot, Double-Blind, Placebo-Controlled, Randomized Clinical Trial on a Combined Product. *High blood pressure & cardiovascular prevention: the official journal of the Italian Society of Hypertension,* 24(3), 283–288. https://doi.org/10.1007/s40292-017-0206-3.

[67] Welch, R. W., Antoine, J. M., Berta, J. L., Bub, A., de Vries, J., Guarner, F., Hasselwander, O., Hendriks, H., Jäkel, M., Koletzko, B. V., Patterson, C. C., Richelle, M., Skarp, M., Theis, S., Vidry, S.,

Woodside, J. V., & International Life Sciences Institute Europe Functional Foods Task Force (2011). Guidelines for the design, conduct and reporting of human intervention studies to evaluate the health benefits of foods. *The British journal of nutrition,* 106 Suppl 2, S3–S15. https://doi.org/10.1017/S0007114511003606.

[68] Andrew, R., & Izzo, A. A. (2017). Principles of pharmacological research of nutraceuticals. *British Journal of Pharmacology,* 174(11), 1177.

[69] Visioli F. (2012). Can experimental pharmacology be always applied to human nutrition?. *International journal of food sciences and nutrition,* 63 Suppl 1, 10–13.

[70] Izzo, A. A., Hoon-Kim, S., Radhakrishnan, R., & Williamson, E. M. (2016). A Critical Approach to Evaluating Clinical Efficacy, Adverse Events and Drug Interactions of Herbal Remedies. *Phytotherapy research: PTR,* 30(5), 691–700.

[71] McParlin, C., O'Donnell, A., Robson, S. C., Beyer, F., Moloney, E., Bryant, A., ... & Norman, J. (2016). Treatments for hyperemesis gravidarum and nausea and vomiting in pregnancy: a systematic review. *JAMA,* 316(13), 1392-1401.

[72] Pittler, M. H., & Ernst, E. (2012). Horse chestnut seed extract for chronic venous insufficiency. *Cochrane database of systematic reviews,* (11).

[73] Karsch-Völk, M., Barrett, B., Kiefer, D., Bauer, R., Ardjomand-Woelkart, K., & Linde, K. (2014). Echinacea for preventing and treating the common cold. *Cochrane Database of Systematic Reviews,* (2).

[74] Leach, M. J., & Page, A. T. (2015). Herbal medicine for insomnia: A systematic review and meta-analysis. *Sleep medicine reviews,* 24, 1-12.

[75] Onakpoya, I., Terry, R., & Ernst, E. (2010). The use of green coffee extract as a weight loss supplement: a systematic review and meta-analysis of randomised clinical trials. *Gastroenterology research and practice,* 2011.

[76] Jurgens, T. M., Whelan, A. M., Killian, L., Doucette, S., Kirk, S., & Foy, E. (2012). Green tea for weight loss and weight maintenance in overweight or obese adults. *Cochrane database of systematic reviews,* (12).

[77] Miroddi, M., Navarra, M., Calapai, F., Mancari, F., Giofrè, S. V., Gangemi, S., & Calapai, G. (2015). Review of clinical pharmacology of Aloe vera L. in the treatment of psoriasis. *Phytotherapy Research,* 29(5), 648-655.

[78] Marcus, D. M. (2016). Dietary supplements: what's in a name? What's in the bottle?. *Drug testing and analysis,* 8(3-4), 410-412.

[79] Ting, Y., Jiang, Y., Ho, C. T., & Huang, Q. (2014). Common delivery systems for enhancing in vivo bioavailability and biological efficacy of nutraceuticals. *Journal of Functional Foods,* 7, 112-128.

[80] Directive 2002/46/EC of the European Parliament and of the Council of 10 June 2002 on the approximation of the laws of the Member States relating to food supplements. *Official Journal L* 183, 12/07/2002, pp. 0051–0057.

[81] Hoffman, F. A. (2001). Regulation of dietary supplements in the United States: understanding the dietary supplement health and education act. *Clinical obstetrics and gynecology,* 44(4), 780-788.

[82] Gulati, O. P., & Ottaway, P. B. (2006). Legislation relating to nutraceuticals in the European Union with a particular focus on botanical-sourced products. *Toxicology,* 221(1), 75-87.

[83] Umegaki, K. (2016). Regulatory Aspects of Nutraceuticals: Japanese Perspective. In *Nutraceuticals* (pp. 933-940). Academic Press.

[84] Rodella, L. F., Favero, G., Foglio, E., Rossini, C., Castrezzati, S., Lonati, C., & Rezzani, R. (2013). Vascular endothelial cells and dysfunctions: role of melatonin. *Frontiers in bioscience* (Elite edition), 5, 119–129.

[85] Grossman, E., Laudon, M., & Zisapel, N. (2011). Effect of melatonin on nocturnal blood pressure: meta-analysis of randomized controlled trials. *Vascular health and risk management,* 7, 577–584.

[86] Scheer, F. A., Morris, C. J., Garcia, J. I., Smales, C., Kelly, E. E., Marks, J., ... & Shea, S. A. (2012). Repeated melatonin

supplementation improves sleep in hypertensive patients treated with beta-blockers: a randomized controlled trial. *Sleep,* 35(10), 1395-1402.

[87] Demidov, L. V., Manziuk, L. V., Kharkevitch, G. Y., Pirogova, N. A., & Artamonova, E. V. (2008). Adjuvant fermented wheat germ extract (Avemar™) nutraceutical improves survival of high-risk skin melanoma patients: a randomized, pilot, phase II clinical study with a 7-year follow-up. *Cancer biotherapy & radiopharmaceuticals,* 23(4), 477-482.

[88] Gupta, R. C. (Ed.). (2016). *Nutraceuticals: efficacy, safety and toxicity.* Academic Press.

[89] Foster, S. (2016). Towards understanding of ginseng adulteration: the tangled web of names, history, trade and perception. *Herbal Gram,* 111, 37-57.

[90] Mathon, C., Bongard, B., Duret, M., Ortelli, D., Christen, P., & Bieri, S. (2013). Analysis of the neurotoxin anisatin in star anise by LC-MS/MS. *Food Additives & Contaminants: Part A,* 30(9), 1598-1605.

[91] Mathews N. M. (2018). Prohibited Contaminants in Dietary Supplements. *Sports health,* 10(1), 19–30. https://doi.org/10.1177/1941738117727736.

In: Nutraceuticals
Editors: A. Kumari and G. Singh
ISBN: 978-1-68507-488-3

Chapter 3

EXTRACTION AND ISOLATION OF NUTRACEUTICALS AND PERSPECTIVE FOR FOOD APPLICATIONS

***Deepika Pawar*[1], *Nidhi Bhardwaj*[2], *Anita Kumari*[1,*] *and Ashwani Kumar*[1]**

[1]Department of Nutrition Biology, School of Interdisciplinary and Applied Life Sciences,
Central University of Haryana, Mahendergarh, Haryana, India
[2]Department of Human Genetics and Molecular Medicine, School of Health Science,
Central University of Punjab, Bathinda, Punjab, India

ABSTRACT

Nutraceuticals are medicinal foods that enhance health, modulate immunity, and thereby prevent and cure specific diseases. Food products like garlic, ginger, echinacea, ginseng, onion, senna, turmeric, kale, etc., contain high content of nutraceuticals. Sources of nutraceuticals other

* Corresponding Author's E-mail: anikum@cuh.ac.in.

than plant tissues are prokaryotes, fungi, algae, and microalgae. To commercialize, nutraceuticals are extracted and isolated. Methods used for extraction are soxhlet extraction, steam distillation extraction, maceration, percolation, digestion, and extraction, infusion extraction, decoction, reflux extraction, tincture extraction, enfleurage, solvent extraction, pressurized liquid extraction, subcritical fluid extraction, supercritical extraction, microwave-assisted extraction, ultrasonic-assisted extraction, enzyme assisted extraction and instant controlled pressure drop-assisted extraction, etc. After extraction, various methods were applied for the isolation of nutraceuticals, such as different types of chromatography (gas chromatography, ion-exchange chromatography, HPLC, TLC, CC, size exclusion chromatography, etc.). In the present scenario, nutraceuticals are used for various medicinal and health benefits to cure certain diseases or health conditions (such as to prevent chronic diseases, improve health, delaying the aging process, increasing life expectancy, or supporting the structure or function of the body). This chapter's main key focus will be on extraction methods and isolation of nutraceuticals via conventional, non-conventional methods and the application of nutraceuticals.

1. Introduction

Nutraceuticals are functional foods having health benefits and medicinal effects. Plant tissues, prokaryotes, fungi, algae, and microalgae are the source of nutraceuticals. "The scientist to be successful must be a sort of Janus looking at the same time to the future and the past"— Francisco Grande. This approach leads to the tremendous development in methods for extraction and isolation of nutraceuticals along with time. Extraction methods are categorized as:

(1) Conventional methods
(2) Non-conventional methods or Modern time methods

In conventional methods, maceration, percolation, digestion, infusion, decoction, steam distillation extraction, soxhlet extraction, solvent extraction, tincture extraction, enfleurage and reflux extraction etc. are included. Whereas in non-conventional methods, pressurized liquid extraction, supercritical fluid extraction, microwave-assisted extraction, ultrasonic-assisted extraction, enzyme-assisted extraction, and pulse electric field-assisted extraction are included.

Chromatography is used to isolate nutraceuticals. Nutraceuticals can be used in the form of medicines and have zero side effects, as they are natural components extracted from living tissues. Commercially available nutraceuticals in the market are tablets, capsules, soft gels, gelcaps, liquids, or powders.

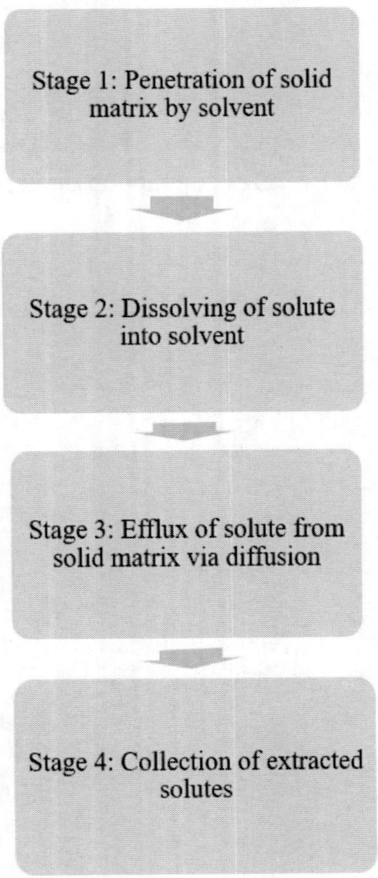

Figure 1. Stages of extraction.

2. METHODS OF EXTRACTION

Extraction of bioactive compounds from the sources of commonly known nutraceuticals is termed nutraceutical extraction. Figure 1 shows four stages of extraction: penetration of solid matrix by solvent, dissolving of solute into the solvent, efflux of solute from solid matrix via diffusion, and collection of extracted solutes.

2.1. Steps of Extraction

Several factors enhance extraction efficiency like properties of extraction solvent, the particle size of materials, time consumed in extraction, the temperature during extraction process, and solvent to solid ratio.

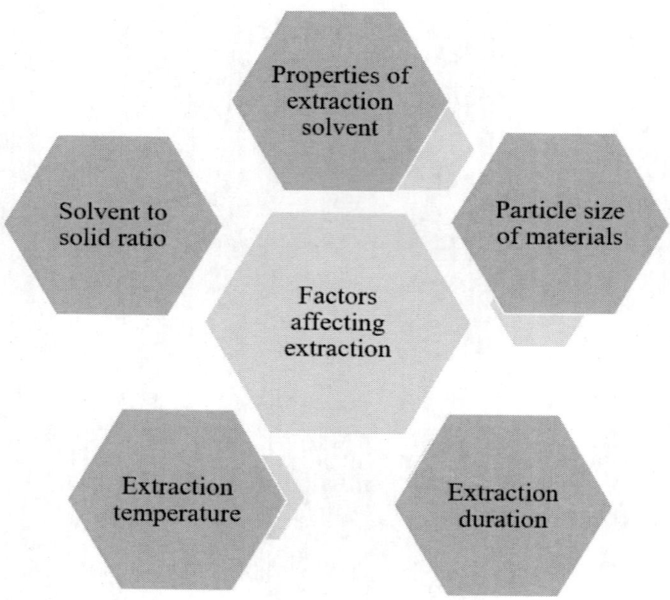

Figure 2. Factors affecting extraction.

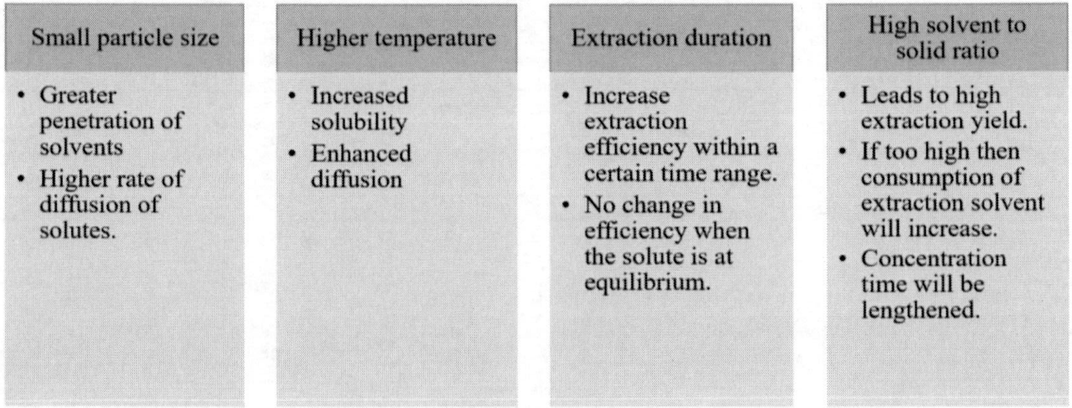

Figure 3. Basis of several techniques.

Factors affecting extraction: Figure 2 explains about factors affecting extraction.

Before selecting a particular solvent, certain elements considered include A) Selectivity, B) Solubility, C) Cost, and D) Safety. Alcohols are universal solvents for extraction processes. Small or fine particle size gives a more significant result in extraction as it allows higher degree of penetration of solvents and solute diffusion. Figure 3 illustrates the basis of several techniques. Solubility and diffusion are enhanced with increased temperature. The high temperature might destroy the resultant extract. A rise in efficiency with increment in extraction time length is a time bound factor. After attainment of equilibrium by solute, efficiency will not be affected by the increase in extraction time. A high solvent to solid ratio leads to more extraction products; beyond the limit, it increases the consumption of extraction solvent [9].

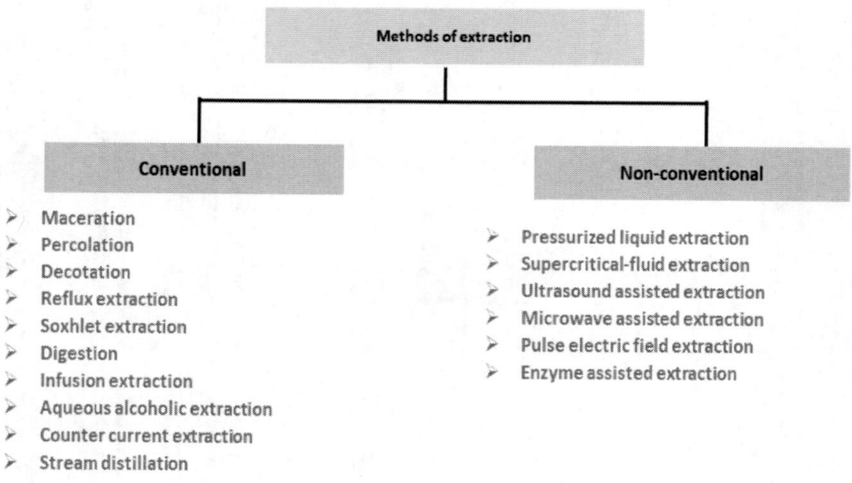

Figure 4. Various methods of extraction of nutraceuticals.

3. Conventional Methods

Various methods of extraction of nutraceuticals are shown in Figure 4. Conventional methods involve standard methods for extraction of nutraceuticals. For the extraction of bioactive compounds from plants and animal sources, conventional methods are used.

3.1. Maceration

Maceration is one of the simplest extraction methods, but it is very time-consuming and less efficient in extraction. The use of liquid makes the tissue softer and broken into pieces. Because of this, the cellular metabolites present in the tissue get leached out in the liquid. Thermolabile compounds are extracted using this method [9].

Nowadays, the microbial maceration technique extracts various bioactive compounds such as phenolics, flavonoids, antioxidants, tannins, and saponins from the fruits, vegetables, legumes, cereals, and pulses. Microbial maceration is very useful because it allows the extraction of the compounds from different sources. It is simpler, low-cost, and energy-efficient [10]. Fruits and vegetables are rich in phytochemicals, nutrients, and other compounds vital for human nutrition [11]. The microbial maceration technique is used to extract these phytochemical products. Industries utilize these phytochemicals and produce other value-added products [11]. Maceration significantly increases the total soluble phenolics, antioxidant, and flavonoids content in *Citrus sinensis*, cabernet sauvignon grapes, tempranillo grapes, kiwifruit, green olive, varicoloured olives, black olives and *Brassica pekinensis* Skeels [12, 13, 14, 15, 16]. Moreover, there is an increase in antioxidant activity of *Citrus sinensis, Basella rubra* [15, 17], flavonoids in cabernet sauvignon grapes, black mulberry [12, 18], pigment composition in *Citrus sinensis*, cabernet sauvignon grapes, tempranillo grapes, black mulberry [12, 18, 15] in comparison to control samples. These wastes obtained from food processing industries are incredibly diverse, novel, natural, and economic sources of numerous phytochemicals [19]. Agro-industrial waste usually contains numerous phytochemicals, which are nutraceuticals.

3.2. Percolation (Per – Through, Collare -to Strain)

Percolation means filtration of any compound. Percolation is more efficient than maceration as it is a continuous process in which the saturated solvent is constantly being replaced by fresh solvent [9]. The vegetable drug is taken in powdered form in the percolator, and the solvent percolates through it. However, some materials (e.g., ginger) can be packed in a dry state into the percolator, which may cause difficulties with other drugs because the addition of solvent may result in swelling of the material. This reduces or blocks the flow of the solvent, thus affecting the extraction process badly [20]. Zhang et al. compared the percolation and refluxing extraction methods to extract *Undariapinnatifida* and found no significant difference in the two methods [21]. Gao developed another percolation method by using the extraction rate of sinomenine and ephedrine hydrochloride; the medicine is soaked for 70% ethanol for 24h and then percolated with times the amount of 70% ethanol. The transfer rates of sinomenine and ephedrine hydrochlorides were 78.23 and 76.92%, respectively [22].

3.3. Decoction

A large number of water-soluble impurities are present in an extract from decoction. It is used to isolate bioactive compounds from complex woody parts such as roots, bark, and stems. In this process, mashing is done to make sure that maximum dissolution takes place. This method is not suitable for the extraction of thermolabile or volatile components [9].

3.4. Reflux Extraction

Reflux extraction is a more efficient and less time-consuming method than percolation or maceration [9]. The highest yield of the natural bio-insecticidal didehydrostemofoline is obtained by refluxing with 70% alcohol from *Stemonacollinsiae* root.

In a study, the extraction efficiency of active ingredients baicalin and puerarin concluded from a TCM compound were compared by using decoction and reflux. The reflux method results are better than decoction as the highest yield is obtained from reflux by using 60% ethanol as the extraction solvent [27].

3.5. Soxhlet Extraction

The Soxhlet extraction method utilizes the principle of reflux and siphoning to continuously extract the herb with fresh solvent. Thus, it is an automatic continuous extraction method. As a result, the efficiency of this method is higher than other methods, and less time and solvent consuming. The high temperature and long extraction time in the Soxhlet extraction will increase the possibilities of thermal degradation [9].

Tea catechins get degraded in Soxhlet extraction because of the high temperature used in the process. Also, the concentrations of both total polyphenols and total alkaloids from the Soxhlet extraction method at 70 °C are decreased [28, 29].

3.6. Digestion

Digestion is similar to maceration; the heat used during this process is gentle and moderate (35C-40°C temperature). The solvent efficiency of the menstruum is thereby increased by this method [30]. Usually, this process opts for the complex parts of plants containing poorly soluble substances.

3.7. Infusion Extraction

These are dilute solutions of the readily soluble constituents of crude drugs [31]. The maceration of crude drugs with cold or boiling water for a short period provides fresh infusions.

3.8. Aqueous Alcoholic Extraction by Fermentation

The fermentation technique is used to prepare some ayurvedic medicines such as asava and arista to extract the active principles. The crude drug (powdered or decoction) is soaked for a specified period to ferment and produce alcohol which facilitates the extraction of active constituents from the plant material. The alcohol thus generated also serves as a preservative.

In large-scale manufacture, wooden vats, porcelain jars, or metal vessels are used in place of earthen vessels. *karpurasava, kanakasava, dasmularista* are prepared using this method; however, Ayurveda is not yet standardized. Nevertheless, this technique is suitable for the production of herbal drug extracts [32].

3.9. Counter-Current Extraction

In counter-current extraction (CCE), a fine slurry is produced by pulverizing the raw materials using toothed disc disintegrators. The material to be extracted moved in one direction within a cylindrical extractor containing extraction solvent in this process. As the starting material moves further, it becomes more concentrated. The quantities of solvent, material, and flow rates are optimized for complete extraction. This process is highly efficient, less time-consuming, and poses no risk from high temperature. The highly concentrated extract comes out at one end of the extractor while the marc (practically free of visible solvent) falls out from the other end.

This extraction process has significant advantages:

i) Minimal solvent is used compared to other extraction methods to extract a unit plant material.
ii) This process is done at room temperature, thus making it most suitable for thermolabile compounds.
iii) The heat produced during comminution is neutralized by the water used for pulverization.
iv) The extraction procedure is more efficient and effective than continuous hot extraction [33].

3.10. Steam Distillation

Distillation is the most popular, widely used and cost-effective method to produce essential oils throughout the world. Distillation of aromatic plants means vaporizing the oils from the plant cellular membranes in the presence of moisture by applying high temperatures and then cooling the vapor mixture. This separates the oil from the water because of the immiscibility and density of the essential oil with respect to water.

The following considerations dictate the choice of a specific process for the extraction of essential oil:

a) Sensitivity of the essential oil to the action of heat and water
b) Volatility of the essential oil
c) Water solubility of the essential oil

The essential oils which are highly soluble in water cannot be extracted using steam distillation. Therefore, the volatile steam nature of oils is a must for this distillation process. Most of the essential oils are steam volatile in nature, heat-stable, and water-insoluble. The components of essential oils are a mixture of various aroma chemicals, monoterpenes, sesquiterpenes, and their oxygenated derivatives, having a boiling point from 150° to 300° C. These oils are released from the plant in the presence of moisture from steam and heat. The latent heat from condensing steam is responsible for changing oil state from liquid to steam. The temperature gradient is necessary to take the latent heat from the condensing steam to vaporize the oil droplet. Thus, the energy from the steam in the form of heat as latent heat of vaporization converts the oil into a vapour. As the boiling point of the oil is higher than that of water, the vaporization takes place with steam based on their relative vapour pressures [34].

4. Non-Conventional Methods

These methods include modern and much advanced techniques to extract nutraceuticals such as pressurized liquid extraction, supercritical fluid extraction, microwave-assisted extraction, ultrasound-assisted extraction, etc.

4.1. Pressurized Liquid Extraction

Pressurized liquid extraction (PLE) is also known as accelerated solvent extraction, enhanced solvent extraction, pressurized fluid extraction, accelerated fluid extraction, or high-pressure solvent extraction. Very high pressure is applied in this extraction, keeping solvents in a liquid state above their boiling point. This results in an increased solubility, increased diffusion rate of lipid solutes in the solvent, and high solvent penetration in the matrix. PLE drastically reduces the extraction time and amount of solvent used [9].

The researchers at University of Macau and other institutes successfully extracted many products such as flavonoids and essential oils from TCM using this technique [35-38]. Maillard reactions occurred when PLE was used at 200 °C to extract antioxidants from grape pomace [39]. Gizir et al. successfully applied PLE to extract an anthocyanin from black carrots and concluded that the degradation rate of anthocyanins is dependent on time [40].

4.2. Supercritical Fluid Extraction (SFE)

In this method, supercritical fluid (SF) is used as the extraction solvent. These supercritical fluids can dissolve many natural products, and they have similar solubility to liquid and similar diffusivity to the gas. Because of small changes in temperature and pressure, the solvation properties change dramatically near critical points. The low critical temperature (31°C), inert and nontoxic nature, and ability to extract thermolabile compounds makes the Supercritical carbon dioxide (S-CO_2) suitable for widespread use in SFE. Lipids and volatile oils are also extracted using S-CO_2 due to its low polarity.

A significantly higher percentage of essential oils and antioxidant activity are obtained by using S-CO_2. [41]. Solvating properties of SFE are enhanced considerably by the use of modifier [9]. S-CO_2 modified with 2% ethanol at 300 bar and 40°C gave higher extracting selectivity of vinblastine (an antineoplastic drug) from *Catharanthusroseus*, which is 92% more efficient for vinblastine extraction compared to traditional extraction methods [42].

4.3. Ultrasound-Assisted Extraction (UAE)

(UAE) is also known as ultrasonic extraction or sonication. Ultrasonic wave energy is used in this extraction. Ultrasound produces cavitation in the solvent, which accelerates the dissolution and diffusion of the solute, and the heat transfer, to improves the efficiency of extraction. The advantages of UAE are low solvent and low energy consumption, less time used for the process. Thermolabile and unstable compounds are extracted using this method [43, 44]. Jovanović et al. obtained a higher yield of polyphenols from *Thymus serpyllum*. Using 50% ethanol as solvent, 1:30 ratio of solid to solvent, 0.3 mm particle size. [45].

4.4. Microwave-Assisted Extraction (MAE)

The ionic conduction and dipole rotation followed by interaction of polar and organic compounds present in the plant matrix is responsible for heat generation in the microwave. Heat and mass transfer occur in the same order in the MAE, producing a synergistic effect to improve the yield. The advantages of MAE are an increase in the extract yield, decreased degradation due to heat, and selective heating of vegetal material. The two MAE methods are solvent solvent-free extraction for volatile compounds, and solvent extraction for non-volatile substances [46, 47].

4.5. Pulse Electric Field (PEF) Extraction

This method significantly increases the extraction yield and decreases the extraction time because the mass transfer increases during extraction by destroying membrane structures. Its effectiveness depends on several parameters, including field strength, specific energy input, pulse number, and treatment temperature. PEF extraction is a non-thermal method and minimizes the degradation of the thermolabile compounds [9].

Bouras studied the antioxidants extracted from Norway spruce bark. They concluded that much higher phenolic content (eight times) and antioxidant activity (30 times) were obtained after the PEF treatment compared to untreated samples [48].

4.6. Enzyme-Assisted Extraction (EAE)

The main barriers to the extraction of natural products are the macromolecules present in their cell membrane, and cell walls. At higher temperature proteins get coagulated and denatured during extraction.

EAE enhances the extraction efficiency due to the hydrolytic action of the enzymes on the components of the cell wall and membrane and the macromolecules inside the cell, which facilitate the release of the natural product. Cellulose, α-amylase, and pectinase are generally used in EAE [9].

The TCM Astragali Radix has polysaccharides as the bioactive component. Chen et al. studied the EAE of polysaccharides from the radix of *Astragalus membranaceus* using various enzymes. They found that glucose oxidase offered better performance in extracting polysaccharides than the other seven enzymes tested (amyloglucosidase, hemicellulase, bacterial amylase, fungal amylase, pectinase, cellulose, and vinozyme). The polysaccharide yield under the optimized EAE condition using glucose oxidase increased more than 250% compared with the non-enzyme treated method [49].

5. Methods of Isolation/Separation

Although nutraceutical extraction provides us beneficial and valuable extracts from foods, these extracts further require purification before being used. This process of purification comes under the second section i.e., isolation of nutraceuticals. This step is necessary because the extracts have several natural compounds complexed with our targeted compounds, and they need to be separated. [9].

5.1. Basis of Isolation

Basis for isolation/purification of extracts received after the extraction process has been mentioned in Figure 5.

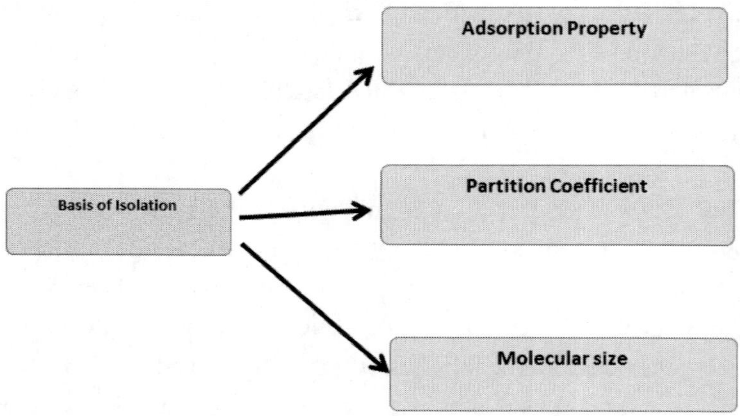

Figure 5. Basis for isolation/purification of extracts received after the extraction process.

5.1.1. On the Basis of Adsorption Properties

Column chromatography is used for separation of natural compounds present in the complex form with our targeted extract. Initially, this chromatography based on adsorption property is used due to its greater capacity, ease or simplicity, greater capacity, and cost-effectiveness. The used adsorbants like silica gel are available at low cost.

5.1.1.1. Principle of Separation

The separation of isolation of the compounds is based on the difference of adsorption properties of each compound to the surface of the adsorbents. For better-isolated products, it is necessary to decide both the stationary and mobile phases wisely. The better the adsorption achieved, the better is separation and maximum retrieval of our targeted compound will occur. Also, in order to avoid irreversible adsorption, additional care is required.

Commonly used adsorbent or stationary phase used includes silica gel due to its low cost and ease of availability. Commonly the silica gel is used for the investigation of phytochemicals. Evidence show that nearly 90% of separation of phytochemicals at the preparative stage was done via silica gel. Silica gel is polar with an attached silanol group. The adsorption on silica gel occurs via dipole-dipole interaction, and the bond formed is a hydrogen bond. Several times the polar compounds undergo irreversible adsorption and make the extraction troublesome. The weakening of the adsorption capacity of silica can be due to pre-isolation water addition to silica gel or the use of mobile phase-containing water. Also, the tailing phenomenon may occur when isolating alkaloids using silica gel; in that case, addition of ammonia may reduce the tailing.[9]

Another example of a strong polar adsorbent is aluminum oxide or Alumina. It is commonly used for better isolation of alkaloids. The Alumina is not preferred much now a days due to its numerous drawbacks like it catalyze dehydration, isomerization or decomposition during the process of isolation.[9]

In the pre-treatment process or standalone system, the adsorptive macroporous resins are used. They do not have ion-exchange groups and structurally consist of macropores. They stand out because they can selectively adsorb almost any natural product. They have several advantages over other adsorbents like high capacity of adsorption, effective cost compared to other adsorbents, and ease of regeneration. The mechanism used for adsorption with this adsorbent includes electrostatic forces, bonding - hydrogen bond, size-sieve action due to macropores between resins and the natural compounds. The influencing factor of resins for isolation are pore size, surface area, and polarity. [50]

5.1.2. On the Basis of Separation Coefficient

When considering partition coefficient, we take into consideration the Partition chromatography (PC). This method of chromatography is based on the following principle:

5.1.2.1. Principle of Isolation

The liquid-liquid extraction is followed in this method. The separation here occurs among two different immiscible fluids based on their relative solubility. The isolation occurs by a) Coating of one liquid phase to a solid matrix as the stationary phase. Materials used for solid matrix includes silica gel, carbon, cellulose, etc., b) The other liquid phase is used as the mobile phase.

The drawback for the technique is the easy removal of stationary phase. As a result, partition chromatography is one of the least used techniques these days. Although, the drawback can be overcome in the bond phase where the liquid stationary phase is bound to the support(inert). This is now used as a stationary phase. For bonded phase, usually, alkyl is taken into consideration. Example includes aryl, cyano, C8, C18, and salines (amino-substituted). They are used widely in the final purification step of various natural compounds [9].

Gravity or centrifugal force are used to hold the liquid stationary phase in counter-current chromatography (CCC). CCC had poor stationary retention, large separation time, so it was rarely used earlier. In the 1980s, significant improvements in CCC were made to develop the modern CCC with hydrodynamic CCC systems with low-pressure drop processes called HSCCC, i.e., hydrostatic counter current chromate-graphy and centrifugal partition chromatography(CPC). HSCCC has one rotating axis with interconnecting chambers for trapping the stationary phase. The HPCCC, i.e., high performance counter-current chromate-graphy is another new generation chromatography. It works similar to the HSCCC, but the g-level of HPCCC is much higher than HSCCC. The HPCCC also benefited less time expenditure as the separation via HSCCC was consuming and used to operate for several hours. On the other hand, HPCCC can work ten times more efficiently than HSCCC [51].

5.1.3. On the Basis of Molecular Size

We know that each component that has been extracted has a different molecular size. This laid the foundation for isolatating or separating these products based on their defined molecular size. This can be achieved by two methods, i.e., MF or membrane filtration and GFC or gel filtration chromatography [9].

5.1.4. Membrane Filtration (MF)

A membrane that is semipermeable is used for this method. The smaller molecules can easily pass through this membrane while the larger ones are retained by it. The pore size of the membrane is a differentiating factor in nanofiltration. Microfiltration and ultrafiltration methodologies are also the kinds of membrane filtration techniques. Membrane filtration is significantly used by the food and pharmaceutical industries to remove the impurities and clarify products.

In case the desired separation/isolation has not occurred using a single semipermeable membrane efficiently, due to similar molecular size of the extracted components or size of the extracts from a range different to the pore size of the used membrane; coupling membrane filtration can be practiced. For example, in a study conducted to isolate the extracts of the olive leaf, a sequential separation was performed as microfiltration, ultrafiltration and nanofiltration, respectively. Each step had a signify-cance like larger than 5KDa of impurities can be removed using microfiltration followed by ultrafiltration. The final stage of nanofi-ltration was responsible for efficient recovery of the polyphenols and flavonoids that were antibacterial.[52]

5.1.5. Gel Filtration Chromatography

It is an often-used laboratory technique with another name called gel permeation or size exclusion chromatography. It is based on the principle that the small molecules will have more retention time than large molecules when gel filtration chromatography is performed. For example, a molecule formed by cross-linking of dextran called as Sephadex was used for the separation of hydrophilic compounds (G-type Sephadex) such as peptides [53] oligosaccharides and polysaccharides. [54]

6. Perspective for Food Applications

The overall health and wellness of human beings mainly depends upon dietary habits and nutritional supplements. Several epidemiological studies proved the health benefits of nutraceuticals, and hence these compounds received more attention of the world and scientific communities. Regular consumption of a diet rich in fruits, vegetables, and whole grains is associated with low incidence and risk of chronic diseases. Estimations suggest that 1/3rd of death due to cancer can be avoided by bringing changes in dietary habits. The chemical compounds obtained from plant-based foods, including antioxidants, primary and secondary metabolites, probiotics, etc. provides long-term benefits to the body and prevent chronic disease, thus increasing life expectancy. More than 50% of the adult population in wesrern countries consume nutraceuticals on a daily basis. Globally, Australia is the one of the top consumers of dietary supplements (nutraceuticals). Following are some of the nutraceutical plants and their medicinal attributes.

6.1. Aloe Vera

This succulent originally from northern Africa is used worldwide as herbal medicine. It is frequently used in the cosmetics industry because of its moisturizing, soothing, and healing effects. It is a common ingredient of lotions, yogurts, laxatives, sunscreen, soaps, and shampoos. Aloe vera is effective in wound healing, inflammation, genital herpes, and psoriasis. The antibacterial and antifungal properties help in the treatment of skin infections. Some studies also claim it to be beneficial in raising blood glucose levels in diabetes. The phytochemicals mannans, anthraquinone, and various lectins make it suitable as a medicinal plant. However, only dose-dependent usage of aloe vera is found to be helpful. Sometimes its excess use leads to toxicity and causes diarrhea, dysfunction of the kidney, and electrolyte imbalance.

6.2. Amaranthus

It is a cosmopolitan plant and can grow in any season. it contains the organic acid squalene and beta-sitosterol and serves as anticarcinogenic and as an antioxidant. In addition, it has cardioprotective effects and also removes stored fats from body tissues.

6.3. Hibiscus Sabdariffa

Hibiscus originated in -world tropics and has been used as a folk medicine to manage cardiac and nerve disorders. This plant has antihypertensive effects and is also used to cure dyspepsia also. In addition, the leaves of this plant are used for crack feet and on boils for a speedy recovery.

6.4. Green Tea

Green tea is obtained from the leaves of *Camellia sinensis*. It is globally used for weight loss management. It contains catechins which seem to promote weight loss and increase in energy expenditure and oxidation of fats. However, the Therapeutic Goods Administration of Australia claimed it to be toxic to the liver.

6.5. Cinnamon

Cinnamon contains catechins, monoterpenes, proanthocyanidins, and sesquiterpenes cinnamonaldehyde, phenols, and trans-cinnamic acids. Meta-analysis claimed the beneficial role of cinnamon in reducing fasting blood glucose levels.

The usage of nutraceuticals in continuously increasing among several countries because of their enormous benefits to the body and fewer side effects. However, further random clinical trials are required to fully understand the safety, efficacy, and mechanism of action of nutraceuticals as an alternative medicine.

CONCLUSION

Medicinal foods or scientifically termed "nutraceuticals," have been in growing demand for decades. The concept of food as an ultimate source to cure almost every disease has been trending ever since the concept of personalized nutrition came to light. As a result, the whole observation leads to an amazing idea of extracting and isolating the particular components of food that cures the disease/disorder by their very own mechanism. These compounds are termed bioactive compounds. The various studies regarding the benefits of these nutraceuticals on health are being carried out regularly. Once the beneficial effect of these nutraceuticals and which compound leads to this is known, we are targeted to extract these compounds and later isolate them to use as functional food, food supplement, or medicinal component. Numerous techniques are used for this with regular advancements and required modifications in them. For example, we have come a long way from earlier used solvent extraction to Ultrasonic assisted or microwave-assisted techniques. Undoubtedly, the effects of these nutraceuticals on human health are doing wonders and providing assurance of good health with no side effects. We no longer see these compounds as a part of food, but they stand out distinctly as a cure. Perhaps one day, we might be able to use these compounds to treat deadly diseases like cancer. The target can only be achieved with the advancements of these techniques for better extraction and isolation of the nutraceuticals.

REFERENCES

[1] https://www.bccresearch.com/market-research/food-and-beverage/nutraceuticals-global-markets.html

[2] Chauhan, B., Kumar, G., Kalam, N. and Ansari, S. H. (2013). Current concepts and prospects of herbal nutraceutical: A review.

Journal of advanced pharmaceutical technology & research, *4*(1): 4.

[3] Zeisel, S. H. (1999). Regulation of nutraceuticals. *Science*. 285: 1853–1855.

[4] Li, P., Xu, G., Li, S. P., Wang, Y. T., Fan, T. P., Zhao, Q. S. and Zhang, Q. W. (2008). Optimizing ultraperformance liquid chromatographic analysis of 10 diterpenoid compounds in Salvia miltiorrhiza using central composite design. *Journal of agricultural and food chemistry*, *56*(4): 1164-1171.

[5] Li, P., Yin, Z. Q., Li, S. L., Huang, X. J., Ye, W. C. and Zhang, Q. W. (2014). Simultaneous determination of eight flavonoids and pogostone in Pogostemon cablin by high performance liquid chromatography. *Journal of Liquid Chromatography & Related Technologies*, *37*(12): 1771-1784.

[6] Yi, Y., Zhang, Q. W., Li, S. L., Wang, Y., Ye, W. C., Zhao, J. and Wang, Y. T. (2012). Simultaneous quantification of major flavonoids in "Bawanghua", the edible flower of Hylocereus undatus using pressurised liquid extraction and high performance liquid chromatography. *Food chemistry*, *135*(2): 528-533.

[7] Zhou, Y. Q., Zhang, Q. W., Li, S. L., Yin, Z. Q., Zhang, X. Q. and Ye, W. C. (2012). Quality evaluation of semen oroxyli through simultaneous quantification of 13 components by high performance liquid chromatography. *Current Pharmaceutical Analysis*, *8*(2): 206-213.

[8] Du, G., Zhao, H., Song, Y., Zhang, Q. and Wang, Y. (2011). Rapid simultaneous determination of isoflavones in Radix puerariae using high-performance liquid chromatography–triple quadrupole mass spectrometry with novel shell-type column. *Journal of separation science*, *34*(19): 2576-2585.

[9] Zhang, Q. W., Lin, L. G. and Ye, W. C. (2018). Techniques for extraction and isolation of natural products: A comprehensive review. *Chinese medicine*, *13*(1): 1-26.

[10] Demain, A. L. and Fang, A. (2000). The natural functions of secondary metabolites. *History of modern biotechnology I*, 69: 1-39. https://doi.org/10.1007/3-540-44964-7_1

[11] Boeing, H., Bechthold, A., Bub, A., Ellinger, S., Haller, D., Kroke, A., and Watzl, B. (2012). Critical review: vegetables and fruit in the prevention of chronic diseases. *European journal of nutrition*, *51*(6): 637-663.

[12] Mayén, M., Mérida, J. and Medina, M. (1995). Flavonoid and non-flavonoid compounds during fermentation and post-fermentation standing of musts from Cabernet Sauvignon and Tempranillo grapes. *American Journal of Enology and Viticulture*, *46*(2): 255-261.

[13] Sun, Y. P., Chou, C. C. and Yu, R. C. (2009). Antioxidant activity of lactic-fermented Chinese cabbage. *Food Chemistry*, *115*(3): 912-917.

[14] Othman, N. B., Roblain, D., Chammen, N., Thonart, P. and Hamdi, M. (2009). Antioxidant phenolic compounds loss during the fermentation of Chétoui olives. *Food Chemistry*, *116*(3): 662-669.

[15] Escudero-López, B., Cerrillo, I., Herrero-Martín, G., Hornero-Méndez, D., Gil-Izquierdo, A., Medina, S. and Fernández-Pachón, M. S. (2013). Fermented orange juice: source of higher carotenoid and flavanone contents. *Journal of agricultural and food chemistry*, *61*(37): 8773-8782.

[16] Li, Y., Ma, R., Xu, Z., Wang, J., Chen, T., Chen, F. and Wang, Z. (2013). Identification and quantification of anthocyanins in Kyoho grape juice-making pomace, Cabernet Sauvignon grape winemaking pomace and their fresh skin. *Journal of the Science of Food and Agriculture*, *93*(6): 1404-1411.

[17] Kumar, S. S., Manoj, P., Shetty, N. P., Prakash, M. and Giridhar, P. (2015). Characterization of major betalain pigments-gomphrenin, betanin and isobetanin from Basella rubra L. fruit and evaluation of efficacy as a natural colourant in product (ice cream) development. *Journal of food science and technology*, *52*(8): 4994-5002.

[18] Pérez-Gregorio, M. R., Regueiro, J., Alonso-González, E., Pastrana-Castro, L. M. and Simal-Gándara, J. (2011). Influence of alcoholic fermentation process on antioxidant activity and phenolic levels from mulberries (*Morus nigra* L.). *LWT-Food Science and Technology*, *44*(8): 1793-1801.

[19] Joshi, V. K., Kumar, A. and Kumar, V. (2012). Antimicrobial, antioxidant and phyto-chemicals from fruit and vegetable wastes: A review. *International Journal of Food and Fermentation Technology*, *2*(2): 123-136.

[20] Singh, J. (2008). Maceration, percolation and infusion techniques for the extraction of medicinal and aromatic plants. *Extraction technologies for medicinal and aromatic plants*, *67*: 32-35.

[21] Zhang, H., Wang, W., FU, Z. M., Han, C. C. and Song, Y. (2014). Study on comparison of extracting fucoxanthin from Undaria pinnatifida with percolation extraction and refluxing methods. *China Food Additives*, 09.

[22] Xiaochun, G., Jianwei, H. and Dai Honglian, X. L. (2009). Study on Optimizing the Technological Condition of Ethanol Percolating Extraction for Goupi Patch [J]. *China Pharmacist*, *10*.

[23] Li, S. L., Lai, S. F., Song, J. Z., Qiao, C. F., Liu, X., Zhou, Y. and Xu, H. X. (2010). Decocting-induced chemical transformations and global quality of Du–Shen–Tang, the decoction of ginseng evaluated by UPLC–Q-TOF-MS/MS based chemical profiling approach. *Journal of Pharmaceutical and Biomedical Analysis*, *53*(4): 946-957.

[24] Zhang, W. L., Chen, J. P., Lam, K. Y. C., Zhan, J. Y. X., Yao, P., Dong, T. T. X. and Tsim, K. W. K. (2014). Hydrolysis of glycol-sidic flavonoids during the preparation of Danggui Buxue Tang: an outcome of moderate boiling of chinese herbal mixture. *Evidence-Based Complementary and Alternative Medicine*, 11. doi: 10.1155/2014/608721

[25] Zhang, Q., Wang, C. H., Ma, Y. M., Zhu, E. Y. and Wang, Z. T. (2013). UPLC-ESI/MS determination of 17 active constituents in two categorized formulas of traditional Chinese medicine, Sanhuang

Xiexin Tang and Fuzi Xiexin Tang: application in comparing the differences in decoctions and macerations. *Biomedical chromatography*, *27*(8): 1079-1088.

[26] Kongkiatpaiboon, S. and Gritsanapan, W. (2013). Optimized extraction for high yield of insecticidal didehydrostemofoline alkaloid in *Stemona collinsiae* root extracts. *Industrial Crops and Products*, *41*: 371-374.

[27] Zhang, L. (2013). Comparison of extraction effect of active ingredients in traditional Chinese medicine compound preparation with two different method. *Heilongjiang Xumu Shouyi*, *9*: 132-133.

[28] Chin, F. S., Chong, K. P., Markus, A. and Wong, N. K. (2013). Tea polyphenols and alkaloids content using soxhlet and direct extraction methods. *World J Agric Sci*, *9*(3): 266-270.

[29] Xu, F. X., Yuan, C., Wan, J. B., Yan, R., Hu, H., Li, S. P. and Zhang, Q. W. (2015). A novel strategy for rapid quantification of 20 (S)-protopanaxatriol and 20 (S)-protopanaxadiol saponins in Panax notoginseng P. ginseng and P. quinquefolium. *Natural Product Research*, *29*(1): 46-52.

[30] http://agritech.tnau.ac.in/horticulture/extraction_techniques%20_medicinal_plants.pdf

[31] Singh, J. (2008). Maceration, percolation and infusion techniques for the extraction of medicinal and aromatic plants. *Extraction technologies for medicinal and aromatic plants*, *67*: 32-35.

[32] Katiyar, C. K. (2008). 6 Aqueous Alcoholic Extraction of Medicinal and Aromatic Plants by Fermentation, *107*.

[33] Handa, S. S. (2008). An overview of extraction techniques for medicinal and aromatic plants. *Extraction technologies for medicinal and aromatic plants*, *1*: 21-40.

[34] Tandon, S. (2008). Distillation Technology for Essential Oils. *Extraction Technologies for Medicinal and Aromatic Plants*, 115.

[35] Yi, Y., Zhang, Q. W., Li, S. L., Wang, Y., Ye, W. C., Zhao, J. and Wang, Y. T. (2012). Simultaneous quantification of major flavonoids in "Bawanghua", the edible flower of Hylocereus

undatus using pressurised liquid extraction and high performance liquid chromatography. *Food Chemistry*, *135*(2): 528-533.

[36] Lv, G. P., Huang, W. H., Yang, F. Q., Li, J. and Li, S. P. (2010). Pressurized liquid extraction and GC-MS analysis for simultaneous determination of seven components in Cinnamomum cassia and the effect of sample preparation. *Journal of separation science*, *33*(15): 2341-2348.

[37] Xu, J., Zhao, W. M., Qian, Z. M., Guan, J. and Li, S. P. (2010). Fast determination of five components of coumarin, alkaloids and bibenzyls in *Dendrobium* spp. using pressurized liquid extraction and ultra-performance liquid chromatography. *Journal of Separation Science*, *33*(11): 1580-1586. Gizir, A. M., Turker, N., & Artuvan, E. (2008). Pressurized acidified water extraction of black carrot [*Daucus carota* ssp. sativus var. atrorubens Alef.] anthocyanins. *European Food Research and Technology*, *226*(3): 363-370.

[38] Xu, F. X., Yuan, C., Wan, J. B., Yan, R., Hu, H., Li, S. P. and Zhang, Q. W. (2015). A novel strategy for rapid quantification of 20 (S)-protopanaxatriol and 20 (S)-protopanaxadiol saponins in *Panax notoginseng, P. ginseng* and *P. quinquefolium*. *Natural product research*, *29*(1): 46-52.

[39] Vergara-Salinas, J. R., Bulnes, P., Zúñiga, M. C., Pérez-Jiménez, J., Torres, J. L., Mateos-Martín, M. L. and Pérez-Correa, J. R. (2013). Effect of pressurized hot water extraction on antioxidants from grape pomace before and after enological fermentation. *Journal of Agricultural and Food Chemistry*, *61*(28): 6929-6936.

[40] Gizir, A. M., Turker, N. and Artuvan, E. (2008). Pressurized acidified water extraction of black carrot [*Daucus carota* ssp. sativus var. atrorubens Alef.] anthocyanins. *European Food Research and Technology*, *226*(3): 363-370.

[41] Conde-Hernández, L. A., Espinosa-Victoria, J. R., Trejo, A., & Guerrero-Beltrán, J. Á. (2017). CO2-supercritical extraction, hydrodistillation and steam distillation of essential oil of rosemary (*Rosmarinus officinalis*). *Journal of food engineering*: *200:* 81-86

[42] Falcão, M. A., Scopel, R., Almeida, R. N., do Espirito Santo, A. T., Franceschini, G., Garcez, J. J. and Cassel, E. (2017). Supercritical fluid extraction of vinblastine from *Catharanthus roseus. The Journal of Supercritical Fluids*: *129*, 9-15.

[43] Barba, F. J., Zhu, Z., Koubaa, M., Sant'Ana, A. S. and Orlien, V. (2016). Green alternative methods for the extraction of antioxidant bioactive compounds from winery wastes and by-products: A review. *Trends in Food Science & Technology*:*49*, 96-109..

[44] Chemat, F., Rombaut, N., Sicaire, A. G., Meullemiestre, A., Fabiano-Tixier, A. S. and Abert-Vian, M. (2017). Ultrasound assisted extraction of food and natural products. Mechanisms, techniques, combinations, protocols and applications. A review. *Ultrasonics Sonochemistry*: *34*, 540-560..

[45] Jovanović, A. A., Đorđević, V. B., Zdunić, G. M., Pljevljakušić, D. S., Šavikin, K. P., Godevac, D. M. and Bugarski, B. M. (2017). Optimization of the extraction process of polyphenols from Thymus serpyllum L. herb using maceration, heat-and ultrasound-assisted techniques. *Separation and Purification Technology*:*179*, 369-380.

[46] Chemat, F. and Cravotto, G. (Eds.). (2012). *Microwave-assisted extraction for bioactive compounds: theory and practice* (Vol. 4). Springer Science & Business Media.

[47] Vinatoru, M., Mason, T. J. and Calinescu, I. (2017). Ultrasonically assisted extraction (UAE) and microwave assisted extraction (MAE) of functional compounds from plant materials. *TrAC Trends in Analytical Chemistry*, *97*: 159-178.

[48] Bouras, M., Grimi, N., Bals, O. and Vorobiev, E. (2016). Impact of pulsed electric fields on polyphenols extraction from Norway spruce bark. *Industrial Crops and Products*, *80*:50-58

[49] Chen, H., Zhou, X. and Zhang, J. (2014). Optimization of enzyme assisted extraction of polysaccharides from Astragalus membranaceus. *Carbohydrate polymers*, *111*: 567-575

[50] Li, J., & Chase, H. A. (2010). Development of adsorptive (non-ionic) macroporous resins and their uses in the purification of

pharmacologically-active natural products from plant sources. *Natural Product Reports*, *27*(10): 1493-1510.

[51] Guzlek, H., Wood, P. L. and Janaway, L. (2009). Performance comparison using the GUESS mixture to evaluate counter-current chromatography instruments. *Journal of Chromatography A*, *1216*(19):4181-4186.

[52] Khemakhem, I., Gargouri, O. D., Dhouib, A., Ayadi, M. A. and Bouaziz, M. (2017). Oleuropein rich extract from olive leaves by combining microfiltration, ultrafiltration and nanofiltration. *Separation and Purification Technology*, *172*: 310-317..

[53] Sila, A. and Bougatef, A. (2016). Antioxidant peptides from marine by-products: Isolation, identification and application in food systems. A review. *Journal of Functional Foods*, *21*:10-26..

[54] Shi, L. (2016). Bioactivities, isolation and purification methods of polysaccharides from natural products: A review. *International journal of biological macromolecules*, *(92)*: 37-48.

In: Nutraceuticals
Editors: A. Kumari and G. Singh
ISBN: 978-1-68507-488-3

Chapter 4

FORMULATION OF FUNCTIONAL FOODS

***Rashim Kumari*[1,2,*] *and Mahesh Gupta*[1]**

[1]Dietetics and Nutrition Division, CSIR-Institute of Himalayan Bioresource Technology, Palampur, Himachal Pradesh, India
[2]Academy of Scientific and Innovative Research (AcSIR), Ghaziabad, Uttar Pradesh, India

ABSTRACT

Nowadays, progress in the field of food and nutrition sciences brings to light that consuming a good diet is important as it can provide nutrients able to modulate body functions, improve health status and wellbeing, and/or reduce the risk of various diseases. Therefore, in recent years much research has been carried out on developing the process and applying technologies to prepare functional foods. Beyond basic nutrition, functional food can bring essential components with healthy benefits. Reliability on functional food products enhances the efficacy of publications. It increases consumer awareness and creates a basis for regulatory approval and development of health claims. The development of functional food includes several different stages, from concept to successful market execution. Acceptability of consumers and providers

[*]Corresponding Author's E-mail: rashimk03@gmail.com.

of health claims showed improved market penetration of functional food, which encourages initiatives for developing new functional food products.

1. INTRODUCTION

Functional food is the food in which a new ingredient(s) (either one or more) has been added to the product so that it has an additional functional property (often regarding health benefits or prevention of disease). In the United States, the Institute of Medicine's Food and Nutrition has defined functional food as "any food and food ingredients that may provide a health benefit beyond the traditional nutrition that it contains." Functional foods may be originated either from animals or from plants.

To know better about functional foods, it is necessary to understandhow the science behind nutrition has changed itself. Nutrition has been developed from the prevention of deficiency nutritionaldue to insufficient intake of required dietary components and the development of dietary guidelines, nutrition standards, and food guides for the benefit of human beings and to minimize the risk of disease.Diet and food are as old as humans have existed on this planet.The term 'nutrition' is quite modern, appearing in the nineteenth century for the first time. In the twentieth century, one of the main contributions of nutrition science is the theory of balanced diet, a required and proper combination of food items which gives, at least, the minimum necessity of all nutrient components required for growth and development, to reduce the risk of certain diseases.

Functional food is not a well-described concept. Additionally, in the future, a wide range of food products can be defined as functional food containing several different nutrient constituents, which have a relation with the various functions of the body, either good state of health or lowering the risk of diseases. Therefore, there is no simple definition of functional food that exists universally.

Functional food originated in Japan. Additionally, since the 1980s, Japan has been at the forefront in organizing the formulation of functional foods and in large-scale research programmes. Further, the Government of Japan also finances them for their regular production analysis to determine physiological regulation of food function and molecular plan. In 1991, as a conclusion of a long making procedure as a national effort and to form a group of potential effective healthy foods to lower the increasing cost of health care, the idea of specified health use foods (FOSHU) was started. Such foods are planned to be used to improve human health with specific health benefit claims. These foods are allowed to be shown and are added as one of the food groups illustrated in the Nutrition Improvement Law as foods for special dietary use.

A functional food should remain in the normal regular food pattern category, as it is not either a drug (pill or capsule) or a dietary supplement. It also must display the particular predictable health effects in total that can likely be taken.

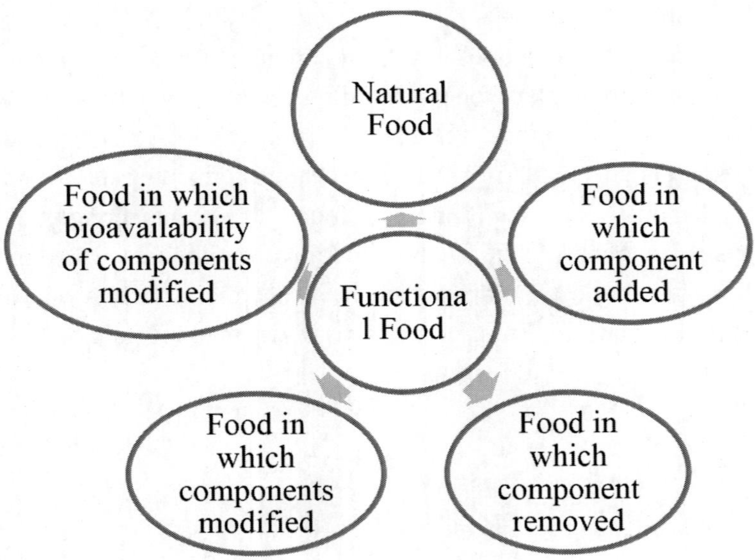

Figure 1. Functional food from a practical point of view.

It is not necessary for all the population members that the functional food will be functional and match the biochemical needs of every individual. Intake of selected food components may become a major job as we progress in understanding the interactions between diet and genes. Figure 1 describes what a functional food can be from a practical point of view.

2. Main Sources and Components of Functional Food

Phytochemicals are the major functional food components derived from plants and may be nutritive or non-nutritive, and have active biological chemicals which show particular preventive function against assured diseases that are non-communicable [1]. Around 900 or more phytochemicals are present in food and having either one portion of vegetables or fruits in our daily diet may provide around 100 various types of phytochemicals[2].

Previously, it was supposed that the components of functional food were found mostly in plant foods, including cereals, vegetables, and fruits. However, probiotics, bioactive peptides, etc. are also present in animal products, especially in dairies such as milk and fermented milk products. Table 1 demonstrates a few functional food components along with their biological functions. Additionally, functional components also showed numerous metabolic activities which have beneficial effects on various diseases in the body[3]. Primary sources of functional food are overviewed below:

Table 1. Components of functional food and their health benefits

Class/Component		Potential Health Benefits
Dietary Fiber	Whole cereal grains	May minimize the threat of coronary heart disease (CHD) and different cancers. Helps in maintaining the level of blood sugar in the body.
	Beta-glucan	May lessen the threat of CHD
	Soluble dietary fiber(SDF)	May cut the risk of different cancers and CHD
	Insoluble dietary fiber(IDF)	May helps in maintaining fit digestive tract. May minimise the threat of different cancers.
Prebiotics	Yeast, Lactobacilli, Bifidobacteria and other beneficial live microorganisms	May contribute in good gut health and overall immunity. Other benefits are related to particular strains
	Inulin, Polydextrose, Fructo-oligosaccharide	May progress the quality of gut health of host and intestinal flora. may improve the absorption of calcium in the body
Flavonoids	Proantocyanidins	May helps in maintaining better health of urinary tract and heart
	Flavonols: e.g., Quercetin, MyricetinKaempferol	Protects from cell damage by neutralizing free radicals Boosts the defense system of cellular antioxidant
	Flavanones: e.g., Naringenina, Hesperetin	Protects from cell damage by neutralizing free radicals; bolster cellular antioxidant defenses
	Flavanols-Catechins, Procyanidins, Epigallocatechin, Epicatechins	May helps in maintaining good health of host heart.
	Anthocyanins: e.g., Cyanidin, Delphinidin, Malvidin	Boosts the defense system of cellular antioxidant Helps in enhancing the functioning of brain
Carotenoids	Beta-carotene	Protects from cell damage by neutralizing free radicals Boosts the defense system of cellular antioxidant In the body helps in making vitamin A
	Lycopene	May helps in reducing various cancer
	Zeaxanthin and Lutein	May helps in maintaining the healthy eye sight

Table 1. (Continued)

Class/Component		Potential Health Benefits
Isothiocyanates	Sulforaphane	Detoxification of unwanted constituents may get improved Boosts the defense system of cellular antioxidant
Phenolic Acids	Ferulic acidand Caffeic acid	Boosts the defense system of cellular antioxidant May contribute in the protection of health of eye sight and heart
Sterols and Plant stanols	Stanol or Sterol esters	May minimize the threat of CHD
	Free Stanols or Sterols	May minimize the threat of CHD
Phyto-strogens	Lignans	May helps in maintaining good health of heart and improves function of immune system
	Isoflavones: e.g., Daidzein, Genistein	May helps in enhancing good bone health, improves function of immune system and brain May helps in maintaining menopausal health in women
Fatty Acids	Conjugated linoleic acid	May helps in enhancing desirable body composition and function of immune system
	Monounsaturated fatty acids	May minimize the threat of CHD
	PUFAs-Omega-3 fatty acids	May minimize the threat of CHD May contribute in maintaining healthy brain function of brain and eye
	Polyunsaturated fatty acids-Omega-3 fatty acids-ALA	May helps in maintaining good health of heart May contribute in maintaining healthy brain function of brain and eye

2.1. Whole Grains

Cereals, particularly barley and oat, are functional food, as they include non-digestible carbohydrates like FOS, trans-GOS, and lactulose, which are beneficial in stimulating the development of bacteria like lactobacilli and bifidobacteria in the large intestines. Cereal has fibers that are water-soluble like arabinoxylan and beta-glucan that delay the gastric emptying, resulting in a reduction of sterols and glucose absorption by the intestine, lowering the LDL, blood serum cholesterol, which results in reducing the threat of atherosclerotic heart disease, post-meal blood glucose, and content of insulin [4]. The fiber in cereal has the ability to reduce the cholesterol level and is believed to result from obvious effects on the GI tract. This is maybe because of the cereal fiber's gel-forming capability [5]. Some of cereal's functional components are also useful in the bakery and dairy industries. Recently functional food research has focused on manufacturing low-fat ice creams and yogurts by using beta-glucan. The addition of beta-glucan during manufacturing of dairy products having low fat changes their taste, appearance and other sensory attributes as like dairy products which have full-fat [6].

2.2. Pulses and Legumes

Legumes, pulses and food derived from soya have isoflavones, which are also known as phytoestrogens. Two main isoflavones are genestein and daidzein. Isoflavones are heterocyclic phenols, and their structure is parallel to estrogenic steroids. As the isoflavones are fragile estrogens, they may proceed as anti-estrogens the endogenous estrogens naturally. This may describe that the people who consume soy food in good quantity have minimum threat diseases due to estrogen like cancer of breast and prostate and intake of isoflavones rich foods also may be concerned with a decrease of diseases. Some evidence suggests that consumption of isoflavones may act defensive and helpful against cancers depending on hormones and against diseases related to age such as cognitive deficit, osteoporosis, and heart disease. Isoflavones are widely digested in the liver and gut of a human. However, soya's effect of lowering the level of cholesterol is the best among reported cardio and vascular protection and may be beneficial for a healthy heart. It is also expected to be intervened by isoflavones through various mechanisms, including artery function effects, estrogen receptor-mediated effects, and cellular effects. Since 2000, the US FDA has given a significant growth in the saleof food products made from soy. The FDA approved the health claim that consumption of low saturates and cholesterol soy protein in the diet may minimize the threat of heart diseases by reducing the level of cholesterol. The claim identifies that an intake of 25g of soy protein is required in a day to show its health benefits. Presently, isoflavones are the bioactive ingredient used mainly for functional foods from soya yogurts to tomato juice to dessert (like pudding). Indeed, as compared to conventional soy yogurts, soya yogurts prepared from germinated soybean extract and lactic acid bacteria have significantly higher amounts of isoflavones, gamma-aminobutyric acid, and free amino acids, therefore may have better health benefits [7].

2.3. Fruits and Vegetables

The benefits provided by functional food and nutraceutical are because of the available phytochemicals constituents and bioactive compounds in vegetables and fruit. They are the key resource of essential nutrients like fiber, vitamins, carotenoids, minerals, folate, phytosterols, and polyphenols. These nutrient components are the major bioactive compounds having the benefits of antioxidants on human health. "The development of degeneration of the central nervous system like Parkinson's and Alzheimer's disease, chronic obstructive pulmonary, rheumatoid arthritis, and other diseases is due to free radicals". Because of antioxidant consumption, these certain diseases, therefore, may be beneficially influenced. Antioxidant plays a vital role in lowering heart disease's threat by inhibiting low-density lipoprotein peroxidation. In contrast, they may also hinder the progression process of other cardiovascular diseases [8]. From various epidemiological studies, it was absorbed that consumption of vegetables and fruits act as defensive agents against various types of human cancers. Because of its compound lycopene, tomatoes have gained much attention in recent years, as the primary carotenoid in tomatoes has antioxidant functions and helps in reducing cancer. The amino acid present in garlic is odorless. On crushing garlic, cloves changed enzymatically by alliinase into allicin followed by decomposition suddenly to form various compounds containing sulfur (diallylsulphide, diallyldisulfide). Epidemiological researches proved that vegetables containing allium may provide a shielding effect on gastrointestinal tract cancers. In tea, specifically, green tea is a rich source of polyphenols. In all polyphenols found in tea, catechins are the primarily and most significantly present. A most essential tea constituent, tea epigallocatechin gallate, is a potential antioxidant and has anti-inflammatory, immuno-stimulatory, and antimicrobial activities. Consumption of green tea reduces the total cholesterol level, lessens the oxidation of lipoprotein, and increases HDL fraction [9]. Epidemiological evidence has linked the regular eating of vegetables such as broccoli, cauliflower, etc. lower the risk of cancer. The high amount of glucosinolates present in all vegetables (like cabbage, turnips, kale, etc.)have been recognized for their anticarcinogenic properties.A plant cell enzyme called myrosinase helps in converting these components to different hydrolysis goods, such assulforaphane, indoles, and isothiocyanates. From various epidemiological studies, it is showed that fruits from the citrus category are helpful against different types of cancers in humans. Although lemons, oranges, grapefruit, and limes are the main source of such vital nutrients, i.e., ascorbic acid, fiber, and folate. Elegbede et al. [10] described that another constituent is accountable for their anticancer action. Citrus fruit is mainly rich in a class of phytochemicals recognized as limonoids. Evident studies have been reported which support the effect of limonene as a cancer-preventative [11]. Non-alcoholic beverages are another important category of functional food segment, which are fortified with vitamins A, C, and E or other nutrients are the drinks healthy for the eyes (including lutein), or drinks for better bone health (including inulin and calcium) [6].

2.4. Prebiotic/Probiotic

In the last two decades, a significant increase in the field of research on different microorganisms and their function in improving people health have been noticed. In the early 1900s, the first study on lactic acid bacteria (LAB) and their existence in the intestine were published [12, 13]. Though, actual advancement was suggested by Metchnikoff in his study, in which he reported enhanced health and long life in humans after consuming fermented dairy products. Nobel Prize was awarded to Metchnikoff (1908) as a symbol of recognition for the significance of these studies. The helpful results of LAB on the health of the human gut have become one of the standards of current awareness on people's nutrition. Numerous studies showed good health effects resulting from the intake of LAB-consisting products and revealed that a few LAB strains provide extra, unique benefits that may have showed a good outcome on the human organism functioning. Such a type of bacteria with these beneficial properties is called 'probiotic'. In 1954, Vergio [14] was the researcher who was first to introduce this term; Although in 1989 Fuller [15] was the first one to define probiotic, he used the term for those live microorganisms of food products, which gives an extra balance advantage of gastrointestinal tract microbial population along with their normal nutritive value benefit. In 2001, Schrezenmeir and de Vrese [16]explained probiotics as the food product consisting of single or more than one live microorganisms culture, which after consumption by human beings in required quantity showed beneficial health effects. According to them, the most adopted definition was provided by WHO, i.e., probiotic are those live microorganisms that while taken in required amounts show health benefits in the living organism. The healthy effect on human beings by these probiotics may include the enhancement of physiological and metabolic processes in the body and the medical condition by minimizing the threat of occurrence of many diseases [17, 18, 19, 20]. The products for enhancing human health may formulate claims for function enhancement, whereas those minimizing the threat of disease occurrence may formulate claims for reduction of disease risk.

3. Main Techniques to Produce Functional Food

For the formulation of functional foods, it is very important to take an incorporated approach that includes various interrelated stepsperformed in synchronization prior tofood production can be effectively launched in the market. The developmental procedure should begin with considerations and thought of the idea and plan of the positioning of the food product, specifically concerning authoritarian positioning, classification. The second step is to recognize the active bioactive components chemically, showing and giving particular health benefits to the host. For instance, the method of modern liquid chromatography (LC/MS) analysis can be used to identify structural target molecules. After that, the next step is to formulate a food product for which possible technologies with suitable diagnostic procedures will be used or formed, if required, to produce a food product with the preferred enhancing health benefit properties [21-23]. Various challenges in the area of functional food products and their possible way outs are proposed, that are harmful or enhancing to eradicate the components which are harmful or increase the amount of indigenous components that are beneficial. To fulfill this target, it is likely to employ processes like enzymatic, supercritical fluid extraction, or membrane separation [24]. Supercritical fluid extraction and membrane separation are the techniques that are also used for harmful components replacement by the beneficial components. In contrast, the microencapsulation technique can be accepted to maximize the bioavailability of beneficial components [25]. Microencapsulation is a modern technology used basically for the packaging of materials in solids, liquids, or gaseous form minutely, and for sealing capsules which in organized rate and below the pressures of particular circumstances can lose their contents. Widely, microencapsulation is a technique used for various food applications, counting giving sustained or maintained release, maintaining the reaction during oxidation, stabilizing the internal matter, maintaining the reaction during oxidation, giving flavors, colors/odors, defensive against loss, and prolonging the life of products [26]. Recently nanotechnology field showed huge potential in enhancing the effectiveness of the release of bioactive compounds and nutraceuticals in functional foods for the benefit of human health [27]. Nano-emulsions, association colloids, nanostructured multiple emulsions, nanofibers, and nanocomposites are the potential nanotechnology-based delivery systems [28]. Processes of foods encapsulation, sphere packaging, and pulsed electric field have been developed, these processes comprise of following features [29]:

- Eradication of those constituents that are recognized as a reason for harmful health effects. For an instance, a protein which is allergenic.
- Maximize the amount of natural bioactive constituents in the food to a level where we can get its desirable effects.
- Adding up the constituent which is absent in most of the food, but only for those whose advantageous results has been confirmed.
- Substitution of constituents, mainly nutrients which are macro, the consumption of those who aremeasured dangerous to wellbeing when taken in large quantity, by a constituent for those whom advantageous results have been expressed.
- Progression in the bioavailability or alteration of food constituents for those whose advantageous results have been confirmed.
- Supervising quantity and effectiveness of bioactive constituents, which are beneficial.

The correct recognition of bioactive compounds is very important to discover the connections between various food components and their benefits on host health. Additionally, deep quantification is required to evaluate the level of intake of diet and safety strategies for possible bioactive compounds essential to obtain desired valuable attributes.To effectively demonstrate the benefits of added value to the consumers, there is a need for rapid analytical methods to check the action of biomarkers on the disease. The four very general steps for the analysis of any analytical method are first doing sampling, then their preparation, shelf life, and determination (partition and recognition) [30]. More than 90% of expansion has its focal point on the last step of investigation in the past few decades. Amazing progress in instrumentation, spectroscopy, and chromatography have shown in speedy improvement in the process for high-throughput partition and recognition of composite multi-component blends with trace amounts of the analytes of interest. The gathering of a representative sample from the whole matrix is the sampling that requires to be examined. The model is taken in such a way that it should characterize the whole population. One of the important parts is sample preservation as the delay between sample collection and its analysis is prevalent.

Accurate preservation of sample ensures that the physical and chemical attributes of sample retain from collection of sample time to its analyses time [30]. Therefore, for sample preservation, enzymatic inactivation is necessary to maintain the correct identification of the sample. Enzymatic inactivation is necessary to maintain the correct identification of the sample and exhibit the benefits of these methods in chemical and metabolic changes during the storage period. Therefore, researchers must preserve samples under suitable conditions and determine the temperature and its effects during storage on the analyte of interest. Preparation of sample consists of steps like drying of sample, sieving, homogenization, extraction, derivatization, and hydrolysis. There are numerous reasons behind sample preparation, such as to elevate the competence of method, to eradicate or lower potential disturbances, to increase the analytical procedure sensitivity by raising the amount of the analyte during the evaluation of mix, and occasionally to modify the interest of analyte to a more appropriate form which without any difficulty can be identified [31]. After completing the preparation of samples, its evaluation will be done by using preferred methods and instruments. Different instruments are used for varied evaluation; it depends upon the knowledge and data to be taken. Even a clearly defined evaluation can be done and drawn in different ways. The aim of measurement of an analytical can either be qualitative or quantitative [30].

4. Current and Future Developments

The food constituents have the potential to alter the target functions in the body. Therefore, inleading the future, functional food science can play a main role in the betterment of human health for further progress and improvement inactive components and/or their role in minimizing the risk of various diseases. In development of functional food, nutritionists and food scientists can present favorable opportunities linked with disease reduction and human health. These types of new and advanced approaches in nutrition are mostly dependent upon various factors *viz.* relevant markers identification, nutrition is mostly dependent upon various factors *viz.* description, measurement, and validation. The steps for analysis of such studies should be done vigilantly with specific references, but in a different way from clinical studies described to help in drug development, not food products. The major target functions that can be modulated by consuming a particular food product should be recognized and described in the body. The science behind these functions, their relation to well-being and health, or a particular pathology process needs to be developed for the necessary scientific base for the formulation and development of new functional foods.

Food regulation concerning the validity guarantee and food safety should be improved. In the twenty-first century, optimizing nutrition is the main challenge in the field of nutritional science. Formulation of functional food is the crucial step of this procedure. Still, a proper explanation of its health claims should be the utmost step of scientific challenge mainly not a promotion one. For the food and the good health of humans, appropriate justification of functional food product claims is condemning for the achievement of functional foods.

In the world market, functional food's potential achievement may be known because the main strength of this area is the diffusion of essential awareness of the specific response of these identified components, which provide much more good benefits to health while consumed than the conventional products. Nutraceutical area has to turn into one of the very vigorous and dynamic sections in the food industry. Numerous companies showed interest and invested in the manufacturing and promotion of nutraceuticals. Presently the need to meet consumer's demands, upcoming challenges and inquiries is increasing day by day, the process of development mainly focused on nutrition optimization and improving the healthy habits of eating also. Multinational companies widely do the manufacturing of functional food. The companies found the chance and understand the urgent need of planning for innovation in tendency strengthening like mode for the conclusion and process consolidation of internationalization, in the direction of network execution and strategies for chain production integration. The soaring of the health of individuals and the diffusion of preventive drugs drags the customers to better their knowledge among food, health, and security. Nowadays, customers are highly conscious and attentive about minimizing the health risk factors and improving their way of life by adopting accurate eating habits. This attentiveness is also because of research done on antioxidants present in food, advantageous components present in vegetables, cereals, fruits, and probiotic microorganisms. As functional foods contain various bioactive compounds, many of them exhibit beneficial properties such as antioxidant, anti-inflammatory, and anti-carcinogenic. Mainly, anti-carcinogenic action of these components is arbitrated by various molecular pathways, such as antioxidant, encouragement of the immune defense system, anti-mutagenic activity, antioxidant enzymes intonation, modulation of hormonal structure, regulation in the growth of cell of genetic expression, and necrobiosis. Though these particles are measures for the prevention of chemo action, their fundamental biochemical role is still uncertain. The nutritional qualities of food are evaluated by the natural compounds present in them and by compounds that formed at the time of development and preservation. Furthermore, the rising demand for fortified foods has increased new challenges concerning the number of nutrients that are micro and ingestion of food supplements related to their effectiveness and safety. A supplementary component of product novelty can be its optimization by nano and microencapsulation. In reality, many functional components are susceptible to ecological factors like humidity, exposure to light, and high temperature. Besides, if applied for the enhancement of Ready to eat products may be spoiled during the various manufacturing steps. The functional component may also combine with another food ingredient and insert aromas, colors, and unwanted savors. Therefore, encapsulation can shield the constituent from any thermal and mechanical loss. More novel elements concern the advancement of alternative procedures to traditional technology for the extraction of solvent. The quality of nutraceuticals and their preservation assessment is one of the critical factors for functional food. It will help to recognize the critical conditions for maintaining phyto-complexes attributes in the product (moisture, temperature, outcome from bioactive compounds interaction with food design). Lastly, the cerebral method of functional foods for subjective awareness and equanimity of weight failure by most individuals is still unknown.

REFERENCES

[1] Murano, P. S. (2003). *Phytochemicals and phytonutrients in understanding food science & technology*. Wadsworth, Bemont, 51–56.

[2] Srividya, A. R., Nagasamy, V. & Vishnuvarthan, V. J. (2010). Nutraceutical as medicine: a review. *Pharmanest, 1*(2), 132–145.

[3] Swanson, J. E. (2003). Bioactive food components. *Encycl Food Cult, Gale Cengage.* http://www.enotes.com/bioactive-food-componentsreference/bioactive-food-components Accessed 18 November 2012.

[4] Charalampopoulos, D., Wang, R., Pandiella, S.S. & Webb, C. (2002). Application of cereals and cereal components in functional foods: A review.*Int. J. Food Microbiol.*, *79*, 131-41.

[5] Brennan, C.S. & Clearyb, L.J. (2005). The potential use of cereal (1/3,1/4)-b-D-glucans as functional food ingredients. *J. Cereal Sci.*, *42*, 1-13.

[6] Siró, I., Kápolna, E., Kápolna, B. & Lugasi, A. (2008). Functional food. Product development, marketing and consumer acceptance-A review. *Appetite*, *51*, 456-67.

[7] Wong, M.C.Y., Emery, P.W., Preedy, V.R. & Wiseman, H. (2008). Health benefits of isoflavones in functional foods? Proteomic and metabonomic advances. *Inflammopharmacology.*, 16, 235-239.

[8] Diplock, A.T., Charleux, J.L. & Crozier-Williet al, G. (1998). Functional food science and defence against reactive oxidative species. *Br. J. Nutr.*, *80*, 77-112.

[9] Ferrari, C.K.B. & Torres, E.A.F.S. (2003). Biochemical pharmacology of functional foods and prevention of chronic diseases of aging. *Biomed Pharmacother.*, *57*,251-260.

[10] Elegbede, J.A., Maltzman, T.H., Elson,C.E. & Gould,M.N. (1993). Effects of anticarcinogenicmonoterpenes on phase II hepatic metabolizing enzymes. *Carcinogenesis.*, *14*, 1221-1223.

[11] Rafter, J.J. (2002). Scientific basis of biomarkers and benefits of functional foods for reduction of disease risk: Cancer. *Br. J. Nutr.*, *88*, 219-224.

[12] Moro, E. (1900). Uber den *Bacillus acidophilus* n. spec. Ein Beitragzur Kenntnis der normalen Darmbakterien des Sauglings (*Bacillus acidophilus* n. spec.). *Jahrbuch fur Kinderheilkunde*, *52*, 38–55. [About the *Bacillus acidophilus* n. Spec. A contribution to the knowledge of the normal intestinal bacteria of the infant (Bacillus acidophilus n. Spec.). *Yearbook of Pediatrics*]

[13] Beijerinck, M.W. (1901). Sur les ferments de lactique de l'industrie. *Des Sciences Exacteset Naturelles*, *6*, 212–243. [On industrial lactic acid bacteria. *Exact and Natural Sciences*]

[14] Vergio, F. (1954). Anti- und Probiotika. *Hippocrates*, 25: 116–119.

[15] Fuller, R. (1989). Probiotics in man and animals. *Journal of Applied Bacteriology*, *66*, 365–378.

[16] Schrezenmeir, J. & de Vrese, M. (2001). Probiotics, prebiotics, and synbiotics – approaching a definition. *American Journal of Clinical Nutrition*, *73*, 361–364.

[17] Saarela, M., Lahteenmaki, L., Crittenden, R., Salminen, S. & Mattila-Sandholm, T. (2002). Gut bacteria and health foods – the European perspective. *International Journal of Food Microbiology*, *78*, 99–117.

[18] Holzapfel, W. H. & Schillinger, U. (2002). Introduction to pre- and probiotics. *Food Research International*, *35*, 109–116.

[19] Saxelin, M., Tynkkynen, S., Mattila-Sanholm, T. & de Vos, W.M. (2005). Probiotic and other functional microbes: from markets to mechanisms. *Current Opinion in Biotechnology*, *16*, 204–211.

[20] Shah, N.P. (2007). Functional cultures and health benefits. *International Dairy Journal*, *17*, 1262–1277.

[21] Diplock, A.T., Aggett, P.J., Ashwell, M., Bornet, F., Fern, E.B. & Roberfroid, M.B. (1999). Scientific concepts of functional foods in Europe: Consensus document. *Br. J. Nutr.*, *81*, 1-27.

[22] Picó, Y. & Barceló, D. (2008). The expanding role of LC-MS in analyzing metabolites and degradation products of food contaminants. *TrAC Trend Anal. Chem.*, *27*, 821-835.

[23] Picó, Y., Farré, M. l., Segarra, R. & Barceló, D. (2010). Profiling of compounds and degradation products from the postharvest treatment of pears and apples by ultra-high pressure liquid chromatographyquadrupole-time-of-flight mass spectrometry. *Talanta*, *81*, 281–293.

[24] Knorr, D. (1999). Novel approaches in food-processing technology: new technologies for preserving foods and modifying function.*Curr. Opin. Biotechnol.*, *10*, 485-91.

[25] Wang, L. & Weller, C.L. (2006). Recent advances in extraction of nutraceuticals from plants. *Trend. Food Sci. Technol.*, *17*, 300-12.

[26] Anal, A.K. & Singh, H. (2007). Recent advances in microencapsulation of probiotics for industrial applications and targeted delivery. *Trend. Food Sci. Technol.*, *18*, 240-251.

[27] Chen, H., Weiss, J. & Shahidi, F. (2006). Nanotechnology in nutraceuticals and functional foods. *Food Technol.*, *60*, 30-36.

[28] Neethirajan, S. & Jayas, D.S. (2011). Nanotechnology for the food and bioprocessing industries. *Food Bioprocess Technol.*, *4*, 39-47.

[29] Korhonen, H. (2002). Technology options for new nutritional concepts. *Int. J. Dairy Sci.*, *55*, 79-88.

[30] Mitra, S. & Brukh, R. (2003). Sample preparation: An analytical perspective, In *Sample Preparation Techniques in Analytical Chemistry*, Eds. by Mitra S.Wiley, New York, 1–36.

[31] Luthria, D.L. (2006). Perspective Significance of sample preparation in developing analytical methodologies for accurate estimation of bioactive compounds in functional foods. *J. Sci. Food Agric.*, *86*, 2266-2272.

In: Nutraceuticals
Editors: A. Kumari and G. Singh

ISBN: 978-1-68507-488-3

Chapter 5

ROLE OF NANOTECHNOLOGY IN VARIOUS ASPECTS OF THE FOOD INDUSTRY

Rajni Sharma, Shiv Kumar Giri and Gulab Singh*
Department of Biotechnology, Maharaja Agrasen University, Baddi, Solan, India

ABSTRACT

The clear relationship between the quality of food we eat, and our health, is demonstrated by various researchers. Gradually, recognition of importance of healthy diet and identification of mechanisms by which foods modulate the metabolism and health contribute to developing functional food concept. Numerous processes and nanomaterials have been developed with the potential to act as functional food. The nanotechnology is playing crucial role in the field of food technology for the production of novel nano-based food materials, delivery of nutrient with the help of nanoemulsion, novel packaging system, implementation of nanotechnology and biosensors to monitor the food quality and waste reduction. Food technologists and engineers are working tirelessly to identify novel ways to re invent food products involving nanotechnology that will be appeal to the worldwide customers. However, consumer

* Corresponding Author's E-mail: gulab24@gmail.com.

health and safety are primary concern. This chapter provides an overview of the developed nanomaterial- based food technologies, delivery vehicles, and new opportunities and challenges for nanotechnology. We also try to cover recent developments and initiatives being adopted to ensure the safe use of nanomaterials based on current predictions.

1. INTRODUCTION

Nanotechnology is defined as the development, manipulation, and utilization of materials at nanometer scales (Fathi et al., 2012). Nanomaterials are usually having a size smaller than 100 nm and behave differently as compared macro scale counterparts. This difference in the physciochemical properties like color, solubility and thermal behavior is mainly due to high surface -to- volume ratio (Zhu et al., 2012; Singh et al., 2017). The difference in physcio- chemical properties helps to improve the sensory qualities such as taste, texture, color of food and also help to increase the self of food materials (Chaudhry, 2008). Since, by the end of the 20th century, there has been a growing realization of the pivotal link between diet and human health. This has led to the development of a new category of foods, the so-called functional foods. Functional food is simply a convenient way to describe foods, or their components, that may provide health benefits beyond nutrition. In other words, functional foods contain a variety of nutrients and non-nutrients components which affect a range of body functions as well as well-being or reduce the chance of infection (Frewer, et al., 2003). The functional components can be enhanced through special growing conditions, or through breeding techniques, e.g., β-carotene rich rice, vitamin-enriched broccoli, and soybeans. Meat, poultry, fish, and eggs can have their composition altered by the animal's diet, e.g., increased levels of conjugated linoleic acid or omega-3 in meat and milk products. The functional foods segment of the food industry is estimated to be worth about 168 billion dollars and is growing at about 9% per annum (Euromonitor International, 2016). Nanotechnology can be applied in all phases of the food cycle – from farm to fork. Application of nanotechnology in the form of utilization of nanocomposites in food packaging is very well accepted by industries as well as common people due to increased shelf-life of packaged food materials. Metal oxide and nano-clays are most commonly utilized nanomaterials in food packaging. However, the number of food products utilizing nanotechnology is relatively small, but still the potential application of nanotechnology such as detecting microorganisms in packaged food, stronger flavor and attractive coloring in the food industry thrilled the manufactures.

Nanotechnology has potential application in all aspects of the food industry like food processing, food packaging, and food monitoring. It has been used to improve the materials properties (improved barrier and mechanical properties, light materials), but also in the development of active and intelligent packaging systems. This chapter will cover the different aspects of nanotechnology in food.

2. Nanotechnology in Food Packaging

Nanotechnology is the science of very small materials that has a big impact in the food industry including packaging. A variety of nanomaterials such as silver nanoparticles, titanium nitride nanoparticles, and nano-titanium dioxide, nano-zinc oxide, and nanoclay are introduced as functional additives to food packaging (Tager, 2014). Food packaging is important for food protection, distribution, transportation from farm and industries to consumer and manipulation, etc (Kuswandi, 2017). In nanotechnology, functional nanomaterial can also be developed which have the properties like antioxidants, antimicrobial, flavor enhancer, nutraceutical and enzymes to protect and improve the shelf life of the food materials (Cha and Chinnan, 2004). Packaging can release nanoscale antimicrobial, antioxidants, flavors, fragrances or nutraceutical into the food or beverages to extend its shelf life or to improve its taste or smell (LaCoste et al., 2005; Nachay, 2007). By 2003, over 90% of nano-packaging was based on nanocomposites, in which nanomaterials were used to improve the barrier function of plastic wrapping for foods, and plastic bottles for beer, soft drinks and juices. Nano-packaging can also be designed to release antimicrobial, antioxidants, enzymes, flavors and nutraceutical to extend shelf life (Cha et al., 2004; LaCoste et al., 2005). Different types of innovative packaging based on nanotechnology are explained below.

2.1. Smart Packaging

Smart packaging is the type of packaging used to indicate the internal condition of the product, such as the shelf life of the product. Nano-sensors are placed on the packaging material of the food product. The sensors sense the conditions and properties of the product and express the results in the form of different types of signals. The sensor expresses the condition of the barrier and properties of the materials (permeability of gases i.e., oxygen). It also detects the microbial load in the food product and explains the shelf life of the product. Nano-sensors are good for the consumer for judging the properties of the product and satisfying their needs (Biji, et al., 2015). Currently, various industries like Nestle, MonoPrix super market and British Airways use chemical sensor- based packaging which indicates deterioration of food via color changes (Pehanich, 2006).

2.2. Active Packaging

In active packaging, a unique system is used to modify the environmental or surrounding conditions for extending shelf life and improving food safety. For example, a very common type of activity is performed to control the oxidation of food products by creating the oxygen- free storage condition. Synthetic antioxidants are widely used in food manufacturing processes (e.g., BHT and BHA common example of antioxidants) (Contini, et al., 2011).

2.3. Antimicrobial Packaging

Food manufacturers have been facing a lots of challenges related to food safety. Food safety means food remains protected from deterioration, provides nutrition to consumers, and will not have any negative effect on consumers health. Many microbial agents (i.e., bacteria) can cause food to spoil when they have perfect conditions to reproduce. Nanotechnology has helped in many ways to keep the microbial agents in food at acceptable levels (Neethirajan and Jayas, 2011; Biji, et al., 2015). In the antimicrobial packaging, the antimicrobial agents are used to control the food- spoiling microbes in food items. Various forms of silver nano-particles are used in large amounts and can control a large range of microbes. *E. coli* is commonly found in contaminated water and can contaminate the product in which water (containing *E. coli*) is used as ingredient. *E. coli* can be controlled by applying the titanium dioxide coating on the packaging material (Chellaram, 2014).

The use of nanoscale particles with antimicrobial properties (e.g., zinc oxide and silver nanoparticles) have been used to extend shelf-life of foods, reducing microbiological growth during storage. This can be very helpful in foods with reduced shelf-life (e.g., fresh meat, poultry, and fish) where the increase in shelf life can bring several advantages for the industry, such as shipping for export purposes (Rao, 2009).

3. Nanotechnology in Food Processing

The nanostructure food ingredients are being developed with the claims that they offer improved taste, texture, and consistency (Cientifica Report, 2006). Nanotechnology increases the shelf-life of different kinds of food materials and also helps bring down the extent of wastage of food due to microbial infestation (Pradhan et al., 2015).

The nanotechnology plays a significant role in improvement of food texture, appearance, taste, nutritional value and shelf life like nano-encapsulation technique is used to improve the flavor (Nakagawa, 2014).

Anthocyanin, which shows various biological activities, is protected by utilizing encapsulation techniques (Zhang et al., 2014). Similarly, edible nano-coating can be used to protect vegetables, fruits, meat, fast food, cheese, bakery and confectionery goods. In addition to protection, nano-encapsulation also prolongs shelf life, provides flavor, enhances color, enzymes, antioxidants and anti-browning compounds to the manufactured products (Azeredo et al., 2009). SiO_2 (E551) and TiO_2 (E171) were approved by the competent authorities for application as food preservative in large quantities (EFSA Scientific Committee, 2009). TiO_2 is used as a coating on doughnuts (Oberdorster et al., 2005). Engineered nanostructure-based aerosols are effective in controlling food borne pathogens like *Listeria, salmonella* and *E. coli* (Pyrgiotakis et al., 2015). Nanotechnology also provides food ingredients as well as water-insoluble food supplements with controlled release and improved dispersibility (McClements et al., 2006).

Production of functional foods such as soft drinks, ice-cream, chocolate and chips are made healthy by reducing fat content, carbohydrates and calorie content of food and also increasing the vitamins, proteins and fibers. Nanotechnology is applied for the modification of food materials as per person needs like taste preference, color of choice, flavor, nutritional properties, allergies, and lower cost of ingredients (Daniells, 2007). Foods which contain nanoscale ingredients and additives are already available on supermarket shelves. Nestle and Unilever developed the ice cream based on nano-emulsion with lower fat content and still retains the same fatty texture and flavor of the ice cream (Renton, 2006). Various nanoparticles are used as food additive such as TiO2, SiO2, and amorphous silica (Gerloff et al., 2009; Uboldi et al., 2012).

The benefits of nanotechnology in food processing include the development of a complete texture for food components; the encapsulation of food additives. The development of new flavors and sensations, and the control of aroma release with increased bioavailability in dietary supplements (Cho and Jones, (2019). Iron oxide (Fe_2O_3) nanoparticles may be utilized in foods as colorants or sources of bioavailable iron (Hilty et al., 2010; Trushina et al., (2011); Zimmermann and Hilty (2011); Wu et al., (2014). The ranges of applications of iron oxide as a food colorant in the United States are highly limited, i.e., up to 0.1 wt % in sausages as part of casings (World Health Organization, 2000). Iron taken in the form of enriched/fortified foods ranges from 10 to 23 mg/day, while that from dietary supplements may range from 10 to 32 mg/day (Fulgoni et al., 2011). TiO_2 particles are used as functional ingredients in certain foods to provide characteristic optical properties such as increased lightness and brightness (Weir et al., 2012). Chewing one piece of chewing gum can result in an intake of 1.5-5.1 mg of TiO_2 nanoparticles (Chen, 2013). Metallic oxide (TiO_2 and SiO_2) is conventionally used in food items mainly for as coloring as well as flavoring agents (Ottaway, 2010). SiO_2 is one of the most commonly used nanomaterial for flavor enhancers or carriers of fragrances (Dekkers et al., 2011).

4. Detection of Deleterious Substances in Food

Nano sensors play an important role in the food industry. Nano sensors are able to detect and quantify pathogens, organic and inorganic contaminants at very low concentrations, as well as also detect the adulteration of different types of materials with high sensitivity and quick response time (Otles et al., 2010). One such application was demonstrated by Valdes et al., (2009) in the case of detection of organophosphate pesticide in fruit and water. Conventional control of microorganisms and toxins are complicated but with the help of nano sensors toxins and microorganisms can be rapidly detected and controlled during storage and processing of food. Different nanotechnology- based sensors are developed by utilizing nanowires and antibodies to detect the toxins, chemical contaminants, adulteration, and pathogens in food materials (McClements, 2015). In supermarkets, nano-electrochemical systems (NEMS) are used to detect hazardous substances. They can be used in quality control for foods as they contain transducers which help in the detection of biochemical as well as chemical signals from food materials (Sozer and Kokini, 2009).

5. Safety Issues

It is increasingly clear that nanotechnology offers a wide spectrum of benefits and can be a game changer for the food industry. The use of nanomaterials in nano-food, nanosensing, packaging incorporating nano-products could leads to migration of nanomaterials and subsequent human exposure through inhalation, skin penetration, and ingestion. Routes of possible exposure to humans include: (i) Leaching of nanomaterials into food from packaging materials. (ii) Ingestion from intake of nano-food and (iii) Disposal of packaging, nanosensors and nanofood in landfills with further release into the environment, air, water and soil (Xia et al., 2014). Because of their small size, high surface area and increased reactivity, nanomaterials pose unique risks and may cause adverse effects on human health and the environment. Upon use, nanomaterials may undergo physicochemical transformations in the local environment (pH, ionic strength, microbial environment, *etc.*) further changing their fate and transport behavior (Garcia et al., 2018). There is growing research related to use of nanomaterials in the field of food. Many research groups are involved in finding the possible negative impact of the nanomaterials from packaging and processing on consumers and environmental health (Bradley et al., 2011; Jain et al., 2016). Nanomaterial interaction with food systems raises concern related to the health of humans and animals. Nano-based products may have negative impact on humans, animals and plants; still there is no standard regulatory rule and regulations are established for nanomaterial applications in the food and agriculture sector. Various regulatory bodies like USFDA and FSANZ are actively participating in the regulation of nanomaterials as food additives in various countries like USA, Australia and New Zealand (Bowman et al., 2006; Cubadda et al., 2016). Although a material is being considered as GRAS (Generally regarded as safe) substance, still additional studies are required as the physicochemical properties of material are completely different in nano-state as compared to the macro-state. Moreover, there is chance of bioaccumulation of nanomaterial in organs and tissues due to its small size (Savolainen et al., 2010). Cushen et al., (2014) developed a model to demonstrate the migration of nanoparticles from packaging material to food. They have studied the migration of silver and copper nanoparticles from nanocomposites and observed and different crucial parameters like particle size, temperature and contact time effect, and the rate of migration of nanomaterials from packing materials to food. Since each nanomaterial has its individual properties, so the toxicity and other impact of nanomaterial should be established on a case-by-case basis (Mahler et al., 2012). Furthermore, regulatory authorities must develop standards rule and regulations for the commercial products to ensure quality of product and health and safety of human and environments.

6. Future Perspectives and Potential Risks of Nanotechnology

There have been tremendous developments in the application of nanotechnology in food science and research. Nanotechnology helps in the detection of pesticides (Fu et al., 2008) pathogens (Hahn et al., 2008), and toxins (Yang and Li (2005), and tracking-tracing-monitoring helps in the maintenance of food quality. One such example is use of carbon nanotubes by incorporating them into packaging materials for the detection of toxic materials, microorganisms and food spoilage (Tully, 2006). Along with the incessant research on the application of nanomaterials, several potential risks and toxicity issues associated with the application of nanomaterials have been revealed, and these concerns must be addressed (Yang et al., 2013; Yang et al., 2005). The effect of these nano-scale particles on human beings, animals, and the environment are unpredictable due to changes over time in their properties. Some nanoparticles can even cross biological barriers, such as the blood-brain barrier, and enter various cells and organs (Su et al., 2004).

Conclusion

The increasing investment and development of nanotechnology is expected to transform food science and food industries. Advances in nanotechnology have brought benefits for the food industries like increased shelf life, enhanced flavor and taste etc and many more applications yet to be realized. Recently, it has been suggested that food nanotechnology can help to develop functional foods and nutraceuticals that provide benefits such as nutrition, health, taste and safety. While nanotechnology provides great benefits still there are emerging concerns that arises from its unknown physicochemical properties of nanomaterials. There are no strict rules and regulations to control the nanotechnology-based food materials with unknown effects of these nanotechnology based food materials. There is an urgent need for the rules and regulations to control the quality and effect of nanotechnology- based food materials on human and environmental health. As the transparency of health and safety impact on human and environmental health should be the prime importance while developing the nanotechnology- based food materials or food development process, and therefore compulsory testing of nanotechnology -based foods is required before they are released to the market.

References

[1] Fathi, M., Mozafari, M. R., Mohebbi, M. (2012) "Nanoencapsulation of food ingredients using lipid based delivery systems," *Trends in Food Science Technology*, vol. 23, no. 1, pp. 13-27.

[2] Zhu, M., Nie, G., Meng, H., Xia, T., Nel, A., Zhao, Y. (2012) "Physicochemical properties determine nanomaterial cellular uptake, transport, and fate," *Accounts of Chemical Research*, vol. 46, no. 3, pp. 622-631.

[3] Singh, T. Shukla S, Kumar P, Bajpai, V. K, Rather I. A. (2017) "Application of nanotechnology in food science: perception and overview," *Frontiers in Microbiology*, vol. 8, p. 1501, 2017.

[4] Chaudhry, Q., Scotter, M., Blackburn, Ross B., Boxall A., Castle L., Aitken R., Watkins R. (2018) Applications and implications of nanotechnologies for the food sector, *Food Additives and Contaminants*, 25:3, 241-258.

[5] IFT. *Functional foods: opportunities and challenges.* (2005) IFT Expert Report Institute of Food Technology; Washington, DC, USA, pp. 7-10.

[6] Frewe, L., Scholderer J., Lambert N. (2003) Consumer acceptance of functional foods: issues for the future. *Br Food J.*; 105:714-731.

[7] Heasman, M., Mellentin, J. (2001). *The Functional foods revolution: healthy people, healthy profits?* 2001 pp. xxii + 313 pp.

[8] Contor L. (2001) Functional food science in Europe. *Nutr Metab Cardiovasc Dis.*; 11:20-23.

[9] Hasler, C. M. (2002) Functional foods: benefits, concerns and challenges - a position paper from the American Council on Science and Health. *J Nutr.*; 132:3772-378.

[10] Bigliardi, B., Galati, F. (2013) Innovation trends in the food industry: the case of functional foods. *Trends Food Sci Technol.*; 31:118-129.

[11] Rao, M. A. (2009) Nanoscale particles in food and food packaging. *J Food Sci.* Nov 9.

[12] Mihindukulasuriya, S. D. F., Lim, L. T. (2014) Nanotechnology Development in Food Packaging: A review. *Trends in Food Science & Technology* 40, 149-167.

[13] Cientifica Report (2006). *Nanotechnologies in the Food Industry, Published August 2006.* Available at: http://www.cientifica.com/www/details.php?id47 (accessed July 24, 2021).

[14] Pradhan, N., Singh, S., Ojha, N., Srivastava, A., Barla, A., Rai, V., Bose S. (2015). Facets of nanotechnology as seen in food processing, packaging, and preservation industry. *BioMed Research Intrernational,* 17 Article ID 365672, 17 pages.

[15] Nakagawa, K. (2014). "Nano- and micro-encapsulation of flavor in food systems," in *Nano- and Microencapsulation for Foods*, Chap. 10, ed. H.-S. Kwak (Oxford: John Wiley & Sons), 249-272.

[16] Zhang, T., Lv, C., Chen, L., Bai, G., Zhao, G., and Xu, C. (2014). Encapsulation of anthocyanin molecules within a ferritin nanocage increases their stability and cell uptake efficiency. *Food Res. Int.* 62, 183-192.

[17] Azeredo, H. M. C., Henrique, C. L. M., Wood, D., Williams, T. G., Roberto, J. A. B., Tara, H. M. (2009) Nanocomposite edible films from mango puree reinforced with cellulose nanofibers. *J. Food Sci.* 74, 31-35.

[18] Naoto, S., Hiroshi O., Mitsutoshi N., Mitsutoshi (2009) micro- and nanotechnology for food processing. (Food safety series) resource: engineering and technology for a sustainable. *World. Am. Soc. Agric. Eng.* 16, 19.

[19] EFSA Scientific Committee (2009). The potential risks arising from nanoscience and nanotechnologies on food and feed safety. *Sci Opin Sci Comm*. https://doi.org/10.2903/j.efsa.2009.958.

[20] Daniells, S. (2007). Thing big, think nano. *Food Navigator.com* Europe 19 December 2007.

[21] Renton, A., (2006) Welcome to the world of nanofoods. *Guardian Unlimited* UK 13 December 2006.

[22] Gerloff, K., Albrecht, C., Boots, A. G., Förster, I., Schins, P. F. R., (2009) "Cytotoxicity and Oxidative DNA Damage by Nanoparticles in Human Intestinal Caco-2 Cells." *Nanotoxicol* 3(4): 355-364.

[23] Uboldi, C., Giudetti, G., Broggi, F., Gilliland, D., Ponti, J., Rossi, F. (2012) Amorphous silica nanoparticles do not induce cytotoxicity, cell transformation or genotoxicity in Balb/3T3 mouse fibroblasts. *Mutat Res*. 14; 745(1-2):11-20.

[24] Oberdörster, G., Oberdörster, E., Oberdörster, J. (2005) Nanotoxicology: an emerging discipline evolving from studies of ultrafine particles. *Environ Health Perspect*.; 113(7):823-839.

[25] Pyrgiotakis, G., et al., 2015. "Inactivation of Foodborne Microorganisms Using Engineered Water Nanostructures (EWNS)." *Environ Sci Technol* 49(6):3737-3745.

[26] Clements, M. C., DeLoid D. J., Pyrgiotakis G., Shatkin, G., Xiao, J. A., Demokritou, H. P., (2016). The role of the food matrix and gastrointestinal tract in the assessment of biological properties of ingested engineered nanomaterials (iENMs): State of the science and knowledge gaps. *NanoImpact*, 3, 47-57.

[27] Cho, Y. H., Jones, O. G. (2019) Assembled protein nanoparticles in food or nutrition applications. *Adv. Food Nutr. Res.*, *88*, 47-84.

[28] Raspopov, R. V., Trushina, E. N., Gmoshinsky, I. V. & Khotimchenko, S. A. (2011) Bioavailability of nanoparticles of ferric oxide when used in nutrition. Experimental results in rats. *Vopr. Pitan.* 80, 25-30.

[29] Zimmermann, M. B., Hilty, F. M. (2011) Nanocompounds of iron and zinc: their potential in nutrition. *Nanoscale* 3, 2390-2398.

[30] Wu, H. H., Yin, J. J., Wamer, W. G., Zeng, M. Y. & Lo, Y. M. (2014) Reactive oxygen species-related activities of nano-iron metal and nano-iron oxides. *J. Food Drug. Anal.* 22, 86-94.

[31] Hilty, F., Arnold, M., Hilbe, M. (2010) Iron from nanocompounds containing iron and zinc is highly bioavailable in rats without tissue accumulation. *Nature Nanotech* 5, 374-380.

[32] WHO (2000) *Safety Evaluation of Certain Food Additives and Contaminants* Vol. 44, 1-2 (World Health Organization, 2000).

[33] Fulgoni, V. L., Keast, D. R., Bailey, R. L., Dwyer, J. (2011) Foods, fortificants, and supplements: where do Americans get their nutrients? *J. Nutr.* 141, 1847.

[34] Weir, A., Westerhoff, P., Fabricius, L., Hristovski, K., von Goetz, N. (2012) Titanium dioxide nanoparticles in food and personal care products. *Environ. Sci. Technol.* 46, 2242-2250.

[35] Chen, X. X., Cheng, B., Yang, Y. X. (2013) Characterization and preliminary toxicity assay of nano-titanium dioxide additive in sugar-coated chewing gum. *Small* (Weinheim an der Bergstrasse, Germany). 9 (9-10):1765-1774.

[36] Ottaway, P. B., (2010) *Nanotechnology in Supplements and Foods – EU Concerns*. Available at: http://www.accessmylibrary.com/coms2/summary_0286-37130259_ITM (accessed July 26, 2021).

[37] Dekkers, S., Krystek, P., Peters, R. J., Lankveld, D. X., Bokkers, B. G., van Hoeven-Arentzen, P. H. (2011) Presence and risks of nanosilica in food products. *Nanotoxicology* 5, 393-405.

[38] Tager, J. (2014) Nanomaterials in food packaging: FSANZ fails consumers gain. *Chain Reaction* 122 (16-17)8-14.

[39] Kuswandi, B., (2017) Environmental friendly food nano-packaging. *Environ Chem Lett* 15:205-221.

[40] Cha, D., Chinnan, M., (2004) Biopolymer-based antimicrobial packaging: A review. *Critic Rev Food Sci Nutrit*; 44:223-237.

[41] LaCoste, A., Schaich, K., Zumbrunnen, D., Yam, K., (2005) Advanced controlled release packaging through smart blending. *Packag Technol Sci*; 18:77-87.

[42] Biji, K. B., Ravishankar C. N., Mohan, C. O., Gopal, T. S. (2015) Smart packaging systems for food applications: a review. *J Food Sci Technol* 52: 6125-6135.

[43] Pehanich, M., (2006) Small gains in processing and packaging. *Food Process* 11: 46-48.

[44] Contini, C., Katsikogianni, M. G., O'Neill F., T., O'Sullivan, M., Dowling, D. P., (2011) Development of active packaging containing natural antioxidants. *Procedia Food Sci* 1: 224-228.

[45] Neethirajan, S., Jayas, D. S. (2011) Nanotechnology for the food and bioprocessing industries. *Food Bioprocess Tech* 4: 39-47.

[46] Biji, K. B., Ravishankar, C. N., Mohan, C. O., Gopal, T. S., (2015) Smart packaging systems for food applications: a review. *J Food Sci Technol* 52: 6125-6135.

[47] Chellaram, C., Murugaboopathi, G., John, A. A., Sivakumar, R., Ganesan, S., Krithika, G. P. (2014) Significance of nanotechnology in food industry. *APCBEE Procedia* 8: 109-113.

[48] Otles, S., Yalcin, B. (2010) "Nano-biosensors as new tool for detection of food quality and safety," *LogForum*, vol. 6(4); 7.

[49] Valdés, M. G., González, A. C. V., Calzón, J. A. G., Díaz-García, M. E. (2009) "Analytical nanotechnology for food analysis," *Microchimica Acta*, vol. 166 (1-2), 1-19.

[50] McClements, D. J., (2015) "Nanoscale nutrient delivery systems for food applications: improving bioactive dispersibility, stability, and bioavailability," *Journal of Food Science*, 80(7), 1602-1611.

[51] Sozer, N., Kokini, J. L. (2009) "Nanotechnology and its applications in the food sector," *Trends in Biotechnology*, 27 (2), 82-89.

[52] Bradley, E. L., Castle, L., Chaudhry, Q. (2011) Applications of nanomaterials in food packaging with a consideration of opportunities for developing countries. *Trends Food Sci. Technol.* 22, 603-610.

[53] Jain, A., Shivendu, R., Nandita, D., Chidambaram, R. (2016) Nanomaterials in food and agriculture: an overview on their safety concerns and regulatory issues. *Crit. Rev. Food Sci. Nutr.* 58.

[54] Bowman, D. M., Hodge, G. A. (2006) Nanotechnology: mapping the wild regulatory frontier. *Futures* 38, 1060-1073.

[55] Cubadda, F., Aureli, F., Raggi, A., Cristina, M., Toscan, B., Mantovani, A. (2016) Nanotechnologies and nanomaterials in the food sector and their safety assessment. *Rapp. ISTISAN* 13, 48.

[56] Xia, Y. N., Rubino, M., Auras, R. (2014) Release of Nanoclay and Surfactant from Polymer-Clay Nanocomposites into a Food Simulant, *Environ. Sci. Technol.*, 48, 13617-13624.

[57] Garcia, C. V., Shin G. H., Kim, J. T. (2018) Metal oxide-based nanocomposites in food packaging: Applications, migration, and regulations, *Trends Food Sci. Technol.*, 82, 21-31.

[58] Savolainen, K., Pylkkänen, K., Norppa, H., Falck, G., Lindberg, H., Tuomi, T., Vippola, M., Alenius, H., Hämeri, K., Koivisto, J., Brouwer, D., Mark, D., Bard, D., Berges, M., Jankowska, E., Posniak, M., Farmer, P., Singh, R., Krombach, F., Bihari, P., Kasper, G., Seipenbusch, M. (2010) Nanotechnologies, engineered nanomaterials and occupational health and safety – A review. *Saf. Sci.* 6, 1-7.

[59] Cushen, M., Kerry, J., Morris, M., Cruz-Romero, M., and Cummins, E. (2014). Evaluasian and simulation of silver and copper nanoparticle migration from polyethylene nanocomposites to food and an associated exposure assessment. *J. Agric. Food Chem.* 62, 1403-1411.

[60] Mahler, G. J., Esch, M. B., Tako, E., Southard, T. L., Archer, S. D., Glahn, R. P., Michael, S. (2012) Oral exposure to polystyrene nanoparticles affects iron absorption. *Nat. Nanotech.* 7, 264-271.

[61] Fu, J., Park, B., Siragusa, G., Jones, L., Tripp, R., Zhao (2008) An Au/Si hetero-nanorod-based biosensor for *Salmonella* detection, *Nanotechnology,* 15, p. e155502.

[62] Hahn, M. A., Keng, P. C., Krauss, T. D. (2008) Flow cytometric analysis to detect pathogens in bacterial cell mixtures using semiconductor quantum dots. *Anal Chem*, 3 864-872.

[63] Yang, L., Li, Y. (2005) Quantum dots as fluorescent labels for quantitative detection of *S. typhimurium* in chicken carcass wash water, *J Food Protect,* 6; 1241-1245.

[64] Tully, E., Hearty, S., Leonard, P., O'Kennedy, R. (2006) The development of rapid fluorescence-based immunoassays, using quantum dot-labelled antibodies for the detection of L. monocytogenes cell surface proteins, *Int. J Biol Macromol,* 1-3 127-134.

[65] Yang, M., Peng, Z., Ning, Y., Chen, Y., Zhou, Q., Deng L. (2013) *Highly specific and cost-efficient detection of Salmonella paratyphi combining aptamers with single-walled carbon nanotubes Sensors,* 13 6865-6881.

[66] Su, S. L., Li, Y. (2004) Quantum dot biolabeling coupled with immunomagnetic separation for detection of *Escherichia coli O157:H7 Anal Chem,* 76 (16).

In: Nutraceuticals
Editors: A. Kumari and G. Singh ISBN: 978-1-68507-488-3

Chapter 6

MECHANISM OF BIOACTIVE COMPOUNDS OF NUTRACEUTICALS

Roshanlal Yadav[1], Munish Siwatch[2,*] and Upasana Yadav[3]

[1,2]Department of Food Technology, Bhaskracharya College of Applied Sciences, University of Delhi, Delhi, India

[3]Department of Nutrition and Dietetics, Faculty of Allied Health Sciences, Manav Rachna International Institute of Research and studies, Faridabad, Haryana, India

ABSTRACT

A positive impact on human health and wellness due to nutraceuticals has exploded globally over the past two decades by becoming a part of the dietary landscape. There has been increasing attention devoted to the possibility that several nutraceuticals may be anticipated to boost short-term well-being leading to additional benefits of long-term defence against certain diseases. Nutraceuticals execute their functionality through their bioactive compounds. These are a

---

[*] Corresponding Author's E-mail: siwatchmunish19@gmail.com.

distinct category of chemicals present in limited measures in plants and animals like fruits, greens, legumes, nuts, oils, vegetables, whole grains, and fatty fish to name a few. These nutraceuticals have been considered for the prevention of cancer, heart disease, Type 2 Diabetes, hypertension, hypercholesterolemia, inflammation, oxidative stress, brain development and age-associated neurodegeneration etc. by incorporating bioactive compounds such as lycopene, resveratrol, dietary fibres, phytoestrogens, plant sterols, probiotics, essential fatty acids, isothiocyanates, organo-sulphur compounds, Inulin-fructo-oligosaccharides, lignin, quercetin, curcumin, gingerol etc. The present chapter will explore the varied sources, biochemical properties, metabolism, health benefits and GRAS status of bioactive ingredients. Significant emphasis will be given to associating the molecular and chemical structures of biologically active components present in foods to their nutritional, pharmacological and functional outcome on human health and wellness.

1. Introduction

Scientifically, the term "bioactive" can be used as an alternative for "biologically active" compounds [1]. In medical terms, bioactive substances are those which exert some effect on living tissues and also cause a biochemical reaction or trigger a response in the living tissue. In association with food, bioactive compounds are defined as food components that can regulate various metabolic and functional processes in humans and also help in the improvement of health [2]. Most of the bioactive compounds include lycopene, carotenoids, resveratrol, dietary fibres, phytoestrogens, plant sterols, probiotics, essential fatty acids, isothiocyanates, organo-sulphur compounds, inulin-fructo-oligosaccharides, lignin, quercetin, curcumin, gingerol etc. These compounds vary extensively in chemical structure and function and are grouped accordingly. Majorly, these bioactive compounds are found largely in fruits, vegetables, legumes and whole grains and can be consumed daily. Additionally, these substances may also present in other live microorganisms, like fungi (mushroom), bacteria and in some animal groups [3]. From various previous studies the beneficial effects in terms of decreasing inflammation, prevention of aging, scavenging free radicals, prevention of cancer, heart disease, type 2 diabetes, hypertension and regulating cell signalling pathways of bioactive compounds have been identified in both cell and animal studies [4, 5]. Nowadays the attention for the consumption of food containing these bioactive compounds has increased because of the safety and lower side effect of these natural compounds. Moreover, under the right circumstances, they can also act as drugs, further enhancing their attractiveness. The bioactive compounds execute their activities through relation with definite biological mechanisms and the interactive effect of other nutrients and food matrices.

Most of the phenolic compounds found in plant cells exhibit antioxidant properties and some research has confirmed positive effects on thrombosis and tumorogenesis. These phenolics are also responsible for the prevention and growth of cancer and cardiovascular diseases (CVD). Different phytoestrogens are present in various food components like flaxseed, soy, whole grains, fruits, and vegetables. These phytoestrogens have antioxidant properties, and several studies established favourable effects on other CVD risk factors, and in animal and cell culture models of cancer. The availability of hydroxytyrosol in olives and resveratrol in nuts and red wine, have antithrombotic, antioxidant and anti-inflammatory properties and also inhibits carcinogenesis. Lycopene, present red colour fruits like tomatoes and watermelon can protect against prostate and other cancers and inhibits the growth of tumour cells in animals. Organosulfur compounds in onions and garlic and, monoterpenes in citrus fruits, cherries, and herbs isothiocyanates in cruciferous vegetables have anticarcinogenic actions as well as cardioprotective effects.

2. Mechanism of Bioactive Compounds

2.1. Spices and Herbs

The spices and culinary herbs have generally been used in foods to enhance flavour and improve their organoleptic properties. Except for proving flavour these spices and herbs have some preservative effects and also have numerous medicinal properties. Various studies have been conducted and it was measured that most of the spices have antioxidant activity which promotes health benefits [6, 7]. In addition to fruits and vegetables, these spices are also a good source of natural antioxidants and it is a part of our diet. The spices chiefly consist of phenolic compounds, flavonoids, tannins, alkaloids, diterpenes and sulphur-containing compounds which have been reported for their antioxidant activity [8, 9]. These compounds revealed diverse antioxidant activities like phenolic acid and flavonoids can capture free radicals and can form complexes with catalytic metal ions rendering them inactive. Previous studies have revealed that herbs and spices such as sage, oregano and rosemary are very good sources of antioxidants because of the availability of high content of phenolic compounds. Antioxidants also can protect lipids and oils available in different food from oxidative degradation. With the addition of these antioxidant compounds, the rancidity can be controlled and retard the development of toxic oxidation products. Moreover, these compounds help in maintaining nutritional quality, and extend the shelf-life of products. Natural antioxidants obtained from edible materials such as spices and herbs have been of increasing interest because of the adverse effect of synthetic antioxidant compounds. The oxidative stress, which is caused by a high concentration of free radicals in cells and tissues, can be induced by various negative factors, such as X-ray, gamma, and UV radiation, pollution, psycho-emotional stress, smoking, intensive physical exertion, drug addiction, and alcoholism. Natural antioxidants contained in spices help to reduce oxidative stress. Chronic oxidative stress has been observed to lead to a variety of heart-related diseases and the generation of various types of cancer cells. Numerous secondary products of lipid oxidation, such as malondialdehyde and 4-hydroxynonenal, can react with biological components such as proteins and DNA. The antioxidant activity of spices is due to the presence of different unique chemical compounds, which majorly consist of biologically active compounds and polyphenolic compounds. Numerous publications have reported that lignans, flavonoids, alkaloids and phenolic acids are the primary biologically active and antioxidants components present in culinary herbs and spices. [10, 11].

Chromatographic technique such as HPLC-MS (High-performance liquid chromatography-mass spectrometry) and HPLC-UV (High-performance liquid chromatography-UV detector) is largely used to determine these bioactive components. The largest amounts of flavonoids have been reported in various herbs like celery, parsley, saffron, oregano, fennel, Tasmanian pepper and dill. In the human diet, the consumption of these culinary herbs and spices may show antioxidant effects. A significant quantity of more than ten flavonoids has been reported in various spices and the largely available component are kaempferol, luteolin and quercetin. The majorly available component in different herbs are as follows: luteolin in celery seeds, apigenin in dried parsley, cyanidin in Tasmanian pepper, luteolin in Mexican oregano, kaempferol and quercetin in capers. Overall all most of the spices and herbs have anti-oxidative properties due to the presence of special kinds of chemical compounds.

2.2. Phytochemicals

Phytochemicals are bioactive components accessible in the form of plant chemicals in fruits, vegetables, grains, and other plant foods that may supply desirable health benefits beyond their basic nutrition which helps to decrease the risk of major chronic diseases [12]. The phytochemicals consist of polyphenols, flavonoids etc. which are described as follows:

2.2.1. Polyphenolic Compounds

Based on the chemical structure there are more than ten classes of polyphenols have been defined and more than 8,000 phenolic compounds have been identified [13]. The chemical structure of some important polyphenolic compounds is embedded in Figure 1. The most common polyphenolic compounds are flavonoids which are present mainly in plant-based food and they can be categorized into 13 (thirteen) classes having more than 5,000 compounds. Very common flavonoids are flavones, flavanols, and their glycosides [13]. Although polyphenols are present in all plant-based foods, their level and quality depend upon the nature of the plant source. Moreover, the most plentiful phenolic compound available in fruits is a flavanol. The primary phenols present in cereals and legumes are phenolic acids, flavonoids and tannins. In wine, major polyphenolic compounds comprise anthocyanins, phenolic acids, flavonoids and tannins. Nuts are rich in tannins [13]. Both phenolic acids and hydrolysable tannins are abundantly available in olive oil. The chief flavonoid present in onions is quercetin glycoside, while in apples and tea it is quercetin-3-rutinoside. Limited information is available about the dietary intake of polyphenolic compounds, which showed that the intake of these phenolics is highly variable among the population groups studied [14]. Different studies have shown that red wine is a very good source of polyphenolic substances and encloses more than 200 individual phenolic which inhibits oxidation of LDL in-vitro [15] and increases the antioxidant capacity of plasma [16]. The main antioxidative compounds recognized in red wine are flavanols, phenolic acids, polymeric-anthocyanidins, and monomeric catechins [17]. Various studies indicated that phenolic compounds have reduced synthesis of prothrombotic, antithrombotic effects and pro-inflammatory mediators that emerge to be the result of reduced susceptibility of platelet aggregation. Furthermore, the previous study indicated that polyphenols present in wine can modulate the production of nitric oxide by the vascular endothelium, resulting in vasorelaxation [18]. Chocolate also has antioxidant activity due to the presence of polyphenol fraction [19].

In a recent study, it was measured that consumption of chocolate caused a decrease in basal plasma oxidation products [20]. The polyphenol present in cacao liquor has been reported to inhibit both superoxide anion production and hydrogen peroxide in the human body, due to the presence of free radical scavengers. In addition to antioxidant effects, the polyphenols of cocoa and chocolate have been shown to inhibit cyclooxygenase activity [21]. The predominant flavonoid present in nuts, seeds, fruits and vegetables is quercetin, which showed a protective effect against cardiovascular disease [22]. Several in vivo and in vitro research have shown that flavonoids have the potential to break diverse stages of the cancer process [23].

p-Coumaric acid

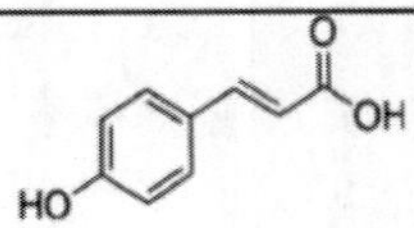

Caffeic acid

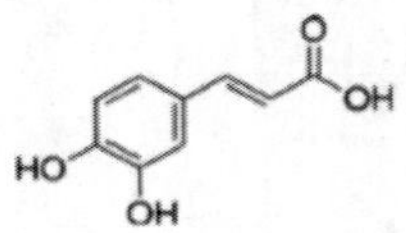

Chlorogenic acid

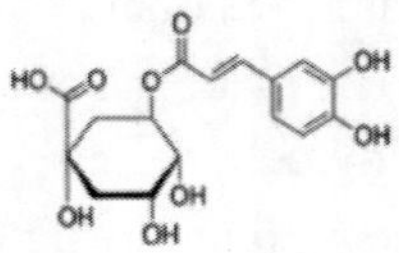

Ferulic acid

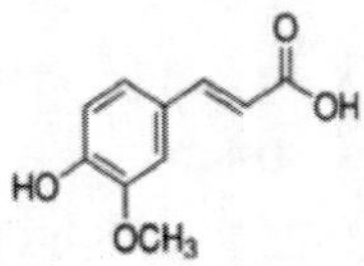

Gallic acid

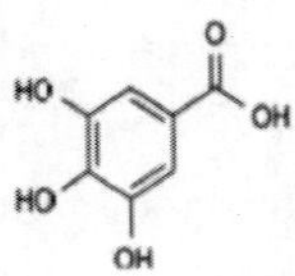

Vanillic acid

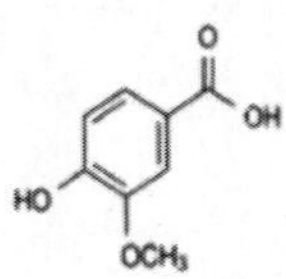

Quercetin

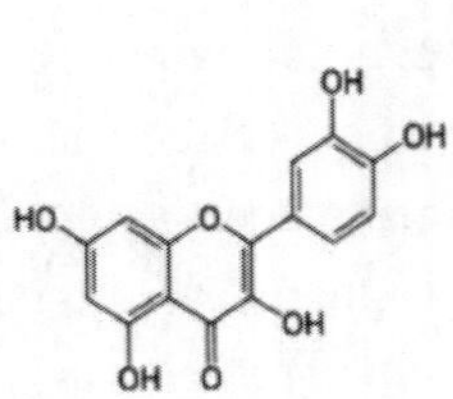

Myricetin

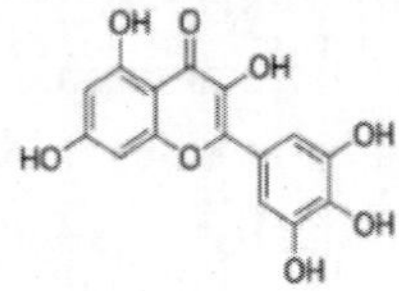

Hesperetin

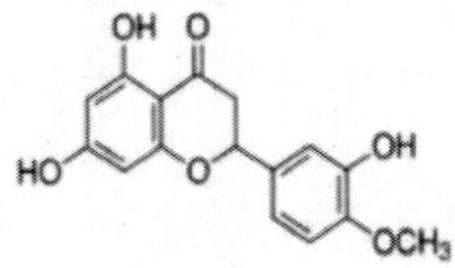

Apigenin

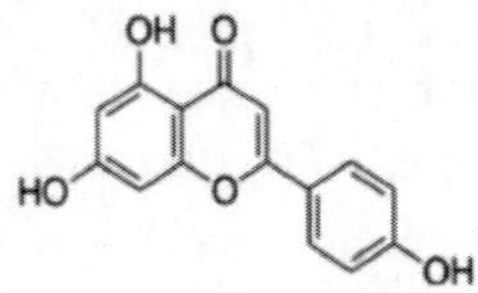

Catechin

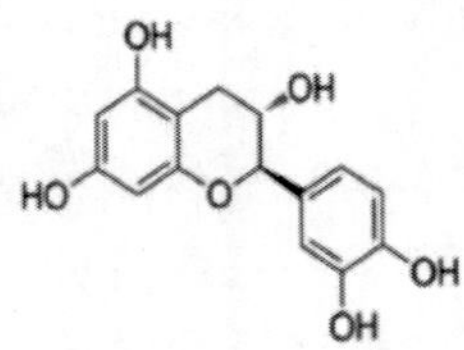

Delphinidin

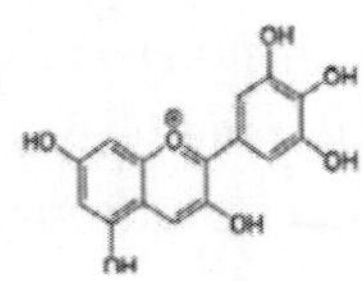

Cyanidin

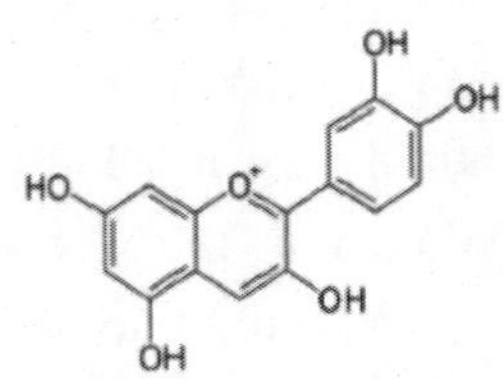

Resveratrol

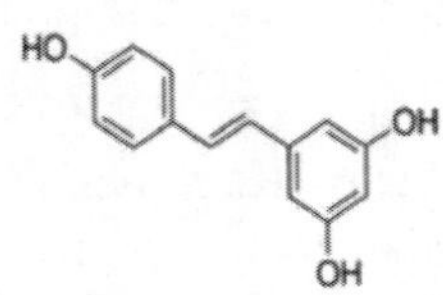

Figure 1. Chemical structure of some important polyphenolic compounds.

Previous studies indicated that these phytochemicals have antioxidant and anticarcinogenic activity [24]. Flavonoids may directly capture some radical species by the chain-breaking antioxidant mechanism. Additionally, flavonoids can suppress lipid peroxidation. Moreover, some flavonoids can chelate pro-oxidant metal ions, such as copper and iron which help in preventing free radical formation [24].

2.2.2. Phytoestrogens

A phytoestrogen is a plant-generated xenoestrogen and these compounds are divided into three main classes i.e., isoflavones, lignans and coumestans [25]. Similar to other phytochemicals, phytoestrogens have antioxidant activity and help in the scavenging of free radicals. In a study, it was reported that phytoestrogen has the potential to suppress tumor promoter–induced hydrogen peroxide and superoxide anion formation [26]. In addition to its antioxidant actions, genistein also enhances the activity of several antioxidant enzymes, including glutathione peroxidase, catalase, superoxide dismutase and glutathione reductase [26]. The di-phenolic compounds are similar to estrogen and, as might be expected, bind to the estrogen receptor. The plant-based phytoestrogens, particularly isoflavones, appear to compete with endogenous estrogen for receptor binding, yet they stimulate weaker estrogenic responses, and at certain concentrations, they exert anti-estrogenic actions [27]. Phytoestrogen action as an estrogen antagonist is dependent on the endogenous estrogen levels, the dietary concentration of these compounds, menopausal status and sex. Phytoestrogens excite sex hormone-binding globulin, which reduces the amount of circulating estrogen, and inhibits gonadotropin output to increase the duration of the menstrual cycle. In soybeans (legumes), a different components such as daidzein, genistein and isoflavones, observed largely [26]. Lignans are the largely available phytoestrogens since they have great involvement in plant cell wall formation, therefore subsist as minor constituents in most of the plants. Flaxseed is the chief dietary source of lignans, whereas it can be available in different concentrations in some plant sources such as whole grains, seaweed, soybeans, fruits, and vegetables [26]. Iso-flavonoids are the most widely calculated phytoestrogens that have the capacity to reduce cardiovascular disease. Soy phytoestrogens have the potential to reduce LDL oxidative susceptibility in humans, decrease thrombin formation and decrease the extent of atherosclerotic lesion formation [28]. It is important to note that isoflavone bioavailability is dependent on gut microflora activity. Thus, isoflavone absorption and its beneficial effects may be highly variable and could explain discrepant study results [29].

2.2.3. Lycopen

Lycopene is an acyclic carotenoid primarily present in tomatoes, watermelon and their products. Other minor sources may include some food products like guava, apricots, watermelon, grapefruit and papaya. The level of lycopene is majorly dependent on the variety and maturity of fruits and vegetables. In a study, it was observed that cooking with fat and oil can promote the extraction of lycopene [30]. There is inadequate proof that dietary supplementation of lycopene lowers LDL cholesterol levels [31] possibly because of an inhibition of cholesterol synthesis and increased LDL degradation. Some reports showed that lycopene intake, as measured by adipose tissue concentrations, is associated with reduced intimal wall thickness and risk of myocardial infarction [32]. The epidemiological studies propose that the consumption of lycopene may protect against different forms of cancer. The anticarcinogenic mechanisms of lycopene remain tentative, but their antioxidative capacities are believed to play a significant role, as oxidative stress is connected to carcinogenesis. It indicates that it may obstruct oxidative damage to lipids, lipoproteins and DNA [33].

2.2.4. Organosulfur Compounds

Organosulfur compounds extracted from garlic have free radical scavenging activity to inhibit lipid peroxidation and they also work as antioxidant compounds. The bioactive component diallyl sulfide may help to suppress the tumor promotion phase of carcinogenesis by reducing polyamine formation using inhibition of ornithine decarboxylase and possibly by stimulating DNA repair [34].

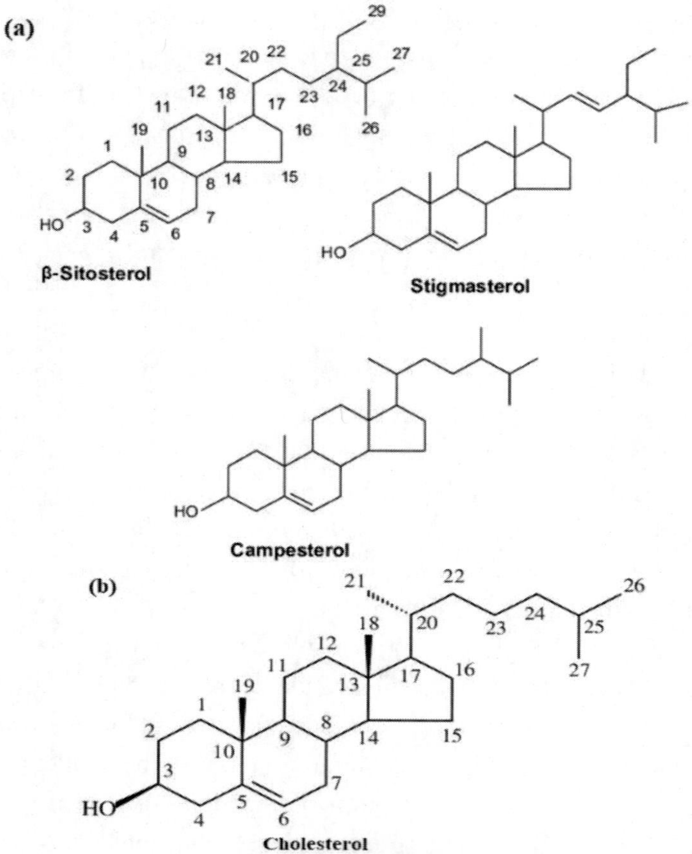

Figure 2. Chemical structure of (a) some plant-based sterols (b) cholesterol.

An additional mechanism by which garlic may suppress carcinogenesis is through a depression in nitrosamine formation [35]. The consumption of garlic, as well as garlic oil, has been shown to decrease triglyceride and LDL levels. The most effective mechanism includes a decrease in cholesterol and fatty acid synthesis and cholesterol absorption [36]. The antioxidant activity was observed from aged garlic extracts, because of some lipid and water-soluble compounds present in garlic extract [37]. In a different study, various organosulfur compounds like lipid-soluble allyl sulfides, allyl amino acid derivatives, saponins and flavonoids were observed in prolonged extraction of fresh garlic at room temperature [38]. The antioxidant activity was estimated from water-soluble organosulfur compounds such as S-allyl mercaptocyteine and S-allylcysteine [29]. Additionally, garlic extract contains lipid-soluble compounds such as triallyl sulfide, diallyl polysulfides, diallyl sulfide and diallyl disulfide also showed antioxidant effects. Moreover, diallyl sulfide, as well as diallyl disulfide bioactive components present in garlic, have anticarcinogenic effects [39]. In a cell culture and animal study, garlic extract showed potent inhibitors of tumorigenesis [40]. However, the epidemiologic study showed no powerful relationship between the intake of garlic and cancer prevention [41]. Allicin is the principal component that has been observed to cause a transient reduction in glutathione (GSH), which was connected with its antiproliferative action [35].

2.2.5. Phytosterol

Plant sterols or phytosterols are a group of substances that are present in the non-saponifiable fraction of plant oils. These compounds are abundantly found in some food groups like vegetable oils, nuts, and seeds. Phytosterols have various medicinal properties and are generally used to reduce blood cholesterol levels and also help to prevent various heart-related diseases. Structurally, plant sterols are analogous to cholesterol except that there always are some substitutions on the sterol side chain at the C24 position (Figure 2). In a study it was found that plant sterol cannot synthesize in human tissue and have very poor absorption in the human body, additionally, faster excretion from the liver was also reported in the same study [42]. The plentiful plant-based phytosterol in diets is β-sitosterol. However, some phytosterol such as stigmasterol, sitosterol and campesterol are considered as primary plant-based sterols in the diet. The phytosterol can be extracted from different plant-based sources, including rice bran oil, shea seed oil, pinewood pulp oil and soybean. It has been investigated that the phytosterols present in oryzanols (a group of ferulate esters of triterpene alcohol sand phytosterols), the unsaponifiable fraction of rice bran oil, decrease cholesterol levels in human plasma [43] and antioxidant activity of tocotrienols (a type of phytosterol) present in rice bran oil was observed [44].

2.2.6. Isothiocyanates

Isothiocyanate is formed by replacing the oxygen with a sulfur in the isocyanate group–N=C=S. The natural isothiocyanates are obtained from plants by enzymatic conversion of metabolites called glucosinolates. Several cruciferous vegetables, including cabbage, brussels sprouts, cauliflower and broccoli are an excellent source of isothiocyanates. Several naturally occurring forms of these phytochemicals include benzyl isothiocyanate, 2-phenethyl isothiocyanate, and sulforaphane [45]. The importance of these compounds has increased due to their noticeable chemopreventive capacity in human as well as animal cell cultures. In a study, it was concluded that various isothiocyanate components like 2-phenethyl isothiocyanate, β-napthyl, α-napthyl, benzyl isothiocyanate, and other arylalkyl isothiocyanates have a protective effect against tumorigenesis in various parts of the body such as liver, lungs, breast, esophagus and stomach [29]. Moreover, the use of 2-phenethyl isothiocyanate has been suggested as a chemo-preventive agent to reduce cancer particularly lung cancer in smokers [46]. Many anticarcinogenic actions were established to occur before or during carcinogen administration but not after [45]. The precise effects of isothiocyanates appear to be dependent on the form of isothiocyanate, treatment schedule, and objective tissue being examined. After thermal treatment, the uptake of isothiocyanates has remarkably decreased because of the inactivation of the myrosinase enzyme. Additionally, it was concluded that most of the isothiocyanates including synthetic and natural have anticarcinogenic activity because they have the potential to decrease activation of carcinogens and increase their detoxification.

2.2.7. Soluble Dietary Fiber

The soluble dietary fiber such as β-Glucan, pectin and psyllium, has been shown to lower total and low-density lipoproteins (LDL) cholesterol levels. The oats, barley, and some yeasts are a very good source of β-Glucan, whereas fruits and vegetables are abundant in pectin content. Moreover, psyllium is derived from the husks of blond psyllium seed and currently is being added to some foods, including cereals and other grain products. In a study, it was found that fiber from 2 servings of oats showed a cholesterol-lowering effect from 2 to 3% beyond what is achieved by serum cholesterol-lowering diet [47]. A comparable cholesterol-lowering effect of pectin, β-glucan and psyllium has been observed in another meta-analysis [48]. The cholesterol-lowering effect is directly related to the quantity of soluble dietary fiber consumed and different foods have varying amounts of these soluble fibers. Thus, soluble fiber has a modest cholesterol-lowering effect that goes beyond what can be achieved by lowering saturated fatty acids and cholesterol. In addition to lowering serum cholesterol levels, a high fiber intake may have other heart-health benefits, such as reducing blood pressure and inflammation. In another study, it was found that soluble fiber is also responsible to prevent the hyper-triglyceridemic response [49].

Vitamins like vitamin C and vitamin E are commonly known as antioxidant vitamins. These vitamins act both synergistically as well as singly for the prevention of oxidative reactions leading to several degenerative diseases including cardiovascular diseases, cancer and cataracts etc. [50]. Fruits and vegetables are rich source of vitamin C and vitamin E. These vitamins exert their protective action by free-radical scavenging mechanisms.

Vitamin C is a water-soluble vitamin. The recommended concentrations of vitamin C are maintained through daily consumption by food, as humans cannot synthesize ascorbic acid de novo. Vitamin C oxidation produces dehydroxy ascorbic acid, which is transferred into our cells via glucose transporters and then reduced back to ascorbic acid for cellular use [51]. McArdle et al. (2002) [52] Studied that supplementation of oral vitamin C leads to an increase in its plasma and skin content. Vitamin C is a powerful free radical scavenger and antioxidant that protects our DNA, tissues and cell membranes from oxidative damage. It also serves as an essential electron donor and cofactor during collagen hydroxylation, encouraging the maturation of extracellular and intracellular collagen [53]. Vitamin C also reduces the malondialdehyde content in the skin, which is an indicator of oxidative stress [52]. The food sources containing greater concentrations of vitamin C are raw green and red peppers, kiwifruit, oranges, strawberries, Brussels sprouts, broccoli and oranges. The Food and Nutrition Board of the Institute of Medicine's recommended dietary allowance of vitamin C is 75 mg a day for women and 90 mg a day for men ages 19 years and older. Scavenging of aqueous radicals by the synergistic effect of ascorbic acid along with tocopherol supplementation is a well-known antioxidant mechanism [54].

Vitamin E is a group of fat-soluble vitamin. The biologically active and most abundant form of vitamin E is alpha-tocopherol, which also is the leading form used in human metabolism [55, 56]. Vitamin E along with tocopherols can scavenge singlet oxygen and transfer hydrogen atoms to other reactive species. This helps in protecting the peroxidation of PUFA within the biological membrane and low-density lipoproteins (LDL). Tocotrienols are more motile within the biological membrane than tocopherols because of the presence of the unsaturated side-chain and hence penetrate tissues with saturated fatty layers, i.e., in liver and brain more efficiently. Alpha-tocopherol acts as a shield to protect the skin from UVB damage by scavenging free radicals, halting the formation of ROS, stabilizing the membranes and surface of cells, minimizing the activation of nuclear factors kappa B and reducing the number of apoptotic cells [53, 57]. Alpha-tocopherol is also accepted to be photoprotective as various researchers studied that alpha-tocopherol along with vitamin C supplementation increases the minimal erythema dose (MED) [55]. UV light decreases the concentration of alpha-tocopherol in the skin and promotes skin aging [55]. Some naturally occurring sources of vitamin E include plant seeds such as peanuts, sunflower seeds, almonds, pecans, walnuts, sesame seeds and pistachios and are also found in lower amounts in fruits and vegetables [58, 59]. The Food and Nutrition Board of the Institute of Medicine's recommended dietary allowances of vitamin E is 15 mg a day for women and men ages 19 years and older. However, the estimated intake might be low because they did not account for the fats added while cooking. They have more recycling ability and are a better inhibitor of liver oxidation [60]. Selenium and vitamin E has a synergistic role against lipid peroxidation. Carotenoids like βcarotene, lycopene, zeaxanthin, lutein are known to be the most efficient singlet oxygen quencher in the biological systems without the production of any oxidizing products. βcarotene traps peroxy free radicals in tissues at low oxygen concentrations. Hence β-carotene complements the antioxidant properties of vitamin E.

2.3. Minerals

2.3.1. Copper

Copper with many innate properties found naturally in the soil as a trace mineral and also acts as an element of fundamental importance for the functioning and formation of several proteins and enzymes, such as Cu/Zn superoxide dismutase and cytochrome C oxidase, which are involved in the processes of respiration, energy metabolism and DNA synthesis [61]. Also, during collagen crosslinking with skin pigmentation with tyrosinase and lysyl oxidase, copper serves as an essential cofactor in enzymatic reactions [62, 63]. Copper promotes fibroblast proliferation and keratinocyte leading to wound repair and skin rejuvenation through its collagen crosslinking properties [64]. Lastly, the anti-bacterial and anti-inflammatory effects of copper make it suitable for topical treatments for wound healing to promote the repair of damaged skin [64]. Nuts, seafood, grains and meat are a good sources of dietary copper. The copper deficiencies are rare in humans and are mostly observed in people with intestinal malabsorption. The Food and Nutrition Board of the Institute of Medicine's recommended dietary allowance (RDA) of copper is 900 lg a day for women and men ages 19 and older. Because of its high redox activity, the therapeutic efficacy of copper coordination compounds is not limited to antiproliferative action. Copper coordination compounds can be highly effective in treating viral infections [65], inflammatory diseases [66], and microbial infections [67] by multiple mechanisms of action.

2.3.2. Selenium

Selenium is an essential mineral and a trace element, found in soil, specific foods and water, the majority of which are highly bioavailable. It exists in various forms, foods of animal origin contain selenocysteine and seleno-methionine and those of plant origin include seleno-methionine. Seleno-methionine is essential to animals and humans, but they can produce selenocysteine from seleno-methionine in the organism. It plays an important role in cell apoptosis, DNA synthesis and repair, and guarding against oxidative damage [53]. Selenium works through thioredoxin reductases and glutathione peroxidases, which remove damaging hydrogen peroxide, lipid hydroperoxides, and peroxy-nitrite formed during oxidative stress, leading to defense against DNA damage and cell membrane stabilization [68, 69]. The concentration of selenium depends upon the concentration within the soil of various land sources, where those plants and animals were produced. The Food and Nutrition Board of the Institute of Medicine's recommended dietary allowances (RDA) of selenium is 55 μg a day for women and men ages 19 years and older. The excess of selenium is excreted in the urine as a seleno-amino sugar in large quantities and it is excreted through the urine may excrete trimethyl-seleniumions, while the respiratory processes as dimethyl-selenide [70].

2.3.3. Zinc

Zinc is an essential mineral that occurs naturally in red meat, seafood, whole grains. Zinc also acts as an essential component for a large number (>300) of enzymes participating in the degradation and synthesis of nucleic acids, proteins, carbohydrates and lipids, as well as in the metabolism of other micronutrients. Zinc helps in stabilization of the molecular structure of membranes and cellular components and in this way contributes to the organ integrity and maintenance of cell. Also, zinc protects against UV-induced cytotoxicity, lipid peroxidation, and oxidative stress induced by ROS made and distributed within the cytosol by macrophages [68, 69]. The maximum concentration of the skin's zinc stores is available in the epidermis, where it acts as an essential element for keratinocyte differentiation and epidermal proliferation [71]. Zinc plays an important role in keratinocyte cell survival and wound healing [71]. It also acts as an anti-inflammatory agent by hindering intercellular adhesion molecule 1, a pro-inflammatory marker of keratinocytes, and reduces the production of nitric oxide [71]. The zinc bioavailability mainly depends upon zinc absorption and reabsorption properties found within the human intestines. The bioavailability of zinc in the body is also influenced by varying concentrations of elements such as copper, iron, phosphate, folate and calcium. Non-milled whole grains, seafood and red meat are the foods having the highest dietary concentrations of zinc. The Food and Nutrition Board of the Institute of Medicine's recommended dietary allowances (RDA) of zinc is 8 mg a day for women and 11 mg a day for men ages 19 years and older.

2.3.4. Amino Acids

N-Acetylcysteine is the prodrug form of L-cysteine and has both tropical and oral bioavailability [72, 73]. N-Acetylcysteine gives rise to glutathione, which is the most abundant endogenous intracellular antioxidant and plays a pivotal role in the body's antioxidant defence [73, 74]. As the human's age increases, the rate of glutathione synthesis decreases, leading to glutathione deficiency, which leaves the body susceptible to oxidative stress [74]. Increased oxidative stress leads to increased aging and DNA damage [74]. A study found that elderly people had significantly lower intracellular glutathione synthesis and concentrations when compared to younger people [74]. After 2 weeks of glycine and cysteine supplementation, elderly people had a significant increase in their cysteine, red blood cell glycine and glutathione concentrations. Also, after supplementation, there was a decline in both plasma markers and oxidative stress for oxidant damage [74]. The Food and Nutrition Board of the Institute of Medicine's recommended dietary allowance (RDA) of N-acetylcysteine has not been established.

2.3.5. Yeast and Fungi

Mushrooms and yeasts are representatives of the taxonomic kingdom of Fungi, which assembles organisms distinct from animals, plants, and bacteria. Mushrooms form multicellular, mostly epigeous, organized structures called fruiting bodies (macrofungi) which are produced by filamentous branched hyphae called mycelia. Practically more than 1,000 mushroom species are edible; approximately 300 are of dietary and medicinal importance [75]. A great majority of edible mushrooms fall into phylum Basidiomycota and some to Ascomycota (as common morel, *Morchellaesculenta,* and winter truffle, *Tuber brumale*); and these two phyla are representatives of the subkingdom of higher fungi (*Dikarya*).

Yeasts are eukaryotic single-cellular organisms, which are not visible to the bare eye. Some yeasts, common in many sectors of the food industry, belong to the order Saccharomycetales, such as baking yeasts (*Saccharomyces cerevisiae*). Therefore, the term "yeasts" is often used as a synonym of *S. cerevisiae.* Other Saccharomycetales, such as *Saccharomyces boulardii,* are utilized for purposes of probiotic food production.

Edible mushrooms possess attractive sensory properties which place. The unique flavor of specialty mushrooms is associated with the presence of glutamate, which gives an oral sensation called umami taste. In fact, glutamate is also found in meat, and the meat-like texture of mushrooms is valued in vegetarian and conventional cooking as a meat substitute. Most mushrooms provide a low caloric value of 100–200 kcal/100g [76]. Despite their low energy contribution, mushrooms are very satiable. The nutritional value of mushrooms is related to having considerable amounts of protein, fiber, vitamins, and minerals and to their low levels of fat, cholesterol, and sodium. Due to having the same group of nutrients, mushrooms are comparable to vegetables but are different in terms of nutrient density. Similar to vegetables, the water content in fresh mushrooms accounts for between 70–95% of its fresh weight (FW), which depends on the environmental conditions and the time of harvest; but protein content, however, is greater than in most vegetables. Proteins in mushrooms range from 15–35% of dry mass (DM) and are less digestible than animal proteins (about 75% versus 90% or more). When compared to the amino acid composition of milk, mushroom proteins can supply approximately half of each of the essential amino acids [77]. Hence, edible mushrooms may be favorable dietary components; and the combination of different mushroom species would provide considerable amounts of amino acids [78]. Mushrooms possess an array of desirable biologically-active substances which are important for reducing the risk and are helpful to the management of certain diseases [79, 80]. Mushrooms demonstrate health-promoting effects, principally immunomodulatory, hypocholesterolemic, hypoglycemic, antitumor, antimicrobial, and hepatoprotective. Therefore, they are interesting materials for the functional food industry. Many of these beneficial functions are associated with the fraction of structural polysaccharides, namely β-glucans, which consist of glucose monomers combined with β-glycosidic bonds. Biologically active glucans may form complexes with proteins. The most active forms of β-glucans from mushrooms are built from D-glucose units joined by 1,3-1,6-β-Dglycosidic bonds. They were isolated from several edible mushrooms, for example, lentinan from shiitake mushroom (*Lentinusedodes*); schizophyllan from *Schizophyllum commune;* or ganopoly from *Ganodermalucidum.* The cholesterol-lowering properties of mushrooms are also related to triterpenes and alkaloid eritadenine isolated from shiitake and button mushrooms. Triterpenes are a group of substances that exhibit not only the cholesterol-lowering potency but also show anti-inflammatory, hypotensive, and antiplatelet activity [81]. Mushrooms are sources of antioxidant compounds, which are active against free radicals, thus protecting tissues from oxidative damage, and the major antioxidants in mushrooms are polyphenols. Another compound demonstrating antioxidant properties in mushrooms is ergosterol. Ergosterol is a precursor of vitamin D2 which can be converted to this vitamin with the participation of ultraviolet (UV) light. In some studies, ergosterol content was positively related to the antioxidant activity of edible fungi [82]. Medicinal mushrooms have their history as immunostimulatory and immunomodulating therapeutics in the management of tumour, viral, and bacterial diseases; but little is known on dietary intakes of commonly consumed mushrooms in association with immune functions in humans. A dietary survey of Chinese women showed that mushroom consumption can decrease the risk of breast cancer by over 60% [83]. This beneficial effect was associated with a daily consumption of at least 10 g of mushrooms, of which the most commonly eaten was the button mushroom. Based on the low-calorie value, low fat [84], and high fiber contents [85], mushrooms can be valuable dietary components for people with high blood cholesterol.

Virtually two representatives of the Saccharomyces genus, *S. cerevisiae* and *S. boulardii,* are utilized in functional food manufacturing processes. Similar to mushrooms, cell walls of S. cerevisiae are sources of water-insoluble β-glucans, polysaccharides consisting of glucose molecules, which are connected with 1,3-1,6-β-D-glycosidic bonds. *Saccharomyces boulardii*are probiotic yeasts, which means they can beneficially affect human health when ingested. *S. boulardii* is far more resistant to the gastrointestinal environment than the probiotic bacteria. It can well tolerate high acidity of gastric juice and the digestion by bile acids in the intestine and can also prevent the growth of harmful microorganisms. *S. boulardii* was used for manufacturing some fermented food products with probiotic activity. An example of such products is a non-dairy probiotic refreshing beverage prepared from organic grains, beans, and seeds. Grainfields Wholegrain Liquid is naturally fermented by the lactic acid bacteria and yeasts *S. cerevisiae* and *S. boulardii,* which allows the body to replenish the supply of probiotic organisms in the intestine.

Yeasts are capable of assimilating large quantities of the trace element selenium, mostly in a form of selenomethionine, which is easily bioavailable for humans [86]. As a safer selenium source compared to inorganic forms, selenium-enriched yeasts were used to manufacture supplements. But currently, selenium yeasts are desirable in functional food for the prevention of diseases of oxidative stress background.

Spirulina is a spiral shaped micro algae that grows naturally in the wild in warm, freshwater lakes. It is a blue green algae *Spirulina plantensis* or *Spirulina maxima* belongs to the family *Oscillatoriaceae*. They are found in fresh water and marine habitats, mostly in warmer regions and are found in hot springs. Spirulina is symbiotic, multicellular and filamentous blue green microalgae with symbiotic bacteria that fix nitrogen from the air. Spirulina is an economically important filamentous cyanobacterium. The annual production of algae is about 10,000 tons which makes it largest microalgal cultivation industry in the world. Spirulina is a protein-rich food product with a relatively low carbohydrate content of around 15% dry weight. It also contains phycocyanin containing phycobiliproteins which are through to be some of spirulina's active ingredients. In addition, spirulina also contains several trace minerals, vitamins, pro and pseudo-vitamins. Phycocyanin is the main active compound in spirulina, it has powerful antioxidant and anti-inflammatory properties. Spirulina may have anti-cancer properties, especially against a type of precancerous lesion called OSMF (*Oralsubmucos fibrosis).* Also, it lowers blood pressure levels. Spirulina supplements have been shown to be very effective against allergic rhinitis, helping to reduce various symptoms.

CONCLUSION

The mechanism of action of bioactive compounds may be multiple, especially in the food matrix. Based on evidence supported in this chapter it can be concluded that the bioactive compounds are majorly available in most of the food components such as spices, fruits, vegetables, legumes, whole grains, living microorganisms and in some animal groups. Plant-based foods mostly have bioactive compounds such as phenolic compounds, flavonoids, tannins, alkaloids, diterpenes and sulfur-containing compounds which can scavenge free radicals in the human body and have antiproliferative action. Because of this the herbs and spices majorly have antioxidant antimicrobial properties. The soluble dietary fiber such as β-Glucan, pectin and psyllium, has the property to reduce LDL cholesterol levels in the human body. Vitamin C and E also have antioxidant properties which act both synergistically as well as singly for the prevention of oxidative reactions leading to several degenerative diseases. Minerals present in different foods have an important role in cell apoptosis, DNA synthesis and repair, and guarding against oxidative damage. Considering their nutritional and biological properties, mushrooms and yeasts are promising candidates for functional food manufacturing, as well as ingredients in healthy food formulations. Overall the bioactive compounds have a mechanism in the body that may promote good health and prevent degenerative diseases as atheroscle-rosis, cardiovascular disease, cancer, diabetes, as well as certain functional disorders. Functional foods were recently redefined as "important sources in the prevention, management and treatment of chronic diseases of the modern age," and the food component that contains bioactive compounds is well placed in this concept.

REFERENCES

[1] Cammack, R., Atwood, T., Campbell, P., Parish, H., Smith, A., Vella, F., and Stirling, J. (2006). *Oxford Dictionary of Biochemistry and Molecular* Biology (2nd Ed). Oxford University Press, 74-75.

[2] Galanakis, C. M. (2017). Nutraceutical and Functional Food Components, *Effects of Innovative Processing Techniques,* 1st Ed. Academic Press.

[3] Elga, Š., Galoburda, R., and Seglioa, D. (2013). Sea Buckthorn Vegetative Parts – A Good Source of Bioactive Compounds. Proceedings of the Latvian Academy of Sciences. Section B. *Natural, Exact, and Applied Sciences,* 2(67): 101-108.

[4] Huang, T. C., Lu, K. T., Wo, Y. Y. P., Wu, Y. J., and Yang, Y. L. (2011). Resveratrol Protects Rats from Aβ-induced Neurotoxicity by the Reduction of iNOS Expression and Lipid Peroxidation. *PLOS ONE,* 6(12): e29102. https://doi.org/10.1371/journal.pone.0029102

[5] Mahmoud, M. F., Hassan, N. A., El Bassossy, H. M., and Fahmy, A. (2013). Quercetin protects against diabetes-induced exaggerated vasoconstriction in rats: effect on low grade inflammation. *PLOS ONE,* 8(5): 63784.

[6] Srinivasan, K. (2013). Dietary spices as beneficial modulators of lipid profile in condition of metabolic disorders and disease. *Food Function.* 4: 503–521.

[7] Baselga-Escudero, L., Souza-Mello, V., Pascual-Serrano, A., Rachid, T., Voci, A., Demori, I., and Grasselli, E. (2017). Beneficial

effects of the Mediterranean spices and aromas on non-alcoholic fatty liver disease. *Trends Food Sci. Technol.* 61: 141–159.

[8] Peter, K. V. (2001). *Handbook of Herbs and Spices*; Woodhead Publishing Limited and CRC Press LLS: Cambridge, UK, Volume 1.

[9] Yesiloglu, Y., Audin, H., and Kilic, I. (2013). In vitro antioxidant activity of various extracts of ginger seed. *Asian J. Chem.* 25: 3573–3578.

[10] AllwynSundarRaj, A., Aaron, S., Seinenbalg, S. S., Tiroutchelvamaa, D., and Ranganathan, T. V. (2014). Review on recent trends in isolation of antioxidants from spices and its biological effects of essential oils. *J. Eng. Res. Appl.* 4: 75–84.

[11] Kaefer, C. M., and Milner, J. A. (2008). The role of herbs and spices in cancer prevention. *J. Nutr. Biochem.* 19: 347–354.

[12] Liu, R. H. (2003). Health benefits of fruit and vegetables are from additive and synergistic combinations of phytochemicals. *The American journal of clinical nutrition,* 78(3): 517S-520S.

[13] Bravo, L. (1998). Polyphenols: chemistry, dietary sources, metabolism, and nutritional significance. *Nutritional Reviews.* 56: 317– 333.

[14] Peterson J, and Dwyer J. (1998). Flavonoids: dietary occurrence and biochemical activity. *Nutr Res.* 18:1995–2018.

[15] Pignatelli, P., Ghiselli, A., Buchetti, B., Carnevale, R., Natella, F., Germano, G., and Violi, F. (2006). Polyphenols synergistically inhibit oxidative stress in subjects given red and white wine. *Atherosclerosis,* 188(1): 77-83.

[16] Kerry, N. L., and Abbey, M. (1997). Red wine and fractionated phenolic compounds prepared from red wine inhibit low density-incomplete lipoprotein oxidation in vitro. *Atherosclerosis*. 135: 93–102.

[17] Rice-Evans, C., Miller, N., and Paganga, G. (1997). Antioxidant properties of phenolic compounds. *Trends in plant science,* 2(4): 152-159.

[18] Schmitt, C. A., and Dirsch, V. M. (2009). Modulation of endothelial nitric oxide by plant-derived products. *Nitric oxide*, 21(2): 77-91.

[19] Afoakwa, E. O., Paterson, A., Fowler, M., and Ryan, A. (2008). Flavor formation and character in cocoa and chocolate: a critical review. *Critical reviews in food science and nutrition*, 48(9): 840-857.

[20] Mayorga-Gross, A. L., and Esquivel, P. (2019). Impact of cocoa products intake on plasma and urine metabolites: A review of targeted and non-targeted studies in humans. *Nutrients,* 11(5): 1163.

[21] Engler, M. B., and Engler, M. M. (2006). The emerging role of flavonoid-rich cocoa and chocolate in cardiovascular health and disease. *Nutrition reviews*, 64(3): 109-118.

[22] Hollman, P. H., and Katan, M. B. (1999). Dietary flavonoids: intake, health effects and bioavailability. *Food and chemical toxicology,* 37(9-10): 937-942.

[23] Chahar, M. K., Sharma, N., Dobhal, M. P., and Joshi, Y. C. (2011). Flavonoids: A versatile source of anticancer drugs. *Pharmacognosy reviews,* 5(9): 1.

[24] Jaradat, N., Hussen, F., and Al Ali, A. (2015). Preliminary phytochemical screening, quantitative estimation of total flavonoids, total phenols and antioxidant activity of Ephedra alata Decne. *J. Mater. Environ. Sci,* 6(6): 1771-1778.

[25] Kurzer, M. S., and Xu. X. (1997). Dietary phytoestrogens. *Annu Rev Nutr.* 17:353–381.

[26] Noguer, M. A., Cerezo, A. B., Navarro, E. D., and Garcia-Parrilla, M. C. (2012). Intake of alcohol-free red wine modulates antioxidant enzyme activities in a human intervention study. *Pharmacological Research,* 65(6): 609-614.

[27] Setchell, K. D. (1998). Phytoestrogens: the biochemistry, physiology, and implications for human health of soy isoflavones. *The American journal of clinical nutrition,* 68(6): 1333S-1346S.

[28] Sargeant, P., Farndale, R. W., and Sage, S. O. (1993). The tyrosine kinase inhibitors methyl 2, 5-dihydroxycinnamate and genistein

reduce thrombin-evoked tyrosine phosphorylation and Ca2+ entry in human platelets. *FEBS letters,* 315(3): 242-246.

[29] Kris-Etherton, P. M., Hecker, K. D., Bonanome, A., Coval, S. M., Binkoski, A. E., Hilpert, K. F., and Etherton, T. D. (2002). Bioactive compounds in foods: their role in the prevention of cardiovascular disease and cancer. *The American journal of medicine*, 113(9): 71-88.

[30] de Alvarenga, J. F. R., Tran, C., Hurtado-Barroso, S., Martinez-Huélamo, M., Illan, M., and Lamuela-Raventos, R. M. (2017). Home cooking and ingredient synergism improve lycopene isomer production in Sofrito. *Food Research International*, 99: 851-861.

[31] Cheng, H. M., Koutsidis, G., Lodge, J. K., Ashor, A., Siervo, M., and Lara, J. (2017). Tomato and lycopene supplementation and cardiovascular risk factors: A systematic review and meta-analysis. *Atherosclerosis,* 257: 100-108.

[32] Arab, L., and Steck, S. (2000). Lycopene and cardiovascular disease. *The American journal of clinical nutrition*, 71(6): 1691S-1695S.

[33] Basu, A., and Imrhan, V. (2007). Tomatoes versus lycopene in oxidative stress and carcinogenesis: conclusions from clinical trials. *European journal of clinical nutrition,* 61(3): 295-303.

[34] Sumiyoshi, H., and Wargovich, M. J. (1989). Garlic (Allium sativum): a review of its relationship to cancer. *Asia Pacific Journal of Pharmacology*. 4:133–140.

[35] Milner, J. A. (2006). Preclinical perspectives on garlic and cancer. *The Journal of nutrition,* 136(3): 827S-831S.

[36] Berthold, H. K., Sudhop, T., and von Bergmann, K. (1998). Effect of a garlic oil preparation on serum lipoproteins and cholesterol metabolism: a randomized controlled trial. *Journal of the American Medical Association,* 279(23): 1900-1902.

[37] Borek, C. (2001). Antioxidant health effects of aged garlic extract. *Journal of Nutrition*, 131: 1010S–1015S.

[38] Katsuki, T., Hirata, K., Ishikawa, H., Matsuura, N., Sumi, S. I., and Itoh, H. (2006). Aged garlic extract has chemopreventative effects

on 1, 2-dimethylhydrazine-induced colon tumors in rats. *The Journal of nutrition,* 136(3): 847S-851S.

[39] Corzo-Martínez, M., Corzo, N., and Villamiel, M. (2007). Biological properties of onions and garlic. *Trends in food science & technology,* 18(12): 609-625.

[40] Yang, C. S., Chhabra, S. K., Hong, J. Y., and Smith, T. J. (2001). Mechanisms of inhibition of chemical toxicity and carcinogenesis by diallylsulfide (DAS) and related compounds from garlic. *The Journal of nutrition*, 131(3): 1041S-1045S.

[41] Jin, Z. Y., Wu, M., Han, R. Q., Zhang, X. F., Wang, X. S., Liu, A. M., and Zhao, J. K. (2013). Raw garlic consumption as a protective factor for lung cancer, a population-based case–control study in a Chinese population. *Cancer prevention research,* 6(7): 711-718.

[42] Brufau, G., Canela, M. A., and Rafecas, M. (2008). Phytosterols: physiologic and metabolic aspects related to cholesterol-lowering properties. *Nutrition Research,* 28(4): 217-225.

[43] Lichtenstein, A. H., Ausman, L. M., Carrasco, W., Jenner, J. L., Ordovas, J. M., Schaefer, E. J. (1994). Hypercholesterolemic effect of dietary cholesterol in diets enriched in polyunsaturated and saturated fat. Dietary cholesterol, fat saturation, and plasma lipids. *ArterisclerThromb.* 14:168–175.

[44] Ozdestan, O., Erol, T., and Acar, A. (2014). Phytosterols in rice bran and usage of rice bran in food industry. In *Proceedings of 9th Baltic Conference on Food Science and Technology "Food for Consumer Well-Being"* FoodBalt (pp. 24-27).

[45] Hecht, S. S. (1999). Chemoprevention of cancer by isothiocya-nates, modifiers of carcinogen metabolism. *Journal of Nutrition,* 129:768S–774S.

[46] Conaway, C. C., Wang, C. X., Pittman, B., Yang, Y. M., Schwartz, J. E., Tian, D., and Chung, F. L. (2005). Phenethyl isothiocyanate and sulforaphane and their N-acetylcysteine conjugates inhibit malignant progression of lung adenomas induced by tobacco carcinogens in A/J mice. *Cancer research,* 65(18): 8548-8557.

[47] Anderson, J. W., Baird, P., Davis, R. H., Ferreri, S., Knudtson, M., Koraym, A., and Williams, C. L. (2009). Health benefits of dietary fiber. *Nutrition reviews*, 67(4): 188-205.

[48] AbuMweis, S. S., Jew, S., and Ames, N. P. (2010). β-glucan from barley and its lipid-lowering capacity: a meta-analysis of randomized, controlled trials. *European journal of clinical nutrition,* 64(12): 1472-1480.

[49] Fried, S. K., and Rao, S. P. (2003). Sugars, hypertriglyceridemia, and cardiovascular disease. *The American journal of clinical nutrition*, 78(4): 873S-880S.

[50] Elliot, J. G. (1999). Application of antioxidant vitamins in foods and beverages. *Food Technology,* 53:46–48.

[51] Subramani, T., Yeap, S. K., Ho, W. Y., Ho, C. L., Omar, A. R., Aziz, S. A., Rahman, N. M., and Alitheen, N. B. (2014). Vitamin C suppresses cell death in MCF-7 human breast cancer cells induced by tamoxifen. *J Cell Mol Med.* 18:305–13.

[52] McArdle, F., Rhodes, L. E., Parslew, R., Jack, C. I. A, Friedmann, P. S., and Jackson, M. J. (2002). UVR-induced oxidative stress in human skin in vivo: effects of oral vitamin C supplementation. *Free RadicBiol Med.* 33:1355–62.

[53] Wu, X., Cheng, J., and Wang, X. (2017). Dietary antioxidants: potential anticancer agents. *Nutr Cancer.* 69:521–33.

[54] Lee, J., Koo, N., and Min, D. B. (2004). Reactive oxygen species, aging and antioxidative nutraceuticals. *CRFSFS*, 3:21–33.

[55] Evans, J. A., and Johnson, E. J. (2010). The role of phytonutrients in skin health. *Nutrients.* 2:903–28.

[56] Jiang, Q. (2014). Natural forms of vitamin E: metabolism, antioxidant and anti-inflammatory activities and the role in disease prevention and therapy. *Free Radic Biol Med.* 72:76–90.

[57] Maalouf, S., El-Sabban, M., Darwiche, N., and Gali-Muhtasib, H. (2002). Protective effect of vitamin E on ultraviolet B light-induced damage in keratinocytes. *Mol Carcinog.* 34:121–30.

[58] McLaughlin, P. J., and Weihrauch, J. L. (1979). Vitamin E content of foods. *J Acad Nutr Diet.* 75:647–65.

[59] Dreher, M. L. (2012). Pistachio nuts: composition and potential health benefits. *Nutr Rev.,* 70:234–40.

[60] Watkins, T. R., Bierenbaurn, M. I., and Giampalala, A. (1999). Tocotrienols: biological and health effects. In: Papas AM, editor. *Antioxidant status, diet, nutrition and health.* Boca Raton: CRC Press; pp. 479–496.

[61] Fengyi, Z., Weifan, W., Wen, L., Li, X., Shilong, Y., Xu-Min, C., Mengyi, Z., Meng, L., Mengtao, M., and Hai-Jun, X. (2018). High anticancer potency on tumorcells of dehydroabietylamine Schiff-base derivatives and a copper(II) complex. *Eur. J. Med. Chem. 146*: 451–459.

[62] Yeowell, H. N., Marshall, M. K., Walker, L. C., Ha, V., and Pinnell, S. R. (1994). Regulation of lysyl oxidase mRNA in dermal fibroblasts from normal donors and patients with inherited connective tissue disorders. *Arch Biochem Biophys.* 308:299–305.

[63] Kanteev, M., Goldfeder, M., and Fishman, A. (2015). Structurefunction correlations in tyrosinases. *Protein Sci.* 24:1360–9.

[64] Driscoll, M. S., Kwon, E. K., Skupsky, H., Kwon, S. Y., and GrantKels, J. M. (2010). Nutrition and the deleterious side effects of nutritional supplements. *Clin Dermatol*. 28:371–379.

[65] Lebon, F., Boggetto, N., Ledecq, M., Durant, F., Benatallah, Z., Sicsic, S. Lapouyade, R., Kahn, O., Mouithys-Mickalad, A., and Deby-Dupont, G. (2002). Metal-Organic Compounds: A New Approach for Drug Discovery: N1-(4-Methyl-2-Pyridyl)-2,3,6-Trimethoxybenzamide Copper (II) Complex as an Inhibitor of Human Immunodeficiency Virus 1 Protease. *Biochem. Pharmacol.* 63: 1863–1873.

[66] Roch-Arveiller, M., Huy, D. P., Maman, L., Giroud, J. P., Sorenson, J. R. J. (1990). Non-Steroidal Anti-Inflammatory Drug-Copper Complex Modulation of Polymorphonuclear Leukocyte Migration. *Biochem. Pharmacol.* 39: 569–574.

[67] Stanila, A., Braicu, C., Stanila, S., and Pop, R. M. (2011). Antibacterial Activity of Copper and Cobalt Amino Acids Complexes. *Not. Bot. Horti Agrobot. Cluj-Na.* 39: 124–129.

[68] Wintergerst, E. S., Maggini, S., and Hornig, D. H. (2007). Contribution of selected vitamins and trace elements to immune function. *Ann Nutr Metab.* 51:301–23.

[69] Pinnell, S. R. (2003). Cutaneous photodamage, oxidative stress, and topical antioxidant protection. *J Am Acad Dermatol.* 48:1–19.

[70] Bendhal, L., and Gammelgaard, B. (2004). Separation and identification of Semethyl-seleno-galactosamine, a new metabolite in basal human urine by HPLC-ICP-MS and CE-nano-ESI-(MS). *Journal of Analytical Atomic Spectrometry,* 19: 950–957.

[71] Ogawa Y, Kawamura T, and Shimada S. (2016). Zinc and skin biology. *Arch Biochem Biophys.* 611:113–9.

[72] Cassidy, P. B., Liu, T., Florell, S. R., Honeggar, M., Leachman, S. A., Boucher, K. M., and Grossman, D. (2017). A phase II randomized placebo-controlled trial of oral Nacetylcysteine for protection of melanocytic nevi against UV-induced oxidative stress in vivo. *Cancer Prev Res* (Phila) 10:36–44.

[73] Kang, S., Chung, J. H., Lee, J. H., Fisher, G. J., Wan, Y. S., Duell, E. A., and Voorhees, J. J.(2003). Topical N-acetyl cysteine and genistein prevent ultraviolet-light-induced signaling that lead to photoaging in human skin in vivo. *J Invest Dermatol.* 120:835–41.

[74] Sekhar, R. V., Patel, S. G., Guthikonda, A. P., Reid, M., Balasubramanyam, A., Taffet, G. E., and Jahoor, F. (2011). Deficient synthesis of glutathione underlies oxidative stress in aging and can be corrected by dietary cysteine and glycine supplementation. *Am J Clin Nutr.* 94:847–53.

[75] Boa, E. (2004). Wild edible fungi: a global overview of their use and importance to people. *NonWood Forest Products*, No. 17, FAO, Forestry Department, Rome, Italy.

[76] Ulziijargal, E., and Mau, J. L. (2011). Nutrient compositions of culinary-medicinal mushroom fruiting bodies and mycelia. *Int J*

Med Mushrooms. 13(4):343-349. doi: 10.1615/intjmedmushr.v13.i4.40.

[77] Kurtzman, Jr., R. H. (2005). Mushrooms: sources for modern western medicine. *MicologíaAplicada International,* 17(2): 21-33.

[78] Ribeiro, B., Guedes de Pinho, P., Andrade, P. B., Baptista, P., and Valentão, P. (2009). Fatty acid composition of wild edible mushrooms species: A comparative study, *Microchemical Journal,* 93: 29-35.

[79] Dubost, N. J., Ou, B., and Beelman, R. B. (2007). Quantification of polyphenols and ergothioneine in cultivated mushrooms and correlation to total antioxidant capacity. *Food Chemistry,* 105:727–735.

[80] Witkowska, A. M., ZujkoMalgorzata, E. and Iwona, M. C. (2011). Comparative study of wild dibble mushrooms as source of Antioxidants. *International Journal of Medical Mushrooms,* 13 (4): 335-341.

[81] Cheung, P. C. K. (2010). The nutritional and health benefits of mushrooms. *Nutrition Bulletin,* 35(4): 292–299.

[82] Shao, S., Hernandez, M., Kramer, J. K. G., Rinker, D. L., and Tsao, R. (2010). Ergosterol profiles, fatty acid composition, and antioxidant activities of button mushrooms as affected by tissue part and developmental stage. *Journal of agricultural and food chemistry,* 58(22): 11616–25.

[83] Zhang, M., Huang, J., Xie, X., and Holman, C. D. J. (2009). Dietary intakes of mushrooms and green tea combine to reduce the risk of breast cancer in Chinese women. *International journal of cancer. Journal international du cancer,* 124(6): 1404–8.

[84] Lee, K. J., Yun, I. J., Kim, K. H., Lim, S. H., Ham, H. J., Eum, W.S., and Joo, J. H. (2011). Amino acid and fatty acid compositions of Agrocybechaxingu, an edible mushroom. *Journal of Food Composition and Analysis,* 24(2): 175–178.

[85] Nitschke, J., Altenbach, H. J., Malolepszy, T., and Mölleken, H. (2011). A new method for the quantification of chitin and chitosan in edible mushrooms. *Carbohydrate research,* 346(11): 1307–10.

[86] Rayman, M. P. (2004). The use of high-selenium yeast to raise selenium status: how does it measure up? *The British journal of nutrition,* 92(4): 557–73.

In: Nutraceuticals ISBN: 978-1-68507-488-3
Editors: A. Kumari and G. Singh

Chapter 7

NUTRACEUTICALS FOR CHRONIC DISEASE PREVENTION, THEIR ADVERSE EFFECTS AND TOXICITY

***Nidhi Bhardwaj*[1], *Deepika Pawar*[2] *and Anita Kumari*[2,*]**
[1]Department of Human Genetics and Molecular Medicine,
School of Health Science,
Central University of Punjab, Bathinda, India
[2]Department of Nutrition Biology,
School of Interdisciplinary and Applied Life Sciences,
Central University of Haryana, India

ABSTRACT

The medicinal food or food products used as a substitute for therapeutic pharmaceuticals to prevent and treat various diseases are known as nutraceuticals. The vitamins, amino acids, minerals, omega-3-fatty acids, and other nutrients present in these foods decrease the risk of chronic disease. Long-term randomized clinical trials, meta-analyses,

---

* Corresponding Author's E-mail: anikum@cuh.ac.in.

and studies have claimed their beneficial role in managing chronic conditions. Their mechanism of action involves various biological processes like anti-oxidant defenses activation, expression of genes associated with cell survival, activation of signal transduction pathways, differentiation and the proliferation of cells. Because of their easy accessibility, enormous health benefits, less cost and fewer side effects, nutraceuticals are used by more than 50% of the adult population in western countries. It is recommended to consult a physician before consuming the nutraceuticals. However, people often overlook this advice and continue self-administrating them. While some nutraceuticals are safe to use, few may have the potential to cause toxicity. Therefore, knowledge of pharmacokinetics and toxicokinetics is necessary for the assessment of toxicity and safety of their use. This chapter discusses some selected nutraceuticals and their mechanism of action, risk factors associated, or clinical safety in managing various chronic diseases, including type 2 diabetes mellitus, inflammatory disease (rheumatoid arthritis, inflammatory bowel disease), hypertension, obesity, and cardiovascular diseases significantly affecting the quality of life.

1. INTRODUCTION

The use of food and food ingredients in managing and treating diseases is from immemorial time. Ayurveda recognized the positive impacts of food and herbal product intake on our bodies thousands of years ago. Acharya Charak defined food as a good diet that provides essential nutrition to maintain the healthy state of the body. The infamous quote by Hippocrates, let the food be thy medicine and medicine be thy food emphasizes the importance of food as medicine. Stefen DeFelic coined the term nutraceuticals and defined it as the food or part of food which prevents disease and provides medicinal benefits along with proper nutrition.

The word nutraceutical is derived from "nutrition" and "pharmaceuticals". Nutraceuticals include dietary supplement (vitamins, amino acids, minerals, herbs), processed foods like cereals, soups, and beverages used for nutrition medicine [1]. They can be obtained from plants and animals also. Because of their multiple therapeutic properties, they provide multipurpose benefits to human health. People often use nutraceuticals to increase life expectancy, delay aging, reducing the risk of chronic diseases giving overall physiological benefits to the body [2]. Both pharmaceuticals and nutraceuticals can be used to cure diseases, but in contrast, pharmaceuticals have patent protection and governmental sanction [3]. Therefore, nutraceuticals can be classified as established and potential nutraceuticals.

- Potential Nutraceuticals – These seem to be a modifiable factor in disease but are not proven efficient by clinical trials evidence.
- Established Nutraceuticals – These proved to be beneficial in disease management by evidence from the clinical trials.

The anti-oxidants, flavonoids, omega-3-fatty acids, polyphenols, and dietary fibers in nutraceuticals help in the cure and prevention of diseases. Green tea, cod liver oil, aloe vera, echinacea, glucosamine, American ginseng, and hibiscus are popular and widely used nutraceuticals in many countries. The herbal nutraceuticals are effective in many life-threatening, hardly curable conditions such as, Alzheimer's disease, cardiovascular, allergies, Parkinson's disease, obesity, and cancer. In recent times nutraceuticals have received significant interest because of their excellent nutritional value, possible therapeutic impacts, promising results for many pathological complications and safety.

2. Role of Nutraceuticals in Management of Chronic Disease

Chronic diseases are progressive degenerative disorders/conditions which last for a long time (years) and require extensive care and medication. Cancer, cardiovascular disorders and diabetes are among the leading causes of death in many countries. Chronic diseases are responsible for more than 55% of the world's burden of conditions and their prevalence is increasing continuously worldwide. Current medicated therapy used to treat and manage chronic diseases is not being fully effective and has many side effects. Figure 1 explains the role of nutraceuticals in management of various chronic diseases.

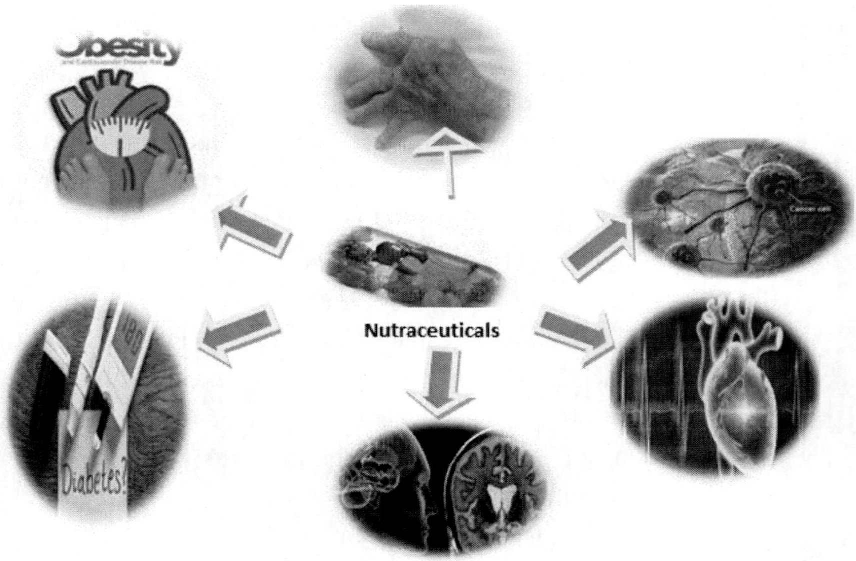

Figure 1. Nutraceuticals for management of various chronic diseases.

The following points should always be considered before using nutraceuticals for chronic disease prevention.

i. Analogs synthesis- synthesis of analog is required for using it as a chemical template for combinatorial synthesis and increasing the efficacy of nutraceuticals.
ii. Molecular target identification- by identifying the molecular target, developing a more refined chemical for explicitly targeting the common shared site is possible.
iii. Synergic effects- by understanding nutraceutical's molecular target and mechanism, it is possible to test the synergic or cumulative effect of two or more nutraceuticals against diseases.

2.1. Cardiovascular Disease

Cardiovascular disease includes a group of disorders related to heart and blood vessels such as heart attack, hypertension, heart failure, cerebrovascular disease/stroke, and heart and many more. Figure 2 illustrates that the risk factors for CVD are modifiable and non-modifiable. Modifiable factors include unhealthy diet, lack of physical activity, hypertension, cigarette smoking, diabetes mellitus, Dyslipidemia (lipid abnormality), and non-modifiable factors, including age, gender, family history, and genetics. Endothelium dysfunction acts as a common reason for developing various cardiovascular diseases. Statins have been previously used to improve the function of the endothelium. Many nutraceutical compounds have been proposed to cause possible positive effects in preventing and curing CVDs by improving endothelial functioning. Dietary factors contribute significantly in the reduction of the associated risk factors [4].

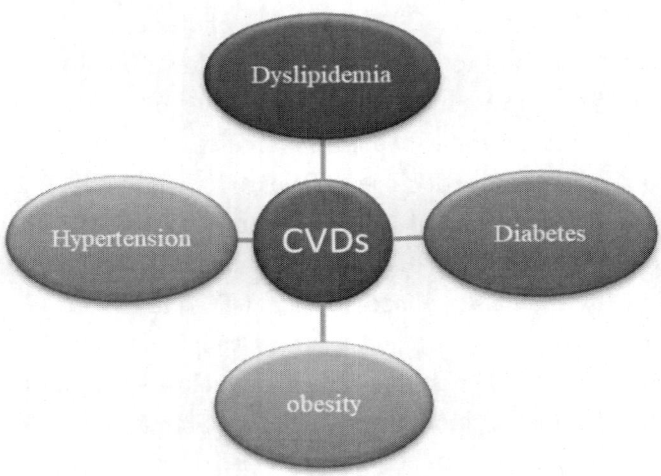

Figure 2. Major modifiable risk factors associated with cardiovascular disease.

Polyphenols include flavonols, stilbenes, anthocyanins, catechins, theaflavins, tannins and phenolic acids [5]. These are widely distributed in grapes and its derivatives, cereals, and beverages like cocoa and tea. Many studies demonstrated that consuming a diet rich in polyphenols lowers the risk of CVDs and has cardioprotective properties by improving endothelial or vascular functions.

2.2. Dyslipidemia

Dyslipidemia is one of the major contributing risk factors for CVDs. The high concentration of total cholesterol (TC) and low-density lipoprotein cholesterol (LDL-C) is a primary reason for abnormal high lipid profiles in cardiovascular diseases. Nutraceuticals have the potential to modify plasma lipid levels that can reduce the burden of CVDs [6]. Consumption of sterols decreases LDL-C by reducing endogenous cholesterol production [7]. The mechanism of action of sterols decreases cholesterol absorption in the intestine. In addition, It upregulates the hepatic LDL receptor, which results in increased uptake of cholesterol by the liver [8]. Metanalysis studies suggested that consumption of statins with sterols lowers the level of total cholesterol and low-density lipoprotein cholesterol to a great extent, thus lowering the risk of CVDs. Dose-dependent consumption of cocca flavanols reduces the level of LDL-C and total glycerides in the body. Both green tea and black tea containing polyphenols also aid in reducing the level of TC and LDL-C significantly. Still HDL-C concentration is not found to be affected by their consumption. Consumption of *Spirullina maxima* significantly alters blood lipid profile by changes in concentration of TC and LDL-C [9].

Armolipid plus is a safe nutraceutical supplement. It contains herbal ingredients such as red yeast rice extract, policosanol, berberine, folic acid, astaxanthin, and ubiquinone/coQ10. Several randomized clinical trials confirmed the positive effects of this by reducing the concentration of plasma TC up to 20%, LDL up to 31%, and triglycerides up to 16.3% [10, 11].

2.3. Hypertension

Hypertension or high blood pressure is another important modifiable factor for CVDs. Studies have proven that the risk of myocardial infarction is reduced by 20-25% for stroke around 40% and 50% for heart failure by lowering the blood pressure [12]. According to a recent randomized clinical trial, consumption of grape seed extract for about six weeks reduces systolic blood pressure by 5.6% and diastolic by 4.7%. Chocolate or cocoa consumption was also found to reduce the diastolic blood pressure and mean arterial blood pressure. Green tea consumption is also found to be associated with a reduced risk of myocardial infarction and stroke. Pomegranate juice, rich in polyphenols, has antihypertensive effects, anti-oxidant, and anti-atherosclerotic properties [13].

2.4. Diabetes Mellitus

Type 2 diabetes is another major complication correlated with CVD. Moreover, people suffering from diabetes and hypertension have a double possibility of developing cardiovascular disorders in comparison to a non-diabetic person [14]. Observational studies suggested that changes in diet might help to prevent and treat diabetes and associated complications. Consumption of red wine and grape seed extract is associated with a significant reduction in risk of diabetes by 30% [15]. Resveratrol supplementation also aids in improving mean hemoglobin A1c, decrease in the level of fasting glucose, lowering insulin and insulin resistance in type 2 diabetes patients. Figure 3 shows mechanism of action of resveratrol.

Administration of potential nutraceuticals in high-risk patients along with medical treatment, increases the effectiveness of the therapy reduces the risk. Studies suggested that these nutraceuticals are well tolerated in patients with almost no side effects. However, clinical studies are required to recognize nutraceuticals with the best clinical effectiveness [16].

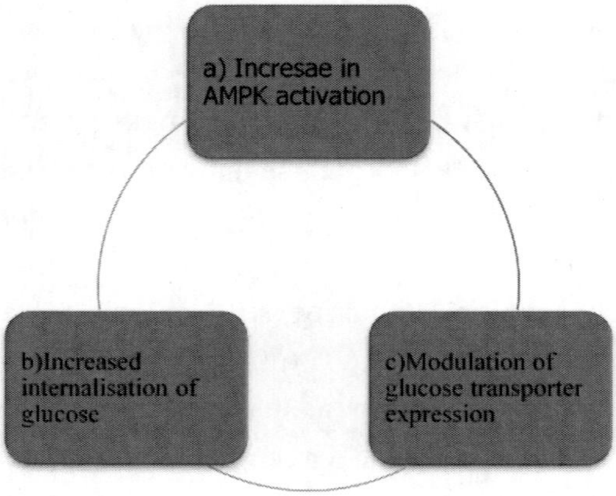

Figure 3. Possible mechanism of action of resveratrol (a) AMPK activation (b) Glucose internalization (c) Glucose transporters.

2.5. Cancer

Cancer development is a long-term step-wise process with many complex factors involved and leads to uncontrolled and abnormal growth of the cells in any part of the body called metastasis [17]. Various anticarcinogenic effects of nutraceuticals on cancerous cells are shown in (Figure 4). A significant increase in the number of reactive oxygen species is also observed in cancerous cells. Cancer is recognized as the second most general cause of mortality in the USA. Sometimes free radicals are produced in mitochondria which ground damage to cell membrane proteins and mutation in DNA, leading to abnormal growth of the cells and oxidative stress. Cancerous cells become resistant to apoptosis because of increased expression of anti-apoptotic proteins such as Bcl-2 and Bcl-xL, IAPs, and FLIP. Simultaneously there is downregulation or decrease in expression of proapoptotic proteins such as Bax, Apaf-1, caspase-8, and death receptors [18]. Epidemiological studies implicated that dietary factors and a healthy lifestyle can prevent carcinogenesis (formation of cancerous cells). Consumption of foods rich in lutein such as spinach, tomatoes, broccoli, oranges, and leafy greens) is associated with a low incidence of colon cancer. Lycopene and beta-carotene can induce apoptosis in prostate cancer cells and malignant lymphoblast cells. These carotenoids destroy the reactive oxygen species from the body and act as anti-oxidants.

Nutraceuticals downregulate the pathways involved in cancer progression to inhibit the proliferation of cells or induce apoptosis. Signaling pathways like IGFR, Wnt/β-catenin signaling, MAPK/ERK pathway, PI3K/Akt/mTOR pathway, EGFR family receptors, sonic hedgehog signaling are suppressed by nutraceutical consumption in subjects [19].

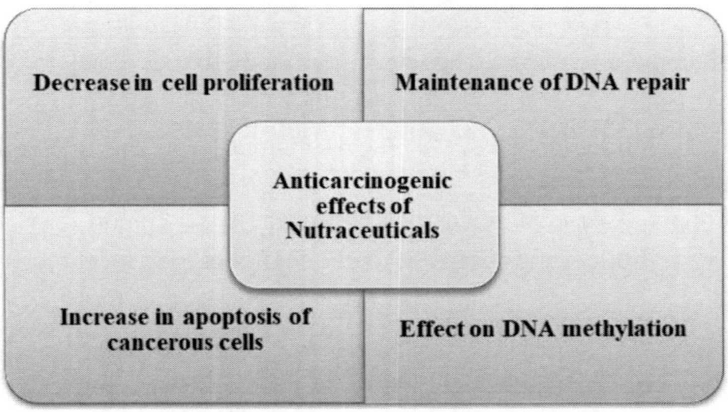

Figure 4. Various anticarcinogenic effects of nutraceuticals on cancerous cells [20, 21].

Indole-3-carbinol, resveratrol, epigallocatechin-3-gallate, curcumin lycopene are some of the widely used nutraceuticals to downregulate these pathways in cancer progression [22]. Garlic has been successfully established as an anticarcinogenic agent. The allyl sulfur compound from garlic has antiproliferation activity and diallyl disulfide compounds induce apoptosis of cancerous cells in lung cancer, breast cancer, and prostate cancer [23].

Often nutraceuticals are co-administrated with a chemotherapeutic drug to control the activity of the drug by regulating the ATP binding Cassettes, namely BCRP, MRP, etc. It also diminishes the potential toxicity of the drug administrated. For, e.g., anthracycline, when consumed alone, induce cardiotoxicity. To avoid this, co-administration of coenzyme Q10 with anthracycline is usually advised. However, because of the poor aqueous solubility of nutraceuticals, their clinical efficacy is constrained. To tackle this problem, nano carrier-based delivery of the nutraceutical approach is considered. [24]. However, to identify the full potential of nano-nutraceuticals, further pre-clinical trials and studies are required on a large scale.

2.6. Inflammatory Disorders

Inflammation refers to the response of body tissues to any infection or injury causing redness, heat, and swelling. Arthritis, inflammatory bowel disease (IBD), celiac diseases, asthma are some prevalent examples of inflammatory disorders.

Crohn's disease and ulcerative colitis are two common forms of IBD associated with diarrhea, nausea, abdominal pain, and bleeding from the rectum [25]. It is one of the most prevalent gastrointestinal disorders in developing nations due to the adoption of western dietary patterns and an inactive way of living. A number of other factors such as genetic susceptibility, immune system dysfunction, nutrition (high content of fat and carbohydrate), and gut micro biota are also responsible for IBDs [26]. There is an imbalance between the inflammatory cytokines such as TNF-α, COX-2, interferon-γ, interleukins (IL-6, IL-1, IL-12), and anti-inflammatory cytokines (IL-4, IL-10, and IL-11), causing tissue damage and inflammation. In addition, the weakening of barrier function of epithelium and increase in permeability of intestine facilitate mucosa inflammation [27].

The current therapeutic drugs available are not capable to completely cure the disease. Amino salicylates, immunosuppressive agentsand corticosteroids are used to regulate the pro-inflammatory cytokines for temporary relief down. Also, there is a high risk of infections and malignancies of IBD [28]. Recent research outcomes using nutraceuticals for treatment suggested that nutraceuticals are beneficial for IBD patients.

They mainly act by interfering within the inflammatory pathways, anti-oxidative properties, and modulating the intracellular signaling mechanisms. Therefore, nutraceuticals have emerged as a promising alternative to conventional therapeutic drugs because of their lower cost and no side effects in the treatment of inflammatory bowel disease. Table 1 summarizes the role of various nutraceuticals in the treatment of IBD.

Probiotics inhibit the growth of pathogenic microorganism by colonizing the colon. They also interact with mucosa epithelial cells, strengthening the barrier and modulating the immune system's response. In addition, prebiotics increase the growth and metabolism of protective commensal gut microorganisms. Large-scale clinical trials showed that VSL#3 consumption (a multi bacterial culture probiotic) for about eight weeks reduces the disease activity and rectal bleeding in ulcerative colitis patients. *Bifidobacteria* and *E. coli* Nissle are also found to be effective in maintaining remission and preventing relapsis in large cohorts of patients [39].

Curcumin, commonly used as a spice, has anti-inflammatory properties. It is obtained from turmeric and also has substantial medicinal value. It reduces the expression of the nuclear factor kappa-light-chain-enhancer of activated B cells (NF-κB) related inflammation pathways. Subsequently, it inhibits the pro-inflammatory cytokines such as TNF-α, IL-12, and IL-2. It is found to be effective and safer in various studies and trials conducted [38]. Although no clinical trials demonstrated aloe vera gel to be effective, it is used in treating UC in many countries as herbal therapy.

Coriolus versicolor, a medicinal mushroom found in China containing polysaccharides such as krestin, lignin, and glucan relieves symptoms by decreasing IgE levels and expression of interferons, TNF-α, IL-4, IL-6, IL-1β in mice models [29]. *Cordyceps militaris*, another medicinal mushroom, prevents damage to the epithelium, migration of inflammatory cells, and shortening the colon by reducing the level of TNF-α, iNOS [30]. The chaga mushrooms rich in anti-oxidants reduce the level of IgE and IgA, IL-4 COX-2, TNF-α, STAT1, and STAT6.

Table 1. Importance of natural and herbal extracts obtained from plants for maintenance of IBD

Natural extract	The main compound acting	Mode of action against IBD
Mushrooms	*Coriolus Versicolor*	Reduction in levels of, IL-6, TNF-α IL-1β [29] Reduces interleukins STAT1 and STAT6
	Cordyceps militaris	Prevent epithelial damage [30] Suppresses mRNA expression of iNOS and TNF- α.[53]
	Ganoderma lucidum	Downregulation of NF-κB and inhibition of phosphorylation MAP Kinase pathways.[31]
Extracts from fruits		
Prunus mume	*Prunus mume extract*	Decrease inflammatory cytokines NF-α and iNOS expressions were also reduced. [32]
Pomegranate	*Polyphenols such as* ellagitannins and ellagic acid	Anti-oxidant effects COX-2, TNF α, STAT6 expressions are decreased [33].
Pomegranate	*Aronia melanocarpa juice*	Reduced NF-κB translocation Anti-oxidant and antiinflammatory [34]
Black check Berry	*Haliotis discus hannai* Ino extract	Prevent damage to colonic tissue [35]
Marine foods	Green algae extract	IFN-γ, IL-4 downregulation
Phytochemicals	Apple polyphenols	Reduced COX-2 and TNF-α Recovered transglutaminase protein by reversing its depletion [36]
	Resveratrol	Reduced expression of TNF-α and NF-κB [37]
	Curcumin	Reduced TNF-α, IL-1β, and MPO [38]

When combined with olive oil and quercitrin, polyunsaturated fatty acids such as EPA and DHA, reduce the iNOS, COX-2, TNF-α, and IL-1β expression levels in colitic rats. Alpha linoleic acid lowers the inflammatory damage by lowering IL-6, COX-2, and TNF-α in mice models. Butyrate, short-chain fatty acid, enhances the production of regulatory T cells and alleviates mucosal inflammation. It also inhibits the activation of NF-κB [39].

Bioactive peptides can also be used as an alternative to conventional therapies in IBDs. For example, alanyl-glutamine treatment in mouse models suppressed the Th17 (helper T cells), cytokines and migration of macrophages towards peritoneal activity to cause a reduction in an inflammatory response.

2.7. Osteoarthritis

It is another inflammatory disease characterized by inflammation of cartilage and synovium, causing stiffness and swelling in joints. Nutraceuticals like fish oil, GAGs (Glucosamine Sulfate, Chondroitin Sulfate, and Hyaluronic acid), olive oil, methionine, and botanical extracts such as avocado/soy unsaponifiable (ASU) components, curcumin, oleoresin from the *Boswellia serrata* tree are found to be effective in the treatment of osteoarthritis [40]. The use of these nutraceuticals inhibits inflammatory pathways and reduces pain in joints.

2.8. Neurological Disorders

Neurological disorder refers to any condition caused due to dysfunction of the brain of nervous system. Neurodegenerative disorder, Alzheimer's disease, Parkinson's disease, psychosis, and depression are some of the neurological disorders with extremely complicated etiology. Protein misfolding is one of the main reasons for neurological disorders. Alzheimer's disease is caused by the misleading of protein Tau and amyloid-β. Misfolded proteins activate the cascade of inflammatory proteins such as NF- κB, iNOS COX, and inflammatory cytokines, which causes neurodegeneration. Their inhibition provides neuroprotective action. Nutraceuticals combined with medicated drug are used to enhance the therapeutic impact and provide excellent neurobiological effects. Most of the commonly used nutraceuticals in the management of neurological disorders are presented in the Figure 5.

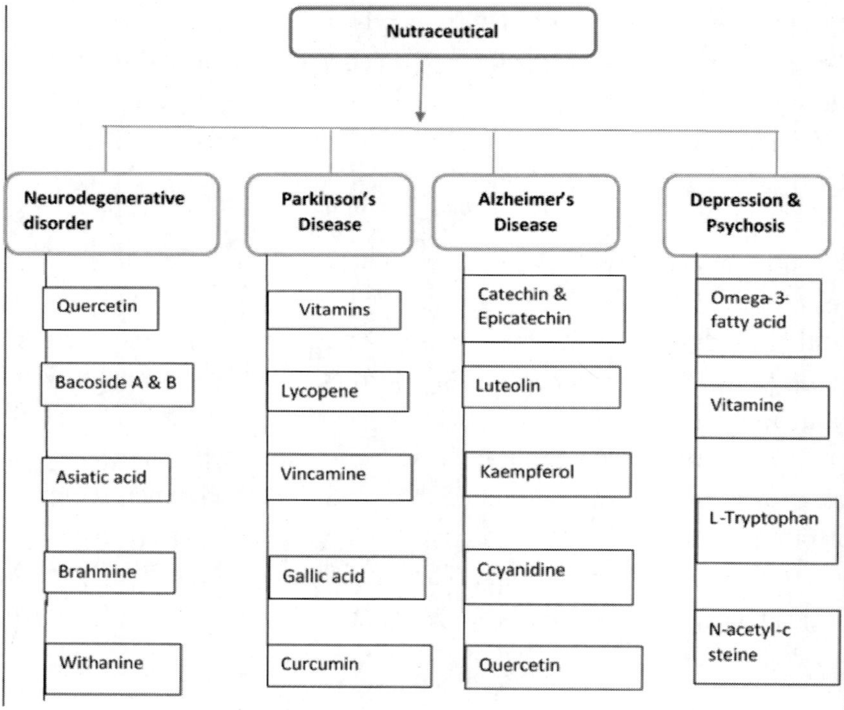

Figure 5. Nutraceutical used for management of various neurological disorders as adjunctive therapy [41, 42].

2.9. Alzheimer's Disease

Primary degenerative types of dementia of Alzheimer's type (PDDAT) or senile dementia of Alzheimer's (SDAT) are two types of Alzheimer's disease. Loss of memory is the most noticeable feature of this disease. Nutraceuticals seem to be helpful in the management of Alzheimer's disease because the aging process and lack of consumption of anti-oxidants results in promoting oxidative stress, thus causes the development of this disease. Food rich in anti-oxidants, PUFAs, saturated and trans fatty acid suppresses neurodegeneration while trans-fat rich diet increases neurodegeneration. Natural compounds such as flavonoids, carotenoids, crocin, luteolin are found to be beneficial in Alzheimer's disease.

The primary polyphenols or flavonoids used in the treatment are catechin, epicatechin, epigallocatechin, and epigallocatechin gallate are obtained from fruits, vegetables wine, and cocoa. The neurological modulating action of flavonoids and their metabolic products is found to interact with the neuronal-glial signaling pathway, responsible for the survival and functioning of neurons [43]. These compounds also inhibit neuropathology by upregulating the activity of anti-oxidants proteins causing synaptic plasticity and neuronal function repairing, β-cryptoxanthin, α and β-carotenes, lycopene are major carotenoids found in the human body. Astaxanthin, a carotenoid derived from seafood, is used because of its anti-oxidant and anti-inflammatory properties. In addition, it has microcirculatory and mitochondrial protective functions implicating it to be a neuroprotective compound. However, in patients with severe complications level of major carotenoids are found to be reduced.

Crocin, a phytoconstituent extracted from saffron, is well known for its antispasmodic, stimulant, improved learning, carminative properties, and expectorant. It alters the level of the oxidative marker hippocampus's oxidative markers level and improves the impacts due to chronic stress [44].

Cyanidin obtained from berries has anti-inflammatory properties by reducing the expression of pro-inflammatory cytokines and thus decreases damage to brain cells. In addition, it inhibits the phospholipaseA2, which is found to play an important role in inflammatory signaling pathways oxidative stress.

2.10. Psychotic Disorder

Psychotic disorders are the mental disorder such as bipolar disease and schizophrenia causing abnormal thinking, perception, and hallucination can also be treated by nutraceuticals. Omega-3- fatty acids and vitamins are commonly used for the treatment of psychotic disorder. Polyunsaturated fatty acids from omega-3-fatty obtained as eicosapentaenoic acid, docosahexaenoic acid are integral components of the cell membrane and cannot be synthesized by the body. So, they must be obtained through diet. EPA and DHA are found to improve dopaminergic and serotoninergic nerve transmission, modulate mitochondrial function and decrease oxidative stress [45]. In addition, they regulate the expression of brain-derived neurotrophic factors and protect against toxicity due to apoptosis.

2.11. Parkinson's Disease

Parkinson's disease is characterized by the drastic depletion of dopamine, oxidative factors, anti-oxidant depletion, and mitochondrial damage. Current anti-Parkinson drugs used for treatment provide temporary relief only and sometimes cause toxicity also. Many natural products are studied widely to be used in the treatment of Parkinson's disease.

3. Nutraceuticals Targeting Mitochondrial Dysfunction and Oxidative Stress

An increase in free radicals with the mitochondrial damage or its abnormal functioning in dopaminergic neurons can cause compromised metabolism of cells and imbalance homeostasis and thus impacts the functioning of brain and neurodegenerative disorder. Ubiquinon /CoQ10 and fish oil are consumed to manage the disease as they are actively involved in the Electron transport chain and ATP production.

Polyphenols protect dopaminergic neurons and decrease free radicals and thus help in improving motor and gait abnormalities in patients. Fish oil has plentiful omega-3-fatty acids. Eicosapentaenoic and docosahexaenoic acids provide protective effects to the neurons. Epigallocatechin-3 gallate (ECCG) obtained from *Camellia sinensis* can surpass blood-brain barrier. The anti-oxidant scavenging activity and iron chelation of phytochemicals significantly enhance motor neurons functioning and DA levels in the brain, thus decreasing neurotoxicity. Vinpocetine and vincamine reduce the synthesis of reactive oxygen species. MitoQ, a synthetic nutraceutical, maintains the functioning of the respiratory chain

4. Nutraceuticals Targeting Endoplasmic Reticulum Stress and Protein Misfolding

Nutraceuticals targeting this mechanism prevent the formation of misfolded proteins and aggregation of proteins. The inability to do so causes apoptosis and neurodegeneration in patients. Vitamin A, beta carotene, CoQ10, glutathione and vitamin E have anti-fibrillogenic properties for Parkinson 's disease treatment. Vitamin A inhibits the intracellular alpha-syn deposition and carotenoids inhibit the proapoptotic factor caspase, involved in apoptosis and decreases the expression of CHOP and (BIP) Grp78.

A flavonoid (Bicalein) found in roots of Scutellaria baicalensis gargi found in Iran. This compound halts the formation of an oligomer of α-Syn and significantly prevents fibrillation. It also decreases the level of inflammatory cytokines, inhibition of apoptosis, and induces autophagy. Resveratrol also seems to be a potent nutraceutical in treatment because of its stability and solubility [54]. Figure 6 summarizes the nutraceuticals used for Parkinson's disease based on the mechanism of action involved.

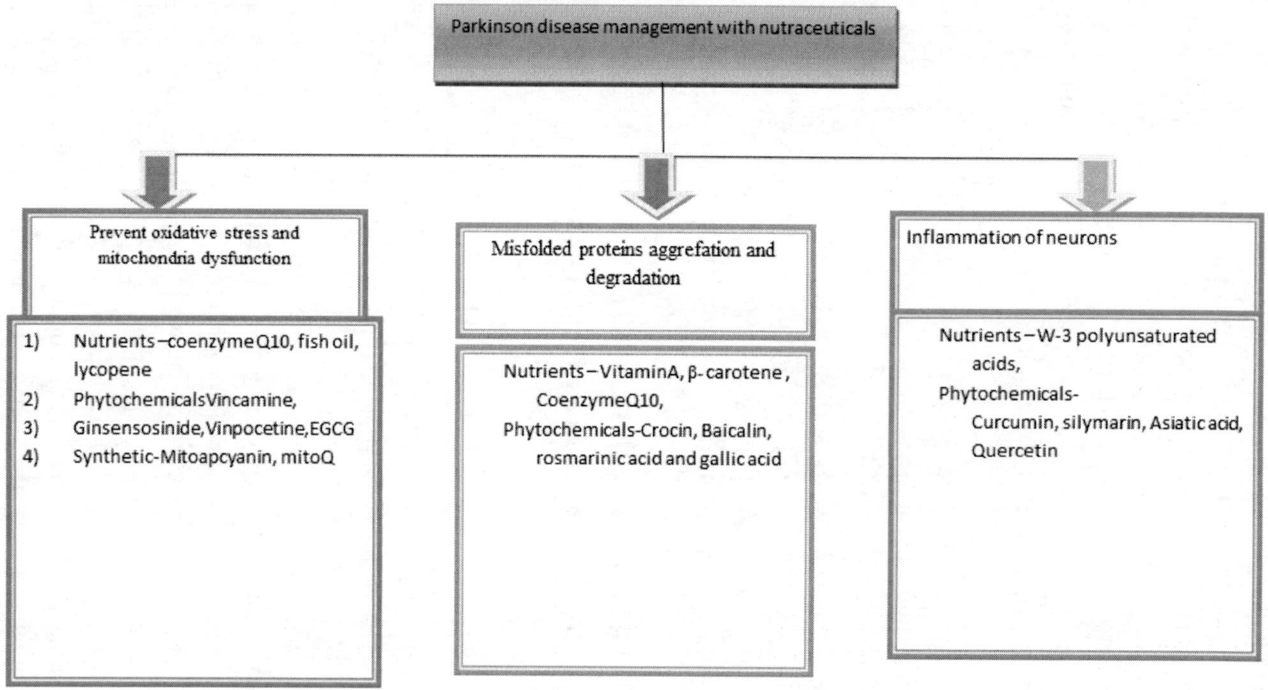

Figure 6. Nutraceuticals used for management of Parkinson's disease based on mechanism of action.

4.1. Depression

Depression refers to the mental condition of any person when they are always in a sad mood, have an impaired routine, and have no interest in doing any social activity. The incidence and prevalence of depression continuously increase among young people in many countries causing a significant burden on society. Moreover, a depressed person is usually less active and productive also; there is a high risk of mortality associated with this mental condition.

Omega-3-fatty acids and folic acids are proved to be beneficial as adjunctive therapy. These fatty acid molecules increase neurogenesis and decrease the synthesis of inflammatory mediators. They also benefit neurotransmission by increasing membrane fluidity and inhibit the reuptake of monoamines. Diet rich in nutrients zinc, folic acids, vitamin D, and other micronutrients have shown positive effects in the management of depression [47].

Zinc amplifies the neurogenesis in the hippocampus and also increases the BNDF. It also modifies the activity of glutamate and NMDA receptors. N- Acetylcysteine having anti-inflammatory and anti-oxidant properties replenish glutathione levels and increases the process of neurogenesis. It also prevents mitochondrial toxicity and modulation of the glutamate pathway, ultimately protecting the person from depression.

S-adenosyl methionine, an essential methyl donor, affects neurotransmitters production and metabolism, affects neurotransmitters production and metabolism and causes an increase in phosphate-dylcholine conversion and decreases the secretion of prolactin.

5. Miscellaneous Benefits of Nutraceuticals

Evidence from various studies supported the use of nutraceuticals to prevent multiple diseases. A good diet and nutritional supplements are important in the health care system and reduce the burden of disease by prevention rather than treatment. Nutraceuticals also protect the pregnant mother and baby against various disorders. Consumption of fish during pregnancy increases the gestation period resulting in higher birth weights. Omega-3- polyunsaturated fatty acids are essential for the development of a healthy nervous system.

A diet rich in anti-oxidants, lutein, and zeaxanthin is proved to be beneficial in age-related macular degeneration (AMD). Consumption of green tea, *Allium* spp., polyphenols, vitamin c and E, lycopene, beta-carotene co-enzyme Q10 have anti-oxidant properties and proved to be effective in AMD [48]. Astaxanthin, a carotenoid isolated from marine organisms, namely sea bream, salmon, trout, and shrimps, is a potent anti-oxidant that protects eyes against ultraviolet light effects and boosts immune system functions. It is also effective in Alzheimer's disease.

Zeaxanthin, lutein is widely used for the management and treatment of eye-related disorders [48]. Lutein is mainly obtained from fruits and vegetables like sweet potatoes, carrots, tomato, corn, mangoes, and leafy green vegetables. Zeaxanthin is also obtained from kiwi, kale, broccoli, green beans, cabbage, lettuce, green vegetables, and sprouts.

Nutraceuticals are also proved to be very useful in improving the functioning of the immune system and disease susceptibility. Herbal extract from the plant genus *Echinacea, Astragalus* plays an important role in developing active immune cells in the bone marrow and lymph [48, 49]. Probiotics are used to maintain balance in pathogenic microflora and commensal microflora of the gut [49]. They are also effective in diarrhea and the establishment of pro and anti-inflammatory cytokines. Phytoestrogens help maintain the hormonal imbalance. Soy isoflavones are used as an alternative to estrogen receptor modulators in hormone replacement therapy.

6. Commonly Used Nutraceuticals

The Structure of commonly used nutraceutical are presented in Figure 7.

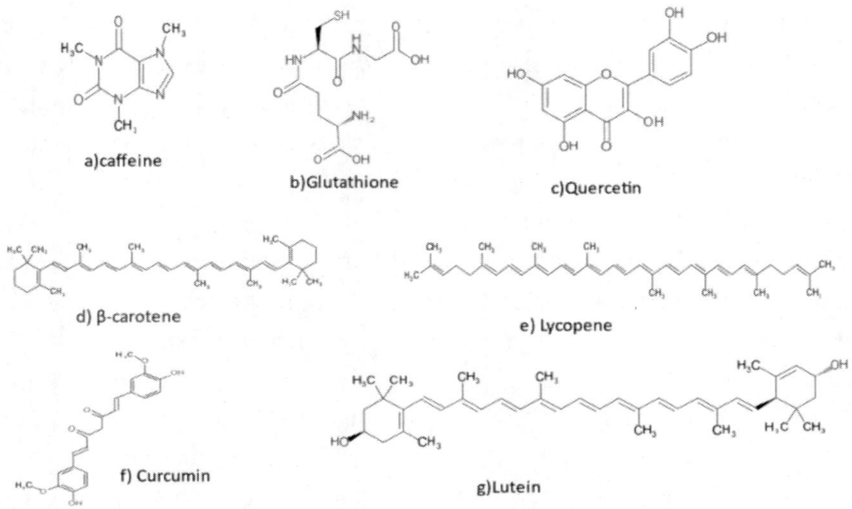

Figure 7. Structure of some commonly used nutraceuticals [48, 55].

7. TOXICITY OF NUTRACEUTICALS

Nutraceuticals such as medicinal plants are used as home remedies and to meet basic nutritional requirements from a long time back. People generally believe that nutraceuticals are entirely safe and often use them without consulting doctors or physicians. Although many nutraceuticals do not cause any harm or toxicity to humans, this is not true for all. There are pieces of evidence from studies conducted in model organisms, but nutraceuticals induced toxicity is very rare in humans. There is no evidence of the delayed effects of nutraceuticals. Only limited nutraceuticals are studied for their safety and toxicity. A large number of nutraceuticals remained poorly understood due to a lack of toxicokinetic studies. Studies of interactions are essential when nutraceuticals and therapeutic drugs are used concurrently.

Green tea is used as a beverage and nutraceuticals all around the world for a long time. It contains more than 2000 chemical compounds. Green tea is beneficial in obesity, inflammatory bowel disease, cancer and cardiovascular diseases. However, a study conducted with mice using green tea in a dose-dependent manner resulted in adverse changes in reproductive organs. Sometimes excessive consumption of green tea increases breast cancer risk [51]. In addition, polycyclic aromatic hydrocarbons act as contaminants for green tea and exaggerate the toxicity potential.

Aloe vera is also known as the 'plant of immortality', is used worldwide for wound healing, anti-aging, anti-inflammatory, laxative, and antibacterial effects. Although there is no major side effect associated with aloe vera but sometimes consuming it orally can cause abdominal cramps, diarrhea and thus decreasing the drug absorption. In addition, studies indicated that aloe-emodin acts as genotoxic in-vivo and induces damage to DNA in lung cancer cells by increasing reactive oxygen species [50].

Nutraceutical-drug interaction can lead to serious outcomes because of the multiple factors involved. Some nutraceuticals influence the enzymes metabolizing drug that is cytochrome P450 and reduces its effects. *Ginkgo biloba* inhibits CYP450 inducers. *Ginseng* acts as an inducer of CYP450. P-glycoprotein acts as a transporter for various lipophilic and amphipathic drugs, toxins, xenobiotics, and carcinogens in the liver, kidney, brain and many organs. Ginseng inhibits the P-glycoprotein. Hypericum increases the expression of P-glycoprotein and significantly modulates the pharmacodynamics of many therapeutic drugs.

Currently, many models are used to determine the safety and toxicity of nutraceuticals. In addition to in-vivo studies, in silico and omics technology bases studies are required to fully determine the potential toxicity [51]. The future of nutraceuticals seems to be very bright due to the increase in the prices of therapeutic drugs and the rising burden of disorders. To promote good and fully effective treatment with fewer side effects, vigorous evaluation of the toxicity of nutraceuticals should be performed. Pharmacists should be well educated about nutraceuticals' positive and negative impact before using nutraceuticals as adjunctive therapy.

Conclusion

The use of therapeutic drugs is associated with very less side effects, so nutraceuticals provide an excellent alternative to these drugs and help in naturally cure the disease by improving the quality of life. The use of nutraceuticals is advantageous in terms of cost and for people who do not prefer to go for chemical therapies. Currently, many nutraceuticals are undergoing research phases, and an increase in the marketing graph of nutraceuticals indicated that people to like to go for them. Nutraceuticals are very convenient in lifestyle as there is an increased craving for a healthy lifestyle. It will play an essential role in the development of future therapeutic essentials. The success of nutraceuticals depends upon the control of purity, safety and efficacy. Effective regulations and legislation are required to ensure the health claims are strongly based on scientific evidence and clinical trials conducted. Nutrigenomics is the newly growing field in which knowledge of genomics, transcriptomics, metabolomics is applied to an individual's genetic makeup for the development of personalized nutraceuticals to prevent chronic diseases.

References

[1] Santini, A., Tenore, G. C. and Novellino, E. (2017). Nutraceuticals: A paradigm of proactive medicine. *European Journal of Pharmaceutical Sciences*, *96*: 53-61.

[2] Shibamoto, T., Kanazawa, K., Shahidi, F., and Ho, C. T. (2008). *Functional food and health: An overview*: 1-6.

[3] Chauhan, B., Kumar, G., Kalam, N. and Ansari, S. H. (2013). Current concepts and prospects of herbal nutraceutical: A review. *Journal of advanced pharmaceutical technology & research*, *4*(1): 4.

[4] Piepoli, M. F., Hoes, A. W., Agewall, S., Albus, C., Brotons, C., Catapano, A. L. and Verschuren, W. M. (2016). Guidelines: Editor's

choice: 2016 European Guidelines on cardiovascular disease prevention in clinical practice: The Sixth Joint Task Force of the European Society of Cardiology and Other Societies on Cardiovascular Disease Prevention in Clinical Practice (constituted by representatives of 10 societies and by invited experts) Developed with the special contribution of the European Association for Cardiovascular Prevention & Rehabilitation (EACPR). *European heart journal*, *37*(29): 2315.

[5] Pandey, K. B. and Rizvi, S. I. (2009). Plant polyphenols as dietary antioxidants in human health and disease. *Oxidative medicine and cellular longevity*, *2*(5): 270-278.

[6] Sahebkar, A., Serban, M. C., Gluba-Brzózka, A., Mikhailidis, D. P., Cicero, A. F., Rysz, J. and Banach, M. (2016). Lipid-modifying effects of nutraceuticals: an evidence-based approach. *Nutrition*, *32*(11-12): 1179-1192.

[7] Wang, P., Chen, Y. M., He, L. P., Chen, C. G., Zhang, B., Xue, W. Q. and Su, Y. X. (2012). Association of natural intake of dietary plant sterols with carotid intima–media thickness and blood lipids in Chinese adults: a cross-section study. *PLoS One*, *7*(3): e32736.

[8] de Jong, A., Plat, J. and Mensink, R. P. (2003). Metabolic effects of plant sterols and stanols. *The Journal of Nutritional Biochemistry*, *14*(7): 362-369.

[9] Parikh, P., Mani, U. and Iyer, U. (2001). Role of Spirulina in the control of glycemia and lipidemia in type 2 diabetes mellitus. *Journal of Medicinal Food*, *4*(4): 193-199.

[10] Mazza, A., Lenti, S., Schiavon, L., Zuin, M., D'Avino, M., Ramazzina, E. and Casiglia, E. (2015). Nutraceuticals for serum lipid and blood pressure control in hypertensive and hypercholesterolemic subjects at low cardiovascular risk. *Advances in therapy*, *32*(7): 680-690.

[11] Barrios, V., Escobar, C., Cicero, A. F. G., Burke, D., Fasching, P., Banach, M. and Bruckert, E. (2017). A nutraceutical approach (Armolipid Plus) to reduce total and LDL cholesterol in individuals

with mild to moderate dyslipidemia: review of the clinical evidence. *Atherosclerosis Supplements*, *24*: 1-15.

[12] Antonakoudis, G., Poulimenos, I., Kifnidis, K., Zouras, C. and Antonakoudis, H. (2007). Blood pressure control and cardiovascular risk reduction. *Hippokratia*, *11*(3): 114.

[13] Asgary, S., Keshvari, M., Sahebkar, A. and Sarrafzadegan, N. (2017). Pomegranate consumption and blood pressure: A review. *Current pharmaceutical design*, *23*(7): 1042-1050.

[14] Arauz-Pacheco, C., Parrott, M. A. and Raskin, P. (2002). The treatment of hypertension in adult patients with diabetes. *Diabetes care*, *25*(1): 134-147.

[15] Guilford, J. M. and Pezzuto, J. M. (2011). Wine and health: A review. *American Journal of Enology and Viticulture*, *62*(4): 471-486.

[16] Sosnowska, B., Penson, P. and Banach, M. (2017). The role of nutraceuticals in the prevention of cardiovascular disease. *Cardiovascular diagnosis and therapy*, *7*(Suppl 1): S21.

[17] Kalimuthu, S. and Se-Kwon, K. (2013). Cell survival and apoptosis signaling as therapeutic target for cancer: marine bioactive compounds. *International journal of molecular sciences*, *14*(2): 2334-2354.

[18] Campbell, K. J. and Tait, S. W. (2018). Targeting BCL-2 regulated apoptosis in cancer. *Open biology*, *8*(5): 180002.

[19] Khare, V., Alam, N., Saneja, A., Dubey, R. D. and Gupta, P. N. (2014). Targeted drug delivery systems for pancreatic cancer. *Journal of biomedical nanotechnology*, *10*(12): 3462-3482.

[20] Surh, Y. J. (2003). Cancer chemoprevention with dietary phytochemicals. *Nature Reviews Cancer*, *3*(10): 768-780.

[21] Gul, K., Singh, A. K. and Jabeen, R. (2016). Nutraceuticals and functional foods: The foods for the future world. *Critical reviews in food science and nutrition*, *56*(16), 2617-2627.

[22] Sarkar, F. H., Li, Y., Wang, Z. and Kong, D. (2010). The role of nutraceuticals in the regulation of Wnt and Hedgehog signaling in cancer. *Cancer and Metastasis Reviews*, *29*(3): 383-394.

[23] Dorant, E., van den Brandt, P. A., Goldbohm, R. A., Hermus, R. J. and Sturmans, F. (1993). Garlic and its significance for the prevention of cancer in humans: a critical view. *British journal of cancer*, *67*(3): 424-429.

[24] Nair, H. B., Sung, B., Yadav, V. R., Kannappan, R., Chaturvedi, M. M. and Aggarwal, B. B. (2010). Delivery of antiinflammatory nutraceuticals by nanoparticles for the prevention and treatment of cancer. *Biochemical pharmacology*, *80*(12): 1833-1843.

[25] Pithadia, A. B. and Jain, S. (2011). Treatment of inflammatory bowel disease (IBD). *Pharmacological Reports*, *63*(3), 629-642.

[26] Ardizzone, S. and Porro, G. B. (2005). Biologic therapy for inflammatory bowel disease. *Drugs*, *65*(16): 2253-2286.

[27] Zhao, L., Wu, H., Zhao, A., Lu, H., Sun, W., Ma, C. and Jia, W. (2014). The in vivo and in vitro study of polysaccharides from a two-herb formula on ulcerative colitis and potential mechanism of action. *Journal of ethnopharmacology*, *153*(1): 151-159.

[28] Horváth, K., Varga, C., Berkó, A., Pósa, A., László, F. and Whittle, B. J. (2008). The involvement of heme oxygenase-1 activity in the therapeutic actions of 5-aminosalicylic acid in rat colitis. *European journal of pharmacology*, *581*(3): 315-323.

[29] Lim, B. O. (2011). Coriolus versicolor Suppresses Inflammatory Bowel Disease by Inhibiting the Expression of STAT1 and STAT6 Associated with IFN-γ and IL-4 Expression. *Phytotherapy Research*, *25*(8): 1257-1261.

[30] Won, S. Y. and Park, E. H. (2005). Anti-inflammatory and related pharmacological activities of cultured mycelia and fruiting bodies of Cordyceps militaris. *Journal of ethnopharmacology*, *96*(3), 555-561.

[31] Hasnat, M. A., Pervin, M., Cha, K. M., Kim, S. K., and Lim, B. O. (2015). Anti-inflammatory activity on mice of extract of Ganoderma lucidum grown on rice via modulation of MAPK and NF-κB pathways. *Phytochemistry*, *114*: 125-136.

[32] Jin, H. L., Lee, B. R., Lim, K. J., Debnath, T., Shin, H. M. and Lim, B. O. (2011). Anti-inflammatory effects of Prunus mume mixture in

colitis induced by dextran sodium sulfate. *Korean Journal of Medicinal Crop Science*, *19*(1): 16-23.

[33] Kamali, M., Tavakoli, H., Khodadoost, M., Daghaghzadeh, H., Kamalinejad, M., Gachkar, L. and Adibi, P. (2015). Efficacy of the Punica granatum peels aqueous extract for symptom management in ulcerative colitis patients. A randomized, placebo-controlled, clinical trial. *Complementary Therapies in Clinical Practice*, *21*(3): 141-146.

[34] Valcheva-Kuzmanova, S., Kuzmanov, A., Kuzmanova, V. and Tzaneva, M. (2018). *Aronia melanocarpa* fruit juice ameliorates the symptoms of inflammatory bowel disease in TNBS-induced colitis in rats. *Food and Chemical Toxicology*, *113*: 33-39.

[35] Debnath, T., Mijan, M. A., Kim, D. H., Jo, J. E., Kim, Y. O., Lee, J. J. and Lim, B. O. (2015). Anti-Inflammatory Effects of H aliotis discus hannai I no on Dextran Sulfate Sodium-Induced Colitis in Mice. *Journal of Food Biochemistry*, *39*(3): 209-217.

[36] D'Argenio, G., Mazzone, G., Tuccillo, C., Ribecco, M. T., Graziani, G., Gravina, A. G. and Romano, M. (2012). Apple polyphenols extract (APE) improves colon damage in a rat model of colitis. *Digestive and Liver Disease*, *44*(7): 555-562.

[37] Martín, A. R., Villegas, I., Sánchez-Hidalgo, M. and De La Lastra, C. A. (2006). The effects of resveratrol, a phytoalexin derived from red wines, on chronic Inflammation induced in an experimentally induced colitis model. *British journal of pharmacology*, *147*(8): 873-885.

[38] Liu, L., Liu, Y. L., Liu, G. X., Chen, X., Yang, K., Yang, Y. X. and Gan, H. T. (2013). Curcumin ameliorates dextran sulfate sodium-induced experimental colitis by blocking STAT3 signaling pathway. *International immunopharmacology*, *17*(2): 314-320.

[39] Al Mijan, M. and Lim, B. O. (2018). Diets, functional foods, and nutraceuticals as alternative therapies for inflammatory bowel disease: Present status and future trends. *World journal of gastroenterology*, *24*(25): 2673.

[40] Kaliora, A. C., Dedoussis, G. V. Z. and Schmidt, H. (2006). Dietary antioxidants in preventing atherogenesis. *Atherosclerosis*, *187*(1): 1-17.

[41] Williams, R. J., Mohanakumar, K. P. and Beart, P. M. (2016). *Neuro-nutraceuticals: Further insights into their promise for brain health.:* 1-3.

[42] Williams, R. J., Mohanakumar, K. P. and Beart, P. M. (2015). *Neuro-nutraceuticals: The path to brain health via nourishment is not so distant:* 1-6.

[43] Glevitzky, I., Dumitrel, G. A., Glevitzky, M., Pasca, B., Otrisal, P., Bungau, S. and Popa, M. (2019). Statistical analysis of the relationship between antioxidant activity and the structure of flavonoid compounds. *Rev. Chim*, *70*: 3103-3107.

[44] Boccardi, V., Tinarelli, C. and Mecocci, P. (2017). Nutraceuticals and Cognitive Dysfunction: Focus on Alzheimer's Disease. *Neuroprotective Effects of Phytochemicals in Neurological Disorders*: 561-579.

[45] Sarris, J., Logan, A. C., Akbaraly, T. N., Amminger, G. P., Balanzá-Martínez, V., Freeman, M. P. and Jacka, F. N. (2015). Nutritional medicine as mainstream in psychiatry. *The Lancet Psychiatry*, *2*(3):271-274.

[46] Makkar, R., Behl, T., Bungau, S., Zengin, G., Mehta, V., Kumar, A. and Oancea, R. (2020). Nutraceuticals in neurological disorders. *International journal of molecular sciences*, *21*(12): 4424.

[47] Van Der Burg, K. P., Cribb, L., Firth, J., Karmacoska, D. and Sarris, J. (2021). Nutrient and genetic biomarkers of nutraceutical treatment response in mood and psychotic disorders: A systematic review. *Nutritional neuroscience*, *24*(4): 279-295.

[48] Iyen, S. I., Ogofotha, O. G., & Oko, O. J. (2019). Advances in nutraceuticals. *International Journal of Medicinal Plants and Natural Products*, *5*: 8-22.

[49] Puneet, G., Andrew, H., Kirschner, B. S. and Guandalini, S. (2000). Is Lactobacillus GG helpful in children with Crohn's disease?

Results of a preliminary, open-label study. *Journal of Pediatric Gastroenterology and Nutrition*, *31*(4): 453-457.

[50] Nasri, H., Baradaran, A., Shirzad, H. and Rafieian-Kopaei, M. (2014). New concepts in nutraceuticals as alternative for pharmaceuticals. *International journal of preventive medicine*, *5*(12): 1487.

[51] Gupta, R. C., Lall, R. and Srivastava, A. (Eds.). (2021). *Nutraceuticals: efficacy, safety and toxicity*. Academic Press.

[52] Gonzalez-Suarez, I., Martin, F., Hoeng, J. and Peitsch, M. C. (2016). Mechanistic network models in safety and toxicity evaluation of nutraceuticals. In *Nutraceuticals* (pp: 287-304). Academic Press.

[53] Han, E. S., Oh, J. Y., & Park, H. J. (2011). Cordyceps militaris extract suppresses dextran sodium sulfate-induced acute colitis in mice and production of inflammatory mediators from macrophages and mast cells. *Journal of ethnopharmacology*, *134* (3): 703-710.

[54] Li, Y., Song, K., Zhang, H., Yuan, M., An, N., Wei, Y. and Gao, Y. (2020). Anti-inflammatory and immunomodulatory effects of baicalin in cerebrovascular and neurological disorders. *Brain Research Bulletin*: 1-3.

[55] Kador, P. F. (2017). Topical applied nutraceutical antioxidant formulation reduces ocular oxidative stress. *Functional Foods in Health and Disease*, *7*(1): 68-87.

In: Nutraceuticals
Editors: A. Kumari and G. Singh
ISBN: 978-1-68507-488-3

Chapter 8

PREBIOTICS AND HEALTH BENEFITS

Heena[1], Neha Verma[2], Shiv Kumar Giri[3] and Anil Kumar[1,*]

[1]Centre for Medical Biotechnology,
Maharshi Dayanand University, Rohtak, Haryana, India
[2]Department of Botany, Baba Mastnath University,
Rohtak, Haryana, India
[3]Department of Biotechnology, Maharaja Agrasen University,
Solan, Himachal Pradesh, India

ABSTRACT

In recent years, human health has become an area of enormous interest for researchers. An individual's health is not only affected by external factors, but also depends on good dietary intake. The latter is required for the smooth functioning of the gut microbiota as they perform a cascade of activities that affect human physiology and health. Dietary components such as prebiotics have received much attention and recognition over the years. They are non-digestible fibers that are fermented to produce short-chain fatty acids (SCFAs) by groups of health-promoting bacteria like Bifidobacterium and Lactobacillus spp.

* Corresponding Author's E-mail: anil.cmbt@mdurohtak.ac.in.

Prebiotics are found naturally in foods such as fruits, vegetables, grains, and pulses. They have also been produced commercially as functional food ingredients and as supplements. The few common examples of prebiotics include inulin, galacto-oligosaccharides (GOS), fructo-oligosaccharides (FOS), polydextrose, and lactitol. Various studies have demonstrated that a gluten-free, less fermented, and low carbohydrates diet affect growth of the gut microbiota and it may result in a plethora of diseases. The primary function of prebiotics is to promote the growth of useful bacteria to limit the progression of colorectal cancer and inflammatory bowel disease. Besides, they also aid in the prevention of obesity and constipation by absorbing several minerals. They have also been reported to modulate the body's immune response by altering the expression of cytokines. This chapter highlights the significance of prebiotics in our dietary intake and their beneficial effects on human health.

1. Introduction

Ecological and evolutionary advancements have taken place all around the globe, and the rise of various industrial sectors has transformed the lifestyle and food habits of peoples. The ailments of untenable eating schedules and the rise of fast-food industries have further changed the global population's eating habits. The food industry has been technically advanced with "ready to eat" versions of food that have made people's lives easier. All these advancements have made a positive impact, but have simultaneously affected public health and their well-being. A strong positive correlation between diet, gut microbiota, and health is widely accepted, and a healthy gut is a sign of whole-body fitness and endurance [1]. In this regard, prebiotics have been very helpful in building up a healthy gut by promoting the growth of beneficial microflora. The term prebiotics was introduced in 1995 by Gibson and Roberfroid, and it was defined as "a non-digestible food component that favorably impacts the host by discriminately promoting the growth/or function of one or a restricted number of microorganisms in the gut, and thus improves host health" [2]. This was reframed in 2004 as the fermented component that allows specific changes, both in composition and/ or activity, in the gastrointestinal microflora that confers benefits upon host betterment and health" [3]. The idea behind prebiotics is to reestablish and maintain microflora, which is beneficial to human health. Prebiotics are food ingredients that fit into the category of functional or medicinal food. Latter can be any fresh or refined food that supplies basic nutrients and has properties to boost a person's health. Prebiotics have become a need of today's food industry and can be added to different foods, including yogurts, cereals, bread, ice cream, and drinks.

2. What Is a Good Prebiotic?

Good prebiotics should possess the below-mentioned properties

1. It should prevent acid formation in the stomach.
2. It should be easy to break down by intestinal enzymes.
3. It should have easy absorption in the upper gastrointestinal (GI) tract.
4. It should not be metabolized by non-probiotic gut microflora such as *Bacteroids* sp. and *E. coli.*
5. It should promote the growth and/or activity of intestinal bacteria, which is associated with human health and their well-being.
6. Furthermore, it should be easy to ferment by the intestinal microflora.

3. Sources and Types of Prebiotics

Prebiotics include low molecular weight indigestible oligosaccharides consisting of 3 to 9 monomer units. Diverse types of oligosaccharides like lactulose, inulin, fructo-oligosaccharide (FOS), galacto-oligosaccharides (GOS) are probable contenders for prebiotics. Both GOS and FOS meet all the criteria to be classified as prebiotics. They can be obtained from different dietary food sources, including sugar beet, *Allium sativum, chicory, Allium cepa*, asparagus, *Jerusalem artichoke*, Triticum, barley, banana, Lycopersicon, peas, beans, honey, rye, soybean, cow's/human's milk, microalgae, and seaweeds. Many food products such as bread, cereal, cookies, and yogurts have also been fortified with prebiotics. However, considering their concentration in natural food sources, they can also be produced artificially at large scales. The purified form of prebiotics like inulin and oligofructose have been produced industrially and are being used as ingredients in many food products. Numerous efforts have been made to hydrolyze natural polysaccharides into non-digestible carbohydrates. Furthermore, the production of prebiotics like lactulose, soybean from oligosaccharides, and raffinose from disaccharides and oligosaccharides also has been attempted using various mechanism including enzymatic and chemical synthesis, direct extraction, and isomerization processes [4]. Various types of prebiotics and their natural, as well as industrial sources, have been discussed below-

3.1. Inulin

It is a mixture of oligosaccharides with a long chain of 2 to 60 fructose molecules. It is commonly found in wheat, onion, asparagus, bananas, garlic, artichokes, and leeks. It remains undigested in the stomach due to lack of specific enzymes there; once it reaches the colon, it favors the growth of *Bfidobacteria* to improve bowel function. Nowadays, the food industries are using inulin as the first-choice food ingredient as it does not have any calories, and also it can be used in margarine and salad dressings as a fat substitute. Inulin can be isolated from chicory roots by differential precipitation with ethanol [5].

3.2. Galacto-Oligosaccharides (GOS)

These are present naturally in both human and cow's milk and offer unique resilience against enzymatic digestion. They could be obtained from microbial fermentation of lactose, or can also be synthesized using the β-galactosidase. These oligosaccharides selectively enhance the count of *Bifidobacterium* and *Lactobacillus* within the human intestinal microbiota [6].

3.3. Lactulose

It is a synthetic disaccharide formed by the isomerization of fructose and galactose, which are linked together with a β-1, 4-glycosidic bond [7]. The lactulose escapes digestion in the mammalian gut due to a lack of enzymes for β-glycosidic bond, and it reaches the stomach and small intestine in the undigested form. Once it enters the colon, it gets digested by lactic acid bacteria into lactate and SCFAs as major end products.

3.4. Lactosucrose

It is a trisaccharide sugar that consists of D-galactose, D-glucose, and D-fructose. It remains undigested in the small intestine and stimulates the intestinal *Bifidobacterium's* population growth, which ensures the normal functioning of the digestive tract. It can either be obtained by a synthesis reaction catalyzed by levansucrase [8] or by a transfructosylation reaction catalyzed by β-fructofuranosidase [9]. Lactosucrose (lactosyl fructoside) can be produced from sucrose and lactose by the action of β-fructofuranosidase, isolated from *Arthrobacter* sp. K-1.

3.5. Fructo-Oligosaccharides (FOS)

These represent short-chain (3-5 units) oligosaccharides composed of D-fructose and D-glucose. These typically contain O- or N-linkages, which are known to be compatible with side chains in proteins or lipids. These can also be used as an alternative to artificial sweeteners. These are present as an ingredient in tomatoes, garlic, asparagus, chicory root, onions, barley, wheat, bananas, and sugarcane juice. At an industrial scale, FOS can be produced in two ways: The first method involves the enzymatic digestion of sucrose using β-fructofuranosidase, and another involves the enzymatic hydrolysis of the polysaccharide (extracted from chicory roots) in a controlled manner [10].

3.6. Soybean Oligosaccharide (SOS)

These are typically found in soybean whey obtained during soy protein production. They are composed of galactose and sucrose units joined by a α-1, 6-linkage. When administered, SOS directly reaches the colon due lack of the α-galactosidase enzyme required for its digestion and elicits its Bifidogenic effect. SOS also include verbascose, raffinose, and stachyose oligosaccharides, which works in the same or less manner [11].

3.7. Xylo-Oligosaccharides

These are composed of a short chain of xylose sugar. Naturally, these are found in fruits, vegetables, milk, wheat bran, honey, and bamboo shoots, however; they can also be produced by (a) enzymatic digestion of lignocellulosic material; (b) conversion of lignocellulosic material to xylan; and (c) by conversion of xylan to xylo-oligosaccharides with steam or diluted mineral acids [12].

3.8. Gentio-Oligosaccharides

These are novel functional oligosaccharides consisting of several glucose residues linked by a β-1,6-glycosidic bond. These can be produced from starch, by their subsequent hydrolysis with either an acid or an enzyme. They can also be obtained from glucose syrup by transglycosylation reaction catalyzed by the β- -glycosidase enzyme [13].

3.9. Isomaltulose or Palatinose

These are natural disaccharides and can be found in honey and sugarcane juice. These are composed of glucose and fructose units linked by a α-1,6- glycosidic bond. These can also be synthesized from sucrose in a two-step process involving enzymatic conversion of (1, 2)-fructoside to a (1, 6)-fructoside by rearrangement within a glycosidic linkage followed by crystallization [14].

3.10. Cyclodextrins

These represent a group of cyclic oligosaccharides composed of the macrocyclic ring where 6 to 8 monomers of glucose are linked by a α-1,4-glycosidic bond. Industrially, these have also been produced from starch using cyclodextrin glycosyltransferases, which catalyze inter and intramolecular transglycosylation and digest the starch using hydrolytic activity [15].

4. Mechanism of Prebiotics Action

Prebiotics can modify the gut environment by modulating the composition and activity of the microbiota. Once prebiotics enter the human intestine, they escape digestion in the small-intestine due to the inability of intestinal enzymes to hydrolyze the polymer bonds. Afterwards, they reach the colon in an undigested form where they get fermented by a group of beneficial bacteria, such as *Lactobacilli* and *Bifidobacteria*. The end products formed by the prebiotic's fermentation are mostly acidic, hence lowers the intestinal gut's pH. It has been observed that even small changes in gut pH, i.e., from 6.5 to 5.5, can modify the population of acid-sensitive *Bacteroides spp;* which stimulates the formation of butyrate [16]. Prebiotics favor the growth of beneficial microorganisms and suppress the growth of harmful microbes by enhancing the synthesis of short-chain fatty acids. Prebiotics such as galacto-oligosaccharides (GOS) function as immunomodulators by influencing the natural killer (NK) cell activity, pro-inflammatory cytokines (IL-8 &10) production, and signaling of toll-like receptor-4 [17]. Prebiotics such as Fructans, e.g.; inulin and fructo-oligosaccharides (FOS), can scavenge the reactive oxygen species (ROS) and stimulate the activity of the antioxidant enzymes to neutralize the burden of the oxidative stress in the body [18].

5. Health Benefits of Prebiotics

The entire human gut hosts numerous microbial species. Once prebiotics reach the colon, they act as a substrate for the useful gut microorganisms and confer various health benefits to humans, as discussed below and shown in Figure 1.

5.1. Prevention from Gastroenteritis

Gastroenteritis disease usually occurs due to the ingestion of pathogenic microorganisms, contaminated food, or water. Usually gastroenteritis is caused by *Shigella, Salmonella, Yersinia enterocolitica, Campylobacter jejuni, Escherichia coli, Vibrio cholera, and Clostridium perfringens.* These pathogens grow and colonize within the gastrointestinal tract, where they dominate the host cell and release the toxins in the food. Contaminated food can disrupt the functions of the intestinal mucosa, causing symptoms like nausea, vomiting, and diarrhea. Prebiotics prevent acute gastroenteritis by increasing the population of beneficial microorganisms in the large intestines [19].

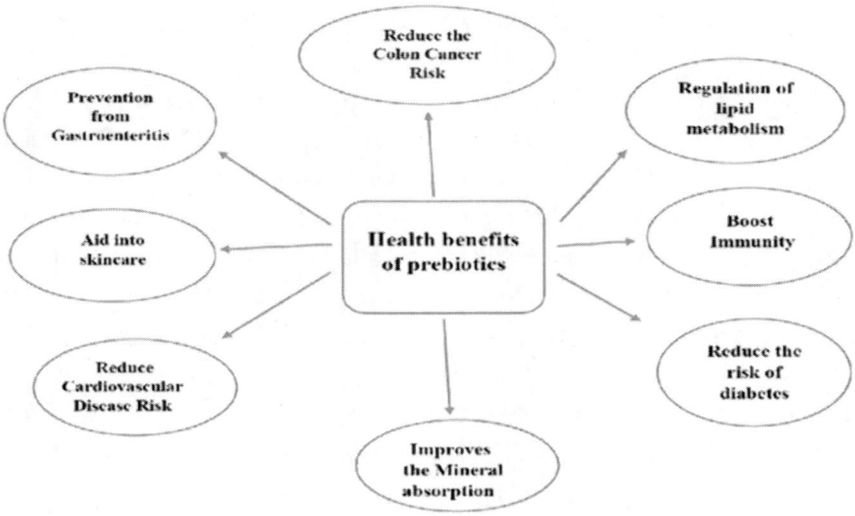

Figure 1. Showing Health benefits of Prebiotics.

5.2. Reduce Colon Cancer Risk

Cancer is the major cause of death all over the world. It occurs due to abnormal cell division and multiplication of cells. The third most common cancer type is colon-rectal cancer, which usually leads to metastatic tumors. Conventional treatments like chemo and radiotherapy may worsen the situation as they make the person immunocompromised. In contrast, feeding a combination of live microorganisms (known as probiotics) along with prebiotics could resist cancer progression by the formation of butyrate-like metabolites [20]. Latter is produced from the fermentation of galacto-oligosaccharides, which help to reduce the progression of tumor cells and induce apoptosis in colon cells. Butyrate is also known to promote the expression of enzymes required to inhibit cancer-causing agents. Clinical studies have showcased the efficient role of synbiotics (combination of prebiotics and probiotics) in reducing the inflammatory state and usage of antibiotics in colon cancer [21].

5.3. Reduce Cardiovascular Disease Risk

The risk of coronary heart disease increases due to elevated levels of low-density lipoproteins (LDL) and blood cholesterol. Due to changes in lifestyle and consumption habits, people have shown keen interest in prebiotics. These can promote the growth of lactic acid bacteria, which can absorb the cholesterol and triglycerides and help to maintain the fats and lipids index in the body. Furthermore, prebiotics like fructo-oligosaccharides lower the synthesis of triglycerides in the body. Consumption of prebiotic dietary fibers helps to decrease LDL and inflammatory elements, while elevating the levels of high-density lipoproteins (HDL) and ferulic acid. In this manner, prebiotics lower the risk of cardiovascular disease and obesity [22].

5.4. Aid into Skincare

The metabolism of aromatic amino acids by gut microbes produces phenol-like compounds that help to reduce the risk of skin allergies and infections such as atopic dermatitis [23]. Prebiotics like GOS have shown the ability to control skin pigmentation, thus aiding in skincare. Moreover, GOS can also boost the skin barrier by elevating the levels of CD 44 and collagen type 1 markers. In females, ingestion of GOS in the presence or absence of probiotic strains such as *Bifidobacterium breve,* can eliminate the deficiency of keratin and water caused by phenols [24].

5.5. Reduce Diabetes Risk

Prebiotics can influence blood glucose and insulin levels by a delay in gastric emptying, intestinal transit time, and absorption of glucose into the bloodstream by producing SCFAs like acetate, butyrate, propionate, etc. The products of prebiotic fermentation-like propionates decrease gluconeogenesis, enhance glycolysis, and stimulate the glucose regulating hormones to balance the insulin and glucose levels in the host [25].

5.6. Prevention of Bone Mineralization

Dietary fibers have positive as well as negative impacts on mineral absorption. Insoluble dietary fiber can interfere with mineral absorption, while the implications of soluble dietary fibers are not much known. The impact of prebiotics on mineral absorption has been demonstrated in clinical studies. Several studies have reported a significant increase in calcium absorption after the intake of 5-20 gm per day doses of lactulose, transgalacto-oligosaccharides, or inulin in combination with oligofructose. Fermentation of inulin-type fructans by probiotics results in colonic mineral absorption, elevates the concentrations of SCFA, and lowers the colon pH. Additionally, the inulin-type fructans also augment calcium accumulation and improve bone mineralization and density in young adults [26].

5.7. Boost the Immune System

Intake of prebiotics plays an important role in improving host immunity by enhancing the population of beneficial microbes. Various studies on animals and humans have reported the prebiotics role in reducing the population of detrimental bacteria with the support of *Lactobacilli* and *Bifidobacteria.* Consumption of D-mannose can decrease the colonization of pathogens such as *Salmonella sp.* by stimulating its adhesion to mannose through type 1 fimbriae. Adhesion of mannose ensures the incompetency of pathogens to bind the epithelial cells [27]. Prebiotics such as oligofructose and inulin aid in amelioration of antibody reactions in response to measles and influenza virus vaccines. Studies have shown that FOS consumption plays a vital role in improving antibody response towards the influenza vaccine and helps to reduce its side-effects. In infants, FOS can reduce diarrhea-associated fever and also helps to reduce the incidence of pyretic seizures [28]. In healthy volunteers, elevated interleukin 4 (IL-4) levels, and modulated immune responses from the toll-like receptor 2 (TLR 2) have been reported on consuming the (2→1) fructans [29]. Intake of prebiotics like GOS can surge the levels of interleukin 8 (IL-8), interleukin 10 (IL-10), and C- reactive proteins into the bloodstream to reduce the inflammatory immune responses from IL-1β. Furthermore, GOS intake also improves the function of NK cells and helps in reducing the menace of atopic dermatitis and eczema in infants.

CONCLUSION

Prebiotics are food supplements usually present in raw food sources and can be synthesized artificially as well. These stimulate the growth of beneficial microorganisms in the gut, and thus play a vital role in maintaining the good health of the host. Prebiotics confer numerous health benefits against gastroenteritis, cancer, diabetes, and cardiovascular diseases. These also play an important role in boosting the host immune system response. However, prebiotics produced industrially are susceptible to various challenges such as critical selection of abundant and cost-effective raw materials, stability at high pH, temperature fluctuations, storage, and functional viability. Despite such challenges, new materials are variously being screened to meet expectations; for example, mushrooms could be used as an invaluable source for cheap, non-dairy-based raw materials. Similarly, cereals and their husks can be a good source for galactose and fructose oligosaccharides. Various experimental studies have demonstrated their role in disease prevention; however, to explore their precise mechanisms of action, further investigations are needed. Further clinical trials should be conducted to gather more information about their other health benefits as well as side effects before considering their role in therapeutics.

REFERENCES

[1] Broekaert, Ilse J., and W. Allan Walker. "Probiotics as flourishing benefactors for the human body." *Gastroenterology Nursing* 29, no. 1 (2006): 26-34.

[2] Gibson, Glenn R., and Marcel B. Roberfroid. "Dietary modulation of the human colonic microbiota: introducing the concept of prebiotics." *The Journal of nutrition* 125, no. 6 (1995): 1401-1412.

[3] Gibson, Glenn R., Hollie M. Probert, Jan Van Loo, Robert A. Rastall, and Marcel B. Roberfroid. "Dietary modulation of the human colonic microbiota: updating the concept of prebiotics." *Nutrition research reviews* 17, no. 2 (2004): 259-275.

[4] Walker, Alan W., Sylvia H. Duncan, E. Carol McWilliam Leitch, Matthew W. Child, and Harry J. Flint. "pH and peptide supply can radically alter bacterial populations and short-chain fatty acid ratios within microbial communities from the human colon." *Applied and environmental microbiology* 71, no. 7 (2005): 3692-3700

[5] Gupta, Anil K., Navdeep Kaur, and Narinder Kaur. "Preparation of Inulin from Chicory Roots." *Journal of scientific and Industrial research* 62, no. 9 (2003): 916-920.

[6] Sako, Tomoyuki, Keisuke Matsumoto, and Ryuichiro Tanaka. "Recent progress on research and applications of non-digestible galacto-oligosaccharides." *International Dairy Journal* 9, no. 1 (1999): 69-80.

[7] Panesar, Parmjit S., Shweta Kumari, and Reeba Panesar. "Biotechnological approaches for the production of prebiotics and their potential applications." *Critical reviews in biotechnology* 33, no. 4 (2013): 345-364.

[8] Takahama, Akihiro, Jiro Kuze, Satoko Okano, Kyoko Akiyama, Toshie NAKANE, Hiroshi Takahashi, and Takeshi Kobayashi. "Production of lactosucrose by Bacillus natto levansucrase and some properties of the enzyme." *Nippon Shokuhin Kogyo Gakkaishi* 38, no. 9 (1991): 789-796.

[9] FUIITA, Koki, Takeshi OSAWA, Katsuhiko MIKUNI, Kozo HARA, Hitoshi HASHIMOTO, and Sumio KITAHATA. "Production of Lactosucrose by β-Fructofuranosidase and Some of Its Physical Properties (Studies on β-Fructofuranosidase from Arthrobacter sp. K-1 Part 3)." *Journal of the Japanese Society of Starch Science* 38, no. 1 (1991): 1-7.

[10] Sangeetha, P. T., M. N. Ramesh, and S. G. Prapulla. "Recent trends in the microbial production, analysis and application of fructooligosaccharides." *Trends in food science & technology* 16, no. 10 (2005): 442-457.

[11] Karr-Lilienthal, L. K., C. T. Kadzere, C. M. Grieshop, and G. C. Fahey Jr. "Chemical and nutritional properties of soybean carbohydrates as related to nonruminants: A review." *Livestock Production Science* 97, no. 1 (2005): 1-12.

[12] Vazquez, M. J., J. L. Alonso, H. Domınguez, and J. C. Parajo. "Xylooligosaccharides: manufacture and applications." *Trends in Food Science & Technology* 11, no. 11 (2000): 387-393.

[13] Crittenden, RG and, and M_J Playne. "Production, properties and applications of food-grade oligosaccharides." *Trends in food science & technology* 7, no. 11 (1996): 353-361.

[14] Lina, B. A. R., D. Jonker, and G. Kozianowski. "Isomaltulose (Palatinose®): a review of biological and toxicological studies." *Food and Chemical Toxicology* 40, no. 10 (2002): 1375-1381.

[15] Munro, I. C., P. M. Newberne, V. R. Young, and A. Bär. "Safety assessment of γ-cyclodextrin." *Regulatory Toxicology and Pharmacology* 39 (2004): 3-13.

[16] Walker, Alan W., Sylvia H. Duncan, E. Carol McWilliam Leitch, Matthew W. Child, and Harry J. Flint. "pH and peptide supply can radically alter bacterial populations and short-chain fatty acid ratios

within microbial communities from the human colon." *Applied and environmental microbiology* 71, no. 7 (2005): 3692-3700.

[17] van der Beek, Christina M., Cornelis HC Dejong, Freddy J. Troost, Ad AM Masclee, and Kaatje Lenaerts. "Role of short-chain fatty acids in colonic inflammation, carcinogenesis, and mucosal protection and healing." *Nutrition reviews* 75, no. 4 (2017): 286-305.

[18] Pasqualetti, Valentina, Annamaria Altomare, Michele Pier Luca Guarino, Vittoria Locato, Silvia Cocca, Sara Cimini, Rossella Palma, Rossana Alloni, Laura De Gara, and Michele Cicala. "Antioxidant activity of inulin and its role in the prevention of human colonic muscle cell impairment induced by lipopolysaccharide mucosal exposure." *PloS one* 9, no. 5 (2014): e98031.

[19] Szajewska, Hania, Alfredo Guarino, Iva Hojsak, Flavia Indrio, Sanja Kolacek, Silvia Salvatore, Raanan Shamir et al. "Use of probiotics for the management of acute gastroenteritis in children: an update." *Journal of pediatric gastroenterology and nutrition* 71, no. 2 (2020): 261-269.

[20] Thilakarathna, W. P. D. Wass, Morgan G. I. Langille, and H. P. Vasantha Rupasinghe. "Polyphenol-based prebiotics and synbiotics: potential for cancer chemoprevention." *Current Opinion in Food Science* 20 (2018): 51-57.

[21] Polakowski, Camila Brandão, Massakazu Kato, Vinicius Basso Preti, Maria Eliana Madalozzo Schieferdecker, and Antonio Carlos Ligocki Campos. "Impact of the preoperative use of synbiotics in colorectal cancer patients: A prospective, randomized, double-blind, placebo-controlled study." *Nutrition* 58 (2019): 40-46.

[22] Thushara, Ram Mohan, Surendiran Gangadaran, Zahra Solati, and Mohammed H. Moghadasian. "Cardiovascular benefits of probiotics: a review of experimental and clinical studies." *Food & function* 7, no. 2 (2016): 632-642.

[23] Davani-Davari, Dorna, Manica Negahdaripour, Iman Karimzadeh, Mostafa Seifan, Milad Mohkam, Seyed Jalil Masoumi, Aydin Berenjian, and Younes Ghasemi. "Prebiotics: definition, types,

sources, mechanisms, and clinical applications." *Foods* 8, no. 3 (2019): 92.

[24] Mohkam, Milad, Navid Nezafat, Aydin Berenjian, Mohammad Ali Mobasher, and Younes Ghasemi. "Identification of Bacillus probiotics isolated from soil rhizosphere using 16S rRNA, recA, rpoB gene sequencing and RAPD-PCR." *Probiotics and antimicrobial proteins* 8, no. 1 (2016): 8-18.

[25] Mahmoudi, Razzagh, Sara Moosazad, and Katayoon Aghaei. "Oral Health by Using Probiotic Products." In *Oral Health by Using Probiotic Products*. IntechOpen, 2019.

[26] Abrams, Steven A., Ian J. Griffin, Keli M. Hawthorne, Lily Liang, Sheila K. Gunn, Gretchen Darlington, and Kenneth J. Ellis. "A combination of prebiotic short-and long-chain inulin-type fructans enhances calcium absorption and bone mineralization in young adolescents–." *The American journal of clinical nutrition* 82, no. 2 (2005): 471-476.

[27] Oyofo, B. A., J. R. DeLoach, D. E. Corrier, J. O. Norman, R. L. Ziprin, and H. H. Mollenhauer. "Prevention of *Salmonella typhimurium* colonization of broilers with D-mannose." *Poultry science* 68, no. 10 (1989): 1357-1360.

[28] Fujiwara, Reiko, Naoki Takemura, Jun Watanabe, and Kei Sonoyama. "Maternal consumption of fructo-oligosaccharide diminishes the severity of skin inflammation in offspring of NC/Nga mice." *British Journal of Nutrition* 103, no. 4 (2010): 530-538.

[29] Shokryazdan, Parisa, Mohammad Faseleh Jahromi, Bahman Navidshad, and Juan Boo Liang. "Effects of prebiotics on immune system and cytokine expression." *Medical microbiology and immunology* 206, no. 1 (2017): 1-9.

In: Nutraceuticals ISBN: 978-1-68507-488-3
Editors: A. Kumari and G. Singh

Chapter 9

PROBIOTICS AND HEALTH BENEFITS

Amita Suneja Dang[1], Neha Verma[2], Shiv Kumar Giri[3] and Anil Kumar[1,*]

[1]Centre for Medical Biotechnology,
Maharshi Dayanand University, Rohtak, Haryana, India
[2]Department of Botany,
Baba Mastnath University, Rohtak, Haryana, India
[3]Department of Biotechnology,
Maharaja Agrasen University, Solan,
Himachal Pradesh, India

ABSTRACT

The scientific community has shown significant interest in probiotics research over the previous few decades because of its potential health benefits. Probiotics are live microbes that enhance the nutritional value of food and also help to maintain the beneficial microflora of the gut. With the increasing demands of functional foods, the probiotic strain has been introduced into a wide range of dairy-based products such as fermented milk, buttermilk, ice cream, milk powder, yogurt, and cheese.

* Corresponding Author's E-mail: anil.cmbt@mdurohtak.ac.in.

In contrast, non-dairy foods include soy-based products, cereals, and a variety of juices. The most frequently used probiotic groups belong to *Lactobacillus, Bifidobacterium, Pediococcus, Lactococcus, Bacillus*, and yeast strains. These probiotic strains have a competitive ability to fight against pathogenic microbes for adhesion sites or act as an antagonist against them. They play a therapeutic role in many disorders like lactose intolerance, gastrointestinal disease, allergies, urogenital infections, diarrhea, and inflammatory bowel syndrome. They also aid in the cholesterol lowering, boosting of the immune response, cancer prevention, and overall improvement of intestinal health. However, the effectiveness of a specific probiotic strain is influenced by short-chain fatty acids and bacitracin production, nutrient competition and low gut pH. Furthermore, overall effectiveness of probiotics depends on their adequate dose. Considering the stress caused by today's lifestyle and consumption habits on human health, in this chapter, we have highlighted the importance of probiotics to better health and quality of life.

1. INTRODUCTION

Probiotics are live microbial feed dietary supplements that confer fitness advantages to the host by balancing its intestinal microbial equilibrium. The gastrointestinal (GI) tract microbiota is described as a complicated and dynamic microbiome. It hosts more than one thousand distinct microbial species, and these are inevitable for homeostatic physiological actions [1]. The microbes that inhabit the intestine make an everlasting diversification and decide the physiological, immune, metabolic, and behavioral improvement. Currently, numerous well-attributed strains of *Lactobacilli* and *Bifidobacteria* are being used as probiotics to decrease the threat of GI infections or cure the same. With the emergence of new sequencing technologies and improvements in revolutionary bioinformatics tools; the classification of the gut microbiota is being upgraded and making our understanding better about the organization and functions of the intestinal microbiota.

To get clinical benefits, knowledge about probiotics, prebiotics, and other food supplements have been translated with great interest into therapeutic strategies to influence the gut microbiota directly or indirectly [2]. Nowadays, consumers have become conscious about the implications of what they eat for their wellbeing; therefore, they look for practical nourishment. In recent years, many studies have reported probiotics' role in balancing gastrointestinal, respiratory, and immunological functions. The release of the bacteriocins and metabolites (lactic and acetic acid) by probiotics describes its protective role against intestinal pathogens [3]. These have shown immense potential as therapeutics alternative chiefly for gastrointestinal illnesses, including acute infectious diarrhea, antibiotic-associated diarrhea, ulcerative colitis, irritable bowel syndrome, and necrotizing enterocolitis [4]. Besides the therapeutic benefits of probiotics, concern has been raised regarding safety issues; for instance, using live microorganisms, especially in sensitive populations. This may be due to the absence of apparent clinical guidelines, the absence of compelling evidence from clinical preliminaries, and restricted studies evaluating the contrasts between viable or non-viable microorganisms in the digestive tract.

1.1. Morphology and Physiology of Probiotic Bacteria

Bifidobacteria are gram-positive, non-motile, anaerobic, and non-spore-forming pleomorphic rods. The cell morphology of probiotic bacteria ranges from regular rods to various branched and club-shaped structures [5]. The multiplying cells in the exponential growth phase depict the development of the intra-cytoplasmic membrane complex portrayed by lamellar, myelinoform, vesicular structures. The nucleoid is localized as the central poly-branched or scattered in the osmophobic zone, and its distribution is dictated by morphogenesis processes such as exobudding and multiseptation.

1.2. Probiotics and Food Products

There is a diverse range of dietary products that contain probiotic strains and are still surging. Mainly dairy-based products are available in the market as fermented milk, ice cream, milk powder, buttermilk, cheese, and yogurts. Latter accounts for the largest share of sales among all dairy products.

Non-dairy food products such as soy-based products, cereals, nutrition bars, and various juices serve as a suitable medium for probiotic delivery to the consumer.

1.3. Desirable Properties of Probiotics

A potential probiotic strain with all its valuable impacts, should possess certain attractive properties, such as-

1. It should show resistance to gastric acid and bile tolerance.
2. It should possess adhesion properties for mucosal and epithelial surfaces.
3. It should destroy pathogenic bacteria.
4. It should be safe for consumption, i.e., non-pathogenic.
5. It should not result in any hypersensitivity reaction.
6. It should be able to affect metabolic activities like lactose intolerance and assimilation of cholesterol.

There are no essential criteria that have been defined for probiotic applications. However, because of in vivo and in vitro disparities, issues of pertinence, and the absence of standard operating procedures need to be considered before their practical use.

Taking everything into account, the level of probiotic dosage should show effectiveness in humans and is expressed as colony-forming unit (CFU) per gram of the product. Still, information on minimum effective concentrations of probiotics is inadequate. A minimum concentration of 10^6 CFU/mL or gram of probiotic microorganisms have been recommended for transfer of health benefits to the consumers. However, probiotic strains should retain their viability under normal storage conditions for maximum benefits [6].

1.4. Probiotic Microorganisms

Various microbial species such as bacteria, yeast, and mold can be used as probiotics (Table 1). However, *Lactobacillus* and *Bifidobacteria* are the most known ones.

2. Mechanism of Probiotic Action

A few mechanisms have been hypothesized concerning probiotics' mode of action, which have been discussed below-

Table 1. List of commonly used probiotics microorganism

Group	Probiotics species	References
Bifidobacterium sp.	*B. infantis, B. animalis, B. adolescentis, B. bifidum, B. breve, Bifidobacterium lactis,*	7-15
Lactobacillus sp.	*L. acidophilus, L. casei, L. crispatus, L. plantarum, L. reuteri, L. acidophilus, L. bulgaricus, L. paracasei, L. rhamnosus, L. fermentum, L. lactis, L. gasseri, L. johnsonii, L. brevis, L. johnsonii, L. salivarius, L. kefir*	
Non-lactic acid bacteria	*Bacillus cereus var, Enterococcus faecalis, E. faecium, E. durans E. coli nissle, Streptococcus thermophilus, Propionibacterium freudenreichii*	
Non-spore forming	*Bacillus. coagulans, B. clausii, B. pumilus, B. licheniformis, B. laterosporus, B. subtilis, B. racemilacticus, Coccobacillus, Leuconostoc, Lactococcus lactis, Pediococcus*	
Non-pathogenic yeast	*Saccharomyces cerevisiae, S. boulardii*	
Mold/Fungus	*Aspergillus niger, A. oryzae, Candida pintolopesii*	

2.1. Production of Short-Chain Fatty Acids (SCFAs)

Probiotics are involved in the production of SFCAs like acetate, propionate, and butyrate that aid in gut health maintenance by acting as key signaling molecules and interacting with peripheral cell receptors. Besides this, these also act as an energy source for enterocytes, hence playing a vital role in energy homeostasis and metabolism regulation. Various studies have demonstrated the role of SCFAs in decreasing fat accumulation.

These also could elevate the intestinal secretion of polypeptide YY and glucagon-like peptides by interacting with G protein-coupled receptors (GPR 41 and 43) [16, 17].

2.2. Production of Bacteriocins

These are positively charged molecules and are composed of ~30–60 amino acids. Bacteriocins have been designated as antimicrobial peptides because of their inherent ability to disrupt the proton motive force of the cytoplasmic membrane of selected pathogens [18]. Probiotics like *Lactobacilli* (*L. plantarum* and *L. acidophilus)* and *Bifidobacteria* have been reported to inhibit the growth of rotaviruses, Helicobacter, and multidrug-resistant Shigella sp. [19].

2.3. Stimulation of Mucosal Barrier

The gastrointestinal mucosa is the initial interface between the exterior and defense systems.

Secretion of mucin by intestinal epithelial cells can prevent adhesion of pathogenic bacteria.

The probiotic *Lactobacillus* sp. facilitates mucous layer secretion and helps to protect against pathogens [20].

2.4. Production of Enzymes

Incomplete lactose absorption and incitement of the intestinal mucosal lactase action has been hypothesized as a potential mechanism against certain sorts of diarrhea. *Lactobacilli* utilized in the fermented milk industry have dynamic β-galactosidase to diminish the lactose concentrations in dairy items that further influence the seriousness of osmotic looseness of the bowels [21].

Most of the known probiotics produce bile salt hydrolase (BSH) which participates in the deconjugation of biliary salts and, in turn, influences cholesterol absorption [22].

2.5. Competitive Exclusion and Colonization

It refers to the condition where one microbial species aggressively competes with another for the receptor sites in the intestinal tract. Few mechanisms like luminal pH reduction, bacteriocin production, and competition for nutritional sources contribute to the competitive exclusion of pathogens [23]. Probiotics can also adopt enzymatic mechanisms such as competitive inhibition to prohibit the colonization of harmful microbes [24].

2.6. Strengthening of Intestinal Barriers

Epithelial integrity is maintained primarily by the intestinal barrier that protects it from pathogenic microorganisms. Consumption of probiotics helps to strengthen the intestinal barrier. One of the possible mechanisms that may fortify intestinal barrier integrity involves enhanced gene expression.

Probiotics like *Lactobacilli* have been found to modulate the expression of several genes encoding adherens junction proteins (E-cadherin and β-catenin) [25].

2.7. Modulation of the Immune Response

The immunomodulatory effects of probiotics result from their interaction with epithelial and immune cells. Latter can activate immune signals to boost the immune response. The probiotic microorganisms regulate the synthesis of secretory IgA immunoglobulin, and the production of cytokines, thus, play a crucial role in mucosal resistance [26].

3. PROBIOTICS AND HEALTH BENEFITS

Most of the well-known probiotics generally belong to *Lactobacillus* and *Bifidobacterium*. The species of these two genera (*B. breve, B. longum, L. fermentum, L. plantarum, L. casei, or L. rhamnosus*) normally reside in the human gut, and are thought to play a crucial role in maintaining human wellbeing [27]. A few of the beneficial health activities of probiotics are discussed below-

3.1. Probiotics and Diarrhea

Probiotics are well-known for their use in diarrheal infections. Various explicit strains, including *Lactobacillus* sp. (*L.GG, L. reuteri and, L. boulardii*), *Bifidobacteria* sp., and others, have shown an enormous advantage for the loose bowels. Lactic acid bacteria release various compounds into the intestinal lumen that synergistically impact processing and manifestations of intestinal malabsorption. Probiotics resist or ameliorate diarrhea by exerting their impact on the immune system. These offer advantages against viral diarrhea, especially in the pediatric, population by increasing the levels of secretory antibodies (IgA). Additionally, probiotic strains compete with pathogenic bacteria or viruses for binding sites on epithelial cells [28]. Probiotics prohibit the development of pathogenic microscopic organisms by releasing the bacteriocins, such as nisin, and promote intestinal mucin secretion, which forestalls the connection of enteropathogens [29].

3.2. Probiotics and Cancer

The progressive aggregation of mutations in the hereditary material of the cell is the primary cause of cancer. It has been reported that 5-10 percent of cancers involve hereditary deformities, while in 90-95 percent, external factors play a significant role [30]. The relationship of probiotics in forestalling, treating, and diminishing the surge of a tumor cell has been well explored. Various studies have demonstrated anti-proliferative or pro-apoptotic activities of probiotics against myeloid leukemia and tumors of breast colon, stomach, and cervix cells [31-34]. The possible mechanisms by which probiotics counteract malignant growth may involve (i) balancing of the gut microbiota; (ii) up gradation of gut capacity; (iii) modulation of immune response; and (iv) protection of intestinal epithelium against DNA damage by the deterioration of possible cancer-causing agents. Probiotics are gaining lots of consideration because of their capacity to balance malignant cell expansion and apoptosis. It has been studied that eating food that is flourished with *Lactobacilli* has a profound effect on the growth rate of colon cancer [35]. Moreover, probiotic-induced stimulation of the immune system, as demonstrated by an increase in T-helper cells and NK cells, might play a significant role in the repression of cancer advancement.

3.3. Anti-Pathogenic Activity of Probiotics

Probiotics are regarded as one of the most advantageous formulations against pathogenic activities. It has been demonstrated that probiotics can inhibit the growth of harmful microorganisms by the synthesis of SCFAs such as acetate propionate, butyrate, and lactate [36].

The SCFA's role is to maintain optimum pH levels in the colonic lumen, which is necessary for the manifestation of various bacterial enzymes and thus aids in the digestion of foreign compounds and cancer-causing agents inside the gut [37]. It has also been suggested that probiotics may show their anti-pathogenic activity by releasing ethanol, acetaldehyde, hydrogen peroxide (H_2O_2), bacteriocins, and peptides. Latter is thought to be involved in the expansion of layer penetrability that results in cell demise by depolarizing the membrane potential [38].

Additionally, the synthesis of H_2O_2 by probiotics causes the oxidation of sulfhydryl groups, bringing about the denaturation of enzymes that participate in lipid peroxidation. Latter results in cell death due to enhanced porosity of the cell membrane [39].

3.4. Anti-Inflammatory Activity of Probiotics

Ulcerative colitis (UC) and Crohn's disease (CD) are among the most severe inflammatory diseases, and together they are called IBD. It is related to hindrance in SCFAs production, especially, acetic acid derivatives, butyrate, and propionate. The SCFAs have been known to play a critical role in maintaining colonic homeostasis and thereby improving colonic capacity [40]. The CD can influence the mucosa, submucosa, and serosa of the GI and may affect it entirely, while UC affects the mucosa and submucosa of the colon. A disparity in the gut microbiota plays a significant role in the pathophysiology of IBD, and its effects can be modified by supplementation with probiotics, prebiotics, and synbiotics [41]. *Lactobacillus, Bifidobacterium, Enterobacter,* and *E. coli* are the most commonly used probiotics in diets that have shown satisfactory results against IBD.

3.5. Role of Probiotics in Allergy

Allergic diseases arise due to the hypersensitivity of the body's immune system in response to foreign particles, i.e., allergens. An allergic reaction
may result from animals, mold, drugs, food, latex, and seasonal changes that result in anaphylaxis, asthma, and contact dermatitis. It has been shown that gut microbiota composition plays a critical role in allergy prevention and treatment.
Therefore, probiotic supplementation beneficially influencing the gastrointestinal microflora is gaining interest as they can reduce allergic complications.

The potential mechanism of probiotics actions against allergic response includes balancing of Th1/Th2 ratio and inhibition of Th17 cells function; development of Tolerogenic regulatory T (Treg) cells; and modification of Toll-like receptors (TLR) [42]. The probiotics' microorganisms such as *B. umbifidum* may incite the T cells to produce more IL-10 [43].

Moreover, probiotic utilization improves skin inflammation, i.e., eczema, by lowering the CD4+, and CD25+ cells count while increasing the CD8+ cells counts.

The probiotic utilization also animates TLR action and may affect the activity of immunoregulatory cytokines, e.g., IL-10 and TGF-β, and different subgroups of Treg cells [42].

3.6. Lactose Intolerance

Lactose intolerance refers to the inability to hydrolyze complex carbohydrates such as lactose into the monosaccharides glucose and galactose due to the deficiency of the enzyme β-galactosidase. After consuming milk or milk products, persons with lactose intolerance develop symptoms such as diarrhea and abdominal uneasiness. It has been demonstrated that traditionally made yogurt preparations using *S.*

thermophilus and *L. delbrueckii sp. Bulgaricus,* shows more efficacy against lactose intolerance, partially because of higher β-galactosidase activity.

The combination of *L. casei* and *B. breve* could also be used for high efficacy against lactose intolerance [44].

3.7. Probiotics and Heart Health

Coronary heart diseases are associated with elevated cholesterol levels in the blood (hypercholesterolemia) and are considered as major risk factors; therefore, lowering cholesterol levels can help to maintain the good health of the heart. The probiotic strains (lactic acid bacteria) play a significant role in lowering down the cholesterol levels. The mechanism by which probiotic strains help in its assimilation involves

1) inhibition of cholesterol synthesis by *de novo* mechanism;
2) lowering of dietary cholesterol absorption in the gut;
3) self-digestion of cholesterol into its metabolites.

Various studies have reported that administration of probiotic microorganisms such as *B. bifidum, B. breve, L. plantarum, L. brevis* LB32, and *L. reuteri* in food products helps in eliminating the risk of heart diseases [45, 46].

3.8. Anti-Diabetic Activities of Probiotics

Firmicutes such as *Megasphaera, Pectinatus and Zymophilus* are Gram-positive bacteria that dominate the human gut.

A recent study has demonstrated the relationship between reduction in firmicutes and type-2 diabetes.

The decreased count of firmicutes is positively related to glucose concentration [47].

Moreover, changes in gut microbiota, also favor the invasion of opportunistic microorganisms that might inhibit the growth of bacteria engaged in SCFAs production [48].

Currently, research has been focused on the management of diabetes by controlling the gut hormones by the intervention of probiotics (*Lactobacillus* and *Bifidobacteria* sp.) and prebiotics (arabinoxylan and its oligosaccharides).

3.9. Probiotics and Blood Pressure Regulation

Blood pressure in hypertensive conditions result from the imbalance of total and low-density lipoprotein (LDL) cholesterol. It has been reported that intake of probiotics and their products might help to regulate blood pressure. Probiotic supplementation containing *Saccharomyces cerevisiae,* S*treptococcus thermophilus, Lactobacillus Spp.* (*L casei, L acidophilus, L helveticus, L bulgaricus, L rhamnosus, L kefiri*), and *Bifidobacteria umbrella* and *longum* could be used as antihypertensive agents [49, 50].

3.10. Probiotics and Urogenital Health Care

Bacterial vaginosis (BV) and urinary tract infection (UTI) are the most common causes of urogenital infections. The BV results from abnormal discharge from the vagina due to the growth of non-endogenous bacteria. Imbalance in the vaginal microenvironment can enhance the incidence of UTI. More than 50 different types of *Lactobacillus Sp.* inhabit the vagina. Probiotic supplements can reassure the balance of *Lactobacillus Sp,* especially the *L. reuteri, L. vaginalis, L. delbrueckii, L. rhamnosus, L. brevis, L. casei,* and *L. salivarius* that are considered as the main regulators of urogenital infections [51].

Conclusion

Probiotics represent a group of microorganisms that improve the health status of the human gut when consumed in adequate doses for appropriate periods. Its mode of action involves production of bacteriocins and volatile fatty acids (SCFAs and BCFAs), modulation of the immune system response, and activity of enzymes involved in the metabolism of various types of toxic substances and carcinogens. The other mechanisms include competitive exclusion, stimulation of mucous production, and maintenance of the intestinal barrier. All these mechanisms work together to favor the growth of beneficial microorganisms while inhibiting the growth of pathogenic ones; thereby resulting in a healthy gut. In the last few decades, a surge in probiotic research-related data has drawn the scientific community's attention to explore further their potential role in therapeutics. However, to consider probiotics as a component of personalized medicine, intensive studies need to be conducted. It will further help us to understand each probiotic microorganism's personality and its overall role in affecting host health. Moreover, technological innovations need to be adopted to ensure that probiotics do not lose their function and useful properties once they enter the gastrointestinal tract.

References

[1] Zhao, Yuhai, Vivian Jaber, and Walter J. Lukiw. "Secretory products of the human GI tract microbiome and their potential impact on Alzheimer's disease (AD): detection of lipopolysaccharide (LPS) in AD hippocampus." *Frontiers in cellular and infection microbiology* 7 (2017): 318.

[2] Dronkers, T. M. G., L. Krist, F. J. Van Overveld, and G. T. Rijkers. "The ascent of the blessed: Regulatory issues on health effects and

health claims for probiotics in Europe and the rest of the world." *Beneficial microbes* 9, no. 5 (2018): 717-723.

[3] Cotter, Paul D., Colin Hill, and R. Paul Ross. "Bacteriocins: developing innate immunity for food." *Nature Reviews Microbiology* 3, no. 10 (2005): 777-788.

[4] Draper, K., C. Ley, and J. Parsonnet. "Probiotic guidelines and physician practice: a cross-sectional survey and overview of the literature." *Beneficial microbes* 8, no. 4 (2017): 507-519.

[5] Leahy, S. C., D. G. Higgins, G. F. Fitzgerald, and D. Van Sinderen. "Getting better with bifidobacteria." *Journal of applied microbiology* 98, no. 6 (2005): 1303-1315.

[6] Sanders, Mary Ellen. "Probiotics: definition, sources, selection, and uses." *Clinical infectious diseases* 46, no. Supplement_2 (2008): S58-S61.

[7] Ziemer, Cherie J., and Glenn R. Gibson. "An overview of probiotics, prebiotics and synbiotics in the functional food concept: perspectives and future strategies." *International Dairy Journal* 8, no. 5-6 (1998): 473-479.

[8] Granato, Daniel, Gabriel F. Branco, Filomena Nazzaro, Adriano G. Cruz, and Jose AF Faria. "Functional foods and nondairy probiotic food development: trends, concepts, and products." *Comprehensive reviews in food science and food safety* 9, no. 3 (2010): 292-302.

[9] Toma, Malda Maija, and Juris Pokrotnieks. "Probiotics as functional food: microbiological and medical aspects." *Acta Universitatis Latviensis* 710 (2006): 117-129.

[10] Taverniti, Valentina, Christian Scabiosi, Stefania Arioli, Diego Mora, and Simone Guglielmetti. "Short-term daily intake of 6 billion live probiotic cells can be insufficient in healthy adults to modulate the intestinal bifidobacteria and *Lactobacilli*." *Journal of Functional Foods* 6 (2014): 482-491.

[11] Chalas, Renata, Magdalena Janczarek, Teresa Bachanek, Elzbieta Mazur, Maria Cieszko-Buk, and Jolanta Szymanska. "Characteristics of oral probiotics–a review." *Current Issues in Pharmacy and Medical Sciences* 29, no. 1 (2016): 8-10.

[12] Pandya, Divya. "Benefits of Probiotics in Oral cavity–A Detailed Review". *Annals of International Medical and Dental Research 2*, no. 5 (2016):1 – 8.

[13] Bayane, Ali, Bréhima Diawara, Robin Dauphin Dubois, Jacqueline Destain, Dominique Roblain, and Philippe Thonart. "Isolation and characterisation of new spore-forming lactic acid bacteria with prospects of use in food fermentations and probiotic preparations." *African Journal of Microbiology Research* 4, no. 11 (2010): 1016-1025.

[14] Adel, M. M., and A. M. Sari. "Probiotic characterization of lactic acid bacteria isolated from local fermented vegetables (makdoos)." *Int J Curr Microbiol Appl Sci* 6, no. 2 (2017): 1673-86.

[15] Sharifi Yazdi, Mohammad Kazem, Abolfazl Davoodabadi, Hamid Reza Khesht Zarin, Maryam Tajabadi Ebrahimi, and Mohammad Mehdi Soltan Dallal. "Characterisation and probiotic potential of lactic acid bacteria isolated from Iranian traditional yogurts." *Italian Journal of Animal Science* 16, no. 2 (2017): 185-188.

[16] Fukuda, Shinji, Hidehiro Toh, Koji Hase, Kenshiro Oshima, Yumiko Nakanishi, Kazutoshi Yoshimura, Toru Tobe et al. "Bifidobacteria can protect from enteropathogenic infection through production of acetate." *Nature* 469, no. 7331 (2011): 543-547.

[17] Canfora, Emanuel E., Johan W. Jocken, and Ellen E. Blaak. "Short-chain fatty acids in control of body weight and insulin sensitivity." *Nature Reviews Endocrinology* 11, no. 10 (2015): 577-591.

[18] Umu, Özgün CO, Knut Rudi, and Dzung B. Diep. "Modulation of the gut microbiota by prebiotic fibres and bacteriocins." *Microbial ecology in health and disease* 28, no. 1 (2017): 1348886.

[19] Kumar, Manesh, Pankaj Dhaka, Deepthi Vijay, Jess Vergis, Vysakh Mohan, Ashok Kumar, Nitin V. Kurkure, Sukhadeo B. Barbuddhe, S. V. S. Malik, and Deepak B. Rawool. "Antimicrobial effects of *Lactobacillus plantarum* and *Lactobacillus acidophilus* against multidrug-resistant enteroaggregative *Escherichia coli*." *International journal of antimicrobial agents* 48, no. 3 (2016): 265-270.

[20] Buck, B. Logan, Eric Altermann, Tina Svingerud, and Todd R. Klaenhammer. "Functional analysis of putative adhesion factors in *Lactobacillus acidophilus* NCFM." *Applied and Environmental Microbiology* 71, no. 12 (2005): 8344-8351.

[21] Macfarlane, George T., and John H. Cummings. "Probiotics and prebiotics: can regulating the activities of intestinal bacteria benefit health?." *Bmj* 318, no. 7189 (1999): 999-1003.

[22] Kumar, Rajesh, Sunita Grover, and Virender Kumar Batish. "Hypocholesterolaemic effect of dietary inclusion of two putative probiotic bile salt hydrolase-producing *Lactobacillus plantarum* strains in Sprague–Dawley rats." *British Journal of Nutrition* 105, no. 4 (2011): 561-573.

[23] Collado, Maria Carmen, Miguel Gueimonde, and Seppo Salminen. "Probiotics in adhesion of pathogens: mechanisms of action." In *Bioactive Foods in Promoting Health*, pp. 353-370. Academic Press, 2010.

[24] Pothoulakis, Charalabos, Ciaran P. Kelly, Manher A. Joshi, Ning Gao, Connor J. O'Keane, Ignazio Castagliuolo, and J. Thomas Lamont. "*Saccharomyces boulardii* inhibits *Clostridium difficile* toxin A binding and enterotoxicity in rat ileum." *Gastroenterology* 104, no. 4 (1993): 1108-1115.

[25] Anderson, Rachel C., Adrian L. Cookson, Warren C. McNabb, Zaneta Park, Mark J. McCann, William J. Kelly, and Nicole C. Roy. "Lactobacillus plantarum MB452 enhances the function of the intestinal barrier by increasing the expression levels of genes involved in tight junction formation." *BMC microbiology* 10, no. 1 (2010): 1-11.

[26] Wang, Haifeng, Kan Gao, Ke Wen, Irving Coy Allen, Guohua Li, Wenming Zhang, Jacob Kocher et al. "*Lactobacillus rhamnosus* GG modulates innate signaling pathway and cytokine responses to rotavirus vaccine in intestinal mononuclear cells of gnotobiotic pigs transplanted with human gut microbiota." *BMC microbiology* 16, no. 1 (2016): 1-14.

[27] Deshpande, Girish, Gayatri Athalye-Jape, and Sanjay Patole. "Paraprobiotics for preterm neonates—The next frontier." *Nutrients* 10, no. 7 (2018): 871.

[28] De Simone, C., B. Bianchi Salvadori, R. Negri, M. Ferrazzi, L. Baldinelli, and R. Vesely. "The adjuvant effect of yogurt on production of gamma-interferon by con A-stimulated human peripheral blood lymphocytes." *Nutrition reports international (USA)* (1986).

[29] Del Giudice, M. Miraglia, and M. G. De Luca. "The role of probiotics in the clinical management of food allergy and atopic dermatitis." *Journal of clinical gastroenterology* 38 (2004): S84-S85.

[30] Anand, Preetha, Ajaikumar B. Kunnumakara, Chitra Sundaram, Kuzhuvelil B. Harikumar, Sheeja T. Tharakan, Oiki S. Lai, Bokyung Sung, and Bharat B. Aggarwal. "Cancer is a preventable disease that requires major lifestyle changes." *Pharmaceutical research* 25, no. 9 (2008): 2097-2116.

[31] Borowicki, Anke, Anke Michelmann, Katrin Stein, Daniel Scharlau, Kerstin Scheu, Ursula Obst, and Michael Glei. "Fermented wheat aleurone enriched with probiotic strains LGG and Bb12 modulates markers of tumor progression in human colon cells." *Nutrition and cancer* 63, no. 1 (2011): 151-160.

[32] Cousin, Fabien J., Sandrine Jouan-Lanhouet, Marie-Thérèse Dimanche-Boitrel, Laurent Corcos, and Gwenael Jan. "Milk fermented by *Propionibacterium freudenreichii* induces apoptosis of HGT-1 human gastric cancer cells." *PloS one* 7, no. 3 (2012): e31892.

[33] Azam, Rosa, Soudeh Ghafouri-Fard, Mina Tabrizi, Mohammad-Hossein Modarressi, Reza Ebrahimzadeh-Vesal, Maryam Daneshvar, Maryam Beigom Mobasheri, and Elahe Motevaseli. "*Lactobacillus acidophilus* and *Lactobacillus crispatus* culture supernatants downregulate expression of cancer-testis genes in the MDA-MB-231 cell line." *Asian Pacific Journal of Cancer Prevention* 15, no. 10 (2014): 4255-4259.

[34] Rossi, Marco, Ali Keshavarzian, and Faraz Bishehsari. "Nutraceuticals in colorectal cancer: a mechanistic approach." *European journal of pharmacology* 833 (2018): 396-402.

[35] Gorbach, SherwoodL, Te-Wen Chang, and Barry Goldin. "Successful treatment of relapsing *Clostridium difficile* colitis with *Lactobacillus* GG." *The Lancet* 330, no. 8574 (1987): 1519.

[36] Tejero-Sariñena, Sandra, Janine Barlow, Adele Costabile, Glenn R. Gibson, and Ian Rowland. "Antipathogenic activity of probiotics against *Salmonella Typhimurium* and *Clostridium difficile* in anaerobic batch culture systems: is it due to synergies in probiotic mixtures or the specificity of single strains?." *Anaerobe* 24 (2013): 60-65.

[37] Kareem, Karwan Yassen, Foo Hooi Ling, Loh Teck Chwen, Ooi May Foong, and Samsudin Anjas Asmara. "Inhibitory activity of postbiotic produced by strains of *Lactobacillus plantarum* using reconstituted media supplemented with inulin." *Gut pathogens* 6, no. 1 (2014): 1-7.

[38] Simova, E. D., D. B. Beshkova, and Zh P. Dimitrov. "Characterization and antimicrobial spectrum of bacteriocins produced by lactic acid bacteria isolated from traditional Bulgarian dairy products." *Journal of Applied Microbiology* 106, no. 2 (2009): 692-701.

[39] Ammor, Salim, Grégoire Tauveron, Eric Dufour, and Isabelle Chevallier. "Antibacterial activity of lactic acid bacteria against spoilage and pathogenic bacteria isolated from the same meat small-scale facility: 1—Screening and characterization of the antibacterial compounds." *Food control* 17, no. 6 (2006): 454-461.

[40] Currò, Diego, Gianluca Ianiro, Silvia Pecere, Stefano Bibbò, and Giovanni Cammarota. "Probiotics, fibre and herbal medicinal products for functional and inflammatory bowel disorders." *British journal of pharmacology* 174, no. 11 (2017): 1426-1449.

[41] Cammarota, Giovanni, Silvia Pecere, Gianluca Ianiro, Luca Masucci, and Diego Currò. "Principles of DNA-based gut microbiota assessment and therapeutic efficacy of fecal microbiota

transplantation in gastrointestinal diseases." *Digestive Diseases* 34, no. 3 (2016): 279-285.

[42] Ozdemir, O. "Prebiotics and probiotics in allergy: Potential mechanisms of prebiotics' and probiotics' actions in allergy-(Part 1)." *MOJ Immunol* 3, no. 1 (2016): 00069.

[43] Miele, Erasmo, Filomena Pascarella, Eleonora Giannetti, Lucia Quaglietta, Robert N. Baldassano, and Annamaria Staiano. "Effect of a probiotic preparation (VSL# 3) on induction and maintenance of remission in children with ulcerative colitis." *Official journal of the American College of Gastroenterology ACG* 104, no. 2 (2009): 437-443.

[44] Almeida, Camila Casuccio, Sônia Letícia Silva Lorena, Célia Regina Pavan, Helena Midori Imamura Akasaka, and Maria Aparecida Mesquita. "Beneficial effects of long-term consumption of a probiotic combination of *Lactobacillus casei* Shirota and *Bifidobacterium breve* Yakult may persist after suspension of therapy in lactose-intolerant patients." *Nutrition in Clinical Practice* 27, no. 2 (2012): 247-251.

[45] Bordoni, Alessandra, Alberto Amaretti, Alan Leonardi, Elisa Boschetti, Francesca Danesi, Diego Matteuzzi, Lucia Roncaglia, Stefano Raimondi, and Maddalena Rossi. "Cholesterol-lowering probiotics: in vitro selection and in vivo testing of bifidobacteria." *Applied Microbiology and Biotechnology* 97, no. 18 (2013): 8273-8281.

[46] Vaishnavi, K., M. Krishma, and P. Rajeswari. "A study on cholesterol degradation by *Lactobacillus*." *Indian J Appl Res* 9 (2016): 12.

[47] Le Barz, Mélanie, Fernando F. Anhê, Thibaut V. Varin, Yves Desjardins, Emile Levy, Denis Roy, Maria C. Urdaci, and André Marette. "Probiotics as complementary treatment for metabolic disorders." *Diabetes & metabolism journal* 39, no. 4 (2015): 291-303.

[48] Hartstra, Annick V., Kristien EC Bouter, Fredrik Bäckhed, and Max Nieuwdorp. "Insights into the role of the microbiome in obesity and type 2 diabetes." *Diabetes care* 38, no. 1 (2015): 159-165.

[49] Rerksuppaphol, Sanguansak, and Lakkana Rerksuppaphol. "A randomized double-blind controlled trial of Lactobacillus acidophilus plus Bifidobacterium bifidum versus placebo in patients with hypercholesterolemia." *Journal of clinical and diagnostic research: JCDR* 9, no. 3 (2015): KC01.

[50] Ekhlasi, Golnaz, Mitra Zarrati, Shahram Agah, Agha Fatemeh Hosseini, Sharieh Hosseini, Shahrzad Shidfar, Seyed Soroush Soltani Aarbshahi, Elham Razmpoosh, and Farzad Shidfar. "Effects of symbiotic and vitamin E supplementation on blood pressure, nitric oxide and inflammatory factors in non-alcoholic fatty liver disease." *EXCLI journal* 16 (2017): 278.

[51] Waigankar, Santosh S., and Vimal Patel. "Role of probiotics in urogenital healthcare." *Journal of mid-life health* 2, no. 1 (2011): 5.

In: Nutraceuticals
Editors: A. Kumari and G. Singh
ISBN: 978-1-68507-488-3

Chapter 10

SYNBIOTICS: MECHANISM AND POTENTIAL HEALTH BENEFITS

***Deepika Pawar*[1], *Nidhi Bhardwaj*[2], *Anita Kumari*[1,*] and *Tejpal Dhewa*[1]**

[1]Department of Nutrition Biology,
School of Interdisciplinary and Applied Life Sciences,
Central University of Haryana, Mahendergarh, Haryana, India
[2]Department of Human Genetics and Molecular Medicine,
School of Health Science, Central University of Punjab,
Bathinda, Punjab, India

ABSTRACT

In the past decade, the keen interest of researchers towards nutraceuticals and functional foods has led to certain advancements in food technology. The combined formulation of probiotics and prebiotics (synbiotics) is one of the significant advancements. The synbiotics have provided countless health benefits such as prevention and treatment of diseases, improved gastrointestinal tract health, and reduced risk of colon

* Corresponding Author's E-mail: anikum@cuh.ac.in.

cancer and mental illness. Furthermore, the introduction of the synbiotic approach has improved and accelerated the growth and implantation of viable probiotics in the gut lining due to the presence of combined prebiotic. In this chapter, the authors covered the fundamental aspect of synbiotics, their possible mechanism of action, and their potential health benefits.

1. Introduction

After hundreds of years of co-evolution, symbiosis has developed between microbes and hosts inhabiting them, such as the gut microbiome. The gastrointestinal tract of humans has numerous microbes inhabiting it and surviving in a symbiotic relationship. Collectively, the human gut microbiota is observed to be composed of nearly 500-1000 bacterial species [1]. A wide range and characteristics of these microbes are kept in the gut. These can be beneficial or harmful for individuals, i.e., both good and bad microbes may be expected in the gut microflora. The key focus in the chapter remains on the beneficial ones, which are supposed to provide countless health benefits such as a healthy gut, strong immunity, improved metabolism, and mental health.

Some of these beneficial microbes are taken via diet and others via pharmaceutical formulas or functional foods. Presently, these beneficial microbes are known as probiotics. Earlier, when probiotic formulations were not developed, food was the primary source of their intake. One such popular probiotic bacteria is *Lactobacillus lacti*; present abundantly in curd, it is proven to reduce the effects of diarrhea [2].

The combination of probiotics and prebiotics in synbiotics is a widely accepted food supplement these days. The WHO (World Health Organization) and FAO (Food and Agriculture organization) define synbiotics as "a supplement that contains both, a prebiotic and a probiotic that works together to promote 'healthy microflora' in the human intestine." The synbiotics are supposed to work according to symbiotic theory, and their excellent mechanism is discussed in this chapter in the later sections.

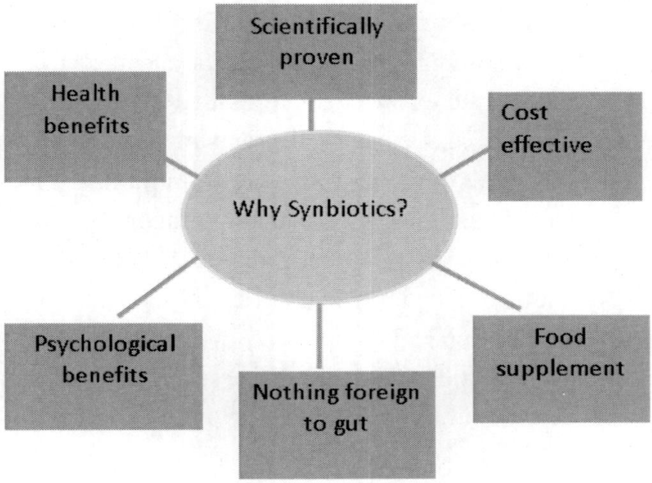

Figure 1. Benefits of synbiotics over usual medications.

Figure 1 has shown how synbiotics are beneficial for the health of mammals as their intervention in several physiological, clinical, and even psychological illnesses has led the individual to better health and even psychology via the gut-brain bidirectional axis. With the advent of synbiotics, life is set free from the bitter taste of drugs; even pharmacological studies potentially formulate these synbiotics in the form of medicines and not solely as foodstuff. The experimentally proven synbiotics are available in the markets at a reasonable rate, unlike expensive medication. Another good part about synbiotics is that it doesn't mean information about something alien to the gut. Its inhabiting microbes are supplemented with superficial dietary fibers with emerging nutritional studies and technology. We are moving towards the betterment of these formulations day by day.

2. Synbiotics: Probiotics and Prebiotics

The supplement of synbiotics (Figure 2) containing combinations of the live bacteria and the non-digestible fructooligosaccharides that stimulate their growth is a tremendous significant milestone in attaining good health. In this combination of probiotics and prebiotics, the latter component enhances the growth of combined probiotic bacteria and flourishes the growth of native bacterial strains in the gastrointestinal tract. The importance of these microbes is visible from the fact that alterations of these microbial strains in early life due to unfavorable conditions have influenced the potential risk for several diseases. Gibson introduced the concept of synbiotics. The term synbiotics alludes to synergism with beneficial effects. The two sub-components of synbiotics are, therefore, a) Probiotics and b) Prebiotics which are discussed briefly in chapters 8 and 9, and general ideas about these are given below:

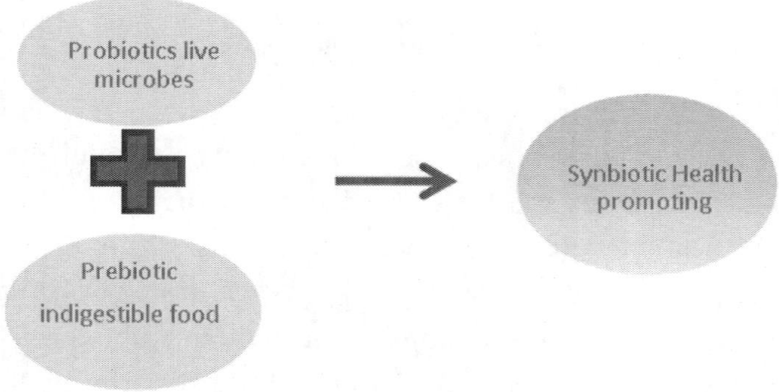

Figure 2. Synbiotics are a combination of probiotics and prebiotics.

1. The enhanced production of beneficial short-chain fatty acids (SCFAs) due to prebiotic consumption: SCFAs can modulate histone acetylation via fermentation. As a result, they enhance the presence of several genes for transcription factors. Example of SCAFs includes butyric acid and propionic acid.
2. Reducing the pH of the gut: Fermentation of these unprocessed foods, main carbohydrates by probiotics like *Bifidobacterium* or *Lactobacillus,* somehow leads to a decrease in pH, making the gut environment more suitable for the survival of inhabiting probiotics (3).

Table 1. Examples of certain prebiotic-influenced health effects/benefits in different studies are clinical disorders

S. No	Administered prebiotic	Outcome of study	Reference
1	OFS	Among individuals facing obesity issues, a reduction in body weight due to oligofructose supplementation was seen, and the control group experienced an increase in the body's weight. In addition, glucose levels were decreased in the oligofructose group and increased in the control group between the initial and final tests.	[5]
2	FOS	No effect of FOS on glucose and lipid metabolism in subjects suffering from type 2 diabetics	[6]
3	AX	Improved glycaemic control in subjects administered AX-rich fibre for clinical condition type 2 diabetes.	[7]
4	OFS	Decreased serum aminotransferases, aspartate aminotransferase, and insulin levels in patients having non-alcoholic fatty liver disease.	[8]
5	GOS, FOS	A high percentage of *Bifidobacterium,* i.e., probiotic growth, enhanced and low level of *E.coli* observed in subjects after administration of GOS, FOS. The study was carried out with bottle-fed infants.	[9]
6	Inulin	Endoscopic and histological inflammation decreased after administration of inulin. In addition, increased intestinal butyrate and lowered pH were also observed, along with a significant reduction in *Bacteroides fragilis*.	[10]
7	GOS, FOS	Allergic symptoms—AD, wheezing, urticaria, and fewer URI were observed less in the subject group given prebiotic than control group individuals in the clinical condition of atopic dermatitis.	[11]
8	GOS	Symptoms for alleviation of lactose intolerance include pain, bloating, cramping, diarrhoea, etc. With the prebiotic administration, it was observed that at least one of these symptoms disappeared or improved in at least 71% of subjects. *Bifidobacteria,* i.e., probiotic bacteria, also significantly increased by 90% among non-lactose tolerant participants after administration of GOS. In addition, lactose fermenting microbes, i.e., *Bifidobacterium, Faecalibacterium,* and *Lactobacillus,* were significantly increased.	[12]
9	Inulin	In human colorectal carcinoma, apoptosis was induced, and growth was inhibited.	[13]

3. The molecular interactions in probiotics and prebiotics: Molecular structure of prebiotics fates their physiological effects. Therefore,

only particular micro-organisms, i.e., probiotic microbes, are the only type of gut microbes to obtain nutrients to flourish their growth (4).

Table 2. Examples of specific probiotic potential carrying bacteria and their health effects in different studies; among individuals with clinical disorders

S. No	Name of the microbes	Health benefit observed	Reference
1.	*E. faecium and 2, S. thermophilus strains*	Decrease in total weight of the body, BP (Systolic), LDL-C, and gain in levels of the fibrinogen in subjects facing obesity.	[14]
2.	*L. acidophilus La5, B. animalis subsp. lactis Bb12*	Decrease in glucose concentration for fasting stage and elevation in HOMA-IR in subjects facing obesity.	[15]
3.	*L. gasseri SBT2055*	Decrease in BMI value and arterial BP (arterial) in subjects facing obesity.	[16]
4.	*Bifidobacterium, Lactobacillus, S. thermophilus*	Better lipid profile and sensitivity for insulin and reduction or lower CRP in subjects facing obesity.	[17]
5.	*L. salivarius Ls-33*	Positive elevation in the ratios of *Prevotellae, Bacteroides,* and *Porphyromonas* in subjects facing obesity.	[18]
6..	*L. casei Shirota*	VCAM-1 level drop in subjects with insulin resistance syndrome.	[19]
7.	*L. plantarum*	Reduction in levels of Glucose and homocysteine subjects with Insulin resistance syndrome	[20]
8.	*L. planatarum A7*	Reduction in the process of methylation, 8-OHDG and SOD in Diabetes type 2 patients	[21]
9.	*L. acidophilus La-5, B. animalis subsp. lactis BB-12*	HDL-C levels gain and reduced LDL-C/HDL-C ratio in Diabetes type 2 patients	[22]
10.	*L. acidophilus La-5, B. animalis subsp. lactis BB-12*	Different changes of HbA1c, TC in groups of subjects for study, and LDL-C in Diabetes type 2 patients	[23]
11.	*L. acidophilus La5, B. lactis Bb12 TC and LDL-C improvement.*	Improved TC and LDL-C in Diabetes type 2 patients	[24]

S. No	Name of the microbes	Health benefit observed	Reference
12.	*L. bulgaris, S. thermophilus*	The reduction was observed for the levels of ALT and γ-GTP among subjects with Non-alcoholic fatty liver disease.	[25]
13.	*Bifidobacterium, Lactobacillus, S. thermophilus*	Fatty liver severity improved, reduced BMI, and more GLP1/aGLP1 among patients with non-alcoholic fatty liver disease.	[26]
14.	*L. rhamnosus GG*	ALT and PG-PS IgAg antibodies reduction among patients having non-alcoholic fatty liver disease.	[27]
15.	*L. reuteri ATCC 55730*	Pain and symptoms associated with intestinal colic disappeared in subjects having Irritable bowel syndrome (IBS).	[28]
16.	*L. rhamnosus HN001*	It decreased the cumulative prevalence of eczema substantially in newborns/infants for the clinical condition of Atopic dermatitis.	[29]
17.	*B. bifidum, B. lactis, L. Lactis*	Reduction of eczema in high-risk for two years due to probiotic administration within three months of birth to infants for clinical condition of Atopic dermatitis.	[30]

Though prebiotic consumption is for the improved growth of probiotics, their availability to target beneficial microbe instead of pathogenic microbes remains an area of concern. Probiotics have developed several characters for the competitive win for nutrients during this case. In addition, prebiotic food products have been shown to provide health benefits, as given in Table 1, to its consumer beyond essential nutrition when added to a regular diet. Thus, it remains an area of interest for various food technologists. Also, some of the administered prebiotic and probiotic bacteria and the health benefits they provide have been mentioned in Tables 1 and 2.

3. Selection Criteria for Synbiotics

While developing a synbiotic formulation, it is important to select an appropriate probiotic and prebiotic. Under this, we check the suggested microbe and indigestible foodstuff we are considering as probiotic and prebiotic have a positive impact on human health with no side-effects either short term or long term when used separately and not in combination. The second thing after the theory search is to look for all the desired attributes for their survival and effective functioning in the gut. Some examples of synbiotic formulas are given in Table 3.

After checking the different potential next steps, look for the effect of the selected prebiotic on the growth of desired probiotic. Prebiotic is limited if it carries specific properties that have a favorable impact on the probiotic. Several lab-based experiments are then carried out where the selected probiotic is administered as a nutrition source to the lab-grown probiotic microbe/bacteria. Later, the pre-preparation is kept for incubation at optimum conditions in an incubator for a probiotic growth period. Now, the effects are seen; either the probiotic will flourish or increase in number, indicating the right choice of probiotic, or the number of probiotics will decline, suggesting the given food doesn't complete the criteria of prebiotic for desired probiotic. Also, the prebiotic will be checked for interaction with other pathogenic microbes present in the gut. In competitive conditions, the prebiotic should stimulate the growth of only desired probiotic, and no pathogenic microbe should be benefitted. Let say we have to check the effect of prebiotic: PRE1 on probiotic: PRO1. We can have two different observations:

3.1. The Probiotic Density Increases

In this case, the number of PRO1 will be increased with PRE1 as the sole source of nutrition for PRO1. It suggests PRE1 as a potential food source for PRO1.

3.2. The Probiotic Density Decreases

In this case, the number of PRO1 will be decreased with PRE1 as the sole source of nutrition for PRO1. It suggests PRE1 as a non-potential food source for PRO1.

Table 3. Several examples of synbiotic formulas (31)

S. No	Genus of Bacteria as Probiotic	Prebiotic combined
1	*Lactobacillus*	Inulin
2	*Lactobacillus, Streptococcus, and Bifidobacterium*	FOS-Fructooligosaccharides
3	*Lactobacillus, Bifidobacterium, Enterococcus*	FOS- Fructo-oligosaccharides
4	*Lactobacillus and Bifidobacterium*	Oligofructose
5	*Lactobacillus and Bifidobacterium*	Inulin

4. MECHANISM USED BY POTENTIAL SYNBIOTICS

The synergistic effect of probiotics and prebiotics lays the foundation for the synbiotics theory. The probiotics aid to our health is nothing new to our knowledge repurchased. Still, the interest is bought back by how these probiotics settling our small and large intestine benefit from the food we consume. This theoretical knowledge was added up with evidence from several empirical studies where researchers observed that prebiotics aided probiotics in tolerating the GI tract's environmental conditions with ease. The requirements included: pH, oxygenation, and temperature. However, how prebiotics provides this added tolerance to probiotics is still into search criteria. The combination of prebiotics and probiotics lead to the formation of dietary supplements carrying viable microbes and assurance to withstand the harsh conditions in the way before it starts its mechanism to provide desired health benefits to host/consumer.

Figure 3 explains the most believed mechanism of synbiotic action is improved viability of combined probiotics for improved shifting [32].

1. Synbiotics lead to enhanced growth and viability of beneficial microflora and elimination of harmful ones.
2. Synbiotics results in the maintenance of the intact intestinal structure and a decrease in concentrations of unwanted metabolites.
3. They also increase the level of useful, beneficial short-chain fatty acids like propionic acid and butyric acid, carbon disulfides,

ketones, and methyl acetates, which influence the health of the host/consumer beneficially [32].

4. Synbiotics have proven therapeutic effects on host health, including prevention and improvement of condition from several health diseases and disorders like diabetes, obesity, osteoporosis, diarrhea, intestinal bowel syndrome, and brain-related disorders.
5. Also, they are anti-cancerogenic, anti-allergic, and anti-bacterial (for pathogenic bacteria inhabiting the gut). All these properties aid synbiotics to act as a potential food supplement with therapeutic benefits.

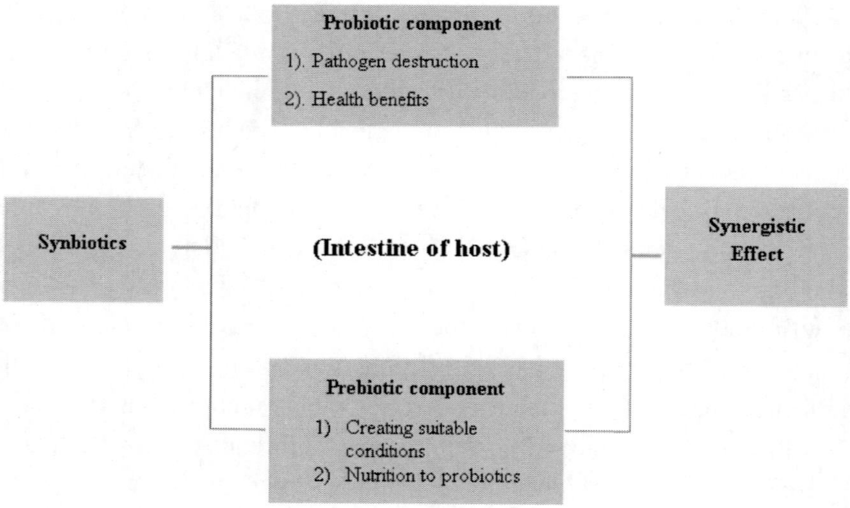

Figure 3. Mechanism of action of synbiotics.

5. HEALTH BENEFITS OF SYNBIOTICS

Over the past few years, the interest of researchers in the synbiotic concept, carrying synergistic effects, has led to many help, valuable findings in the field. The synbiotics may seem just a food supplement, but their benefits are beyond our imagination several times. The proven health/ clinical benefits of synbiotics are given in Figure 4.

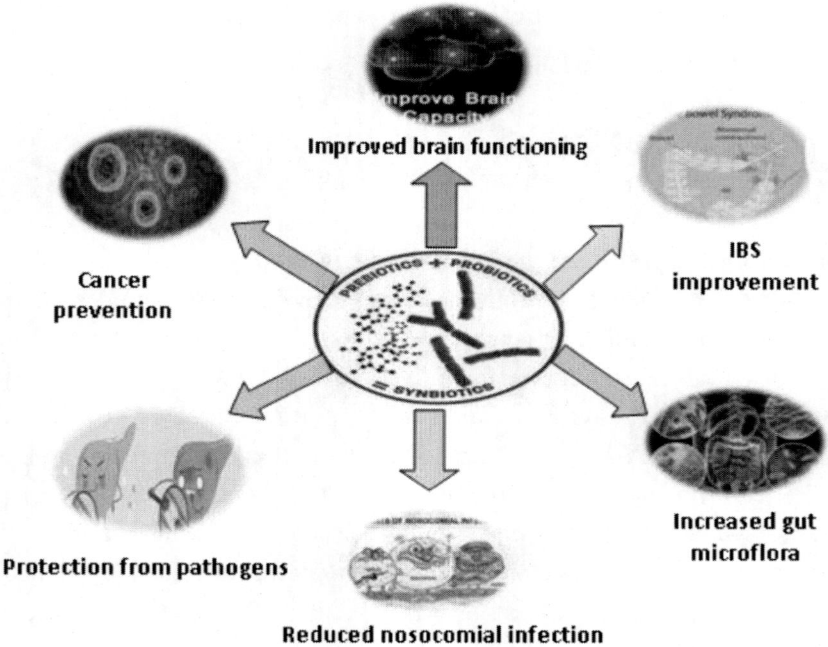

Figure 4. Health benefits of synbiotics.

5.1. Increased Ratio of Healthy Gut Microflora

The healthy gut microflora majorly accounts for the bacteria of genus *Lactobacillus* and *Bifidobacterium*. Administrations of synbiotics have led to a boost in the density of these bacteria by many folds. The increased number of these beneficial microbes serves as stimulation to several health benefits like improved efficiency of the gut, resistance to several diseases, and enhanced work efficiency. These microbes are even sometimes associated with brain functioning and psychology. The concept of the gut-brain axis and its efficiency to modulate our mood leading us to the path of happiness and away from psychological diseases like stress, depression, and anxiety is an approach towards achieving good mental health; when most individuals these days can be seen suffering from stress-related disorders in our day to day lives [33].

5.2. Improvement in the Functioning of Organs in the Human Body

The benefits of Synbiotics are not limited only to the gut but also to the organs such as the liver and brain.

5.2.1. Liver/Hepatic Functioning Improvement

The end product of probiotic bacteria metabolism includes ethanol, SCFA (Short-chain fatty acids), and LPSs (Lipopolysaccharides). These end products can quickly efficiently translocate and penetrate the liver. Inside the liver, the synthesis and storage of hepatic triacylglycerol carrier stimulated by SCAFs. This leads to hepatic detoxification resulting in the storage of triacylglycerol in the liver and intensifying liver steatosis. LPSs also play a crucial role in developing resistance to insulin and increasing inflammatory cell uptake in the case of non-alcoholic fatty liver disease (NAFLD). This is stimulated due to induction of TNF or tumor necrosis factor by LPSs [34].

5.2.2. Improvement in the Brain's Functioning

The Enteric Nervous System (ENS), Central Nervous System (CNS), and subsequently the brain is benefitted via the administration of synbiotics carrying psychological effects; commonly known as psychobiotics. Considerable large-scale improvement in concentration, thinking efficiency, memory, and decision-making ability has been observed in several studies targeting psychobiotics. The reason for the benefits observed is proper maintenance and functioning of the bi-directional brain-gut or gut-brain axis. Diet intake has shown great importance in this regard [35].

5.3. Reduced Chances of Nosocomial Infection

Post-surgical wounds have always been an area of concern. They are most susceptible to infections caused by several pathogenic microorganisms. In places like hospitals, multiple patients taking potential pathogens are current, and the chances of infection increase many folds. Administration of synbiotics in such cases has shown a significant decrease in nosocomial infections by preventing bacterial translocation in post-surgical procedures.

5.4. Aid in Curing/Prevention of Health Diseases

Improvement in health conditions during several illnesses has been observed in several diseases. The conditions discussed in this chapter are Cancer and Intestinal Bowel Syndrome.

5.4.1. Cancer

Cancer is a disease caused by the uncontrolled growth and proliferation of cells in the body. Cancer may either be benign or malignant depending on the limit of its effect on an organ or the whole body. Synbiotics are means of manipulating the gut microbiota to provide therapeutic effects in preventing and curing cancer. Onco-suppressive agents have shown a positive impact on controlling proliferation and enhancing immunomodulation. Synbiotics have demonstrated that they act as adjuncts in cancer treatment during several studies. Even after the disease has occurred, the post effects of its treatment like chemotherapy, radiotherapy, and antibiotics have shown to be minimized as a result of synbiotics administration [36]. In a study, the health effect of a synbiotic administration (where probiotic strains considered were *Lactobacillus rhamnosus GG* and *Bifidobacterium animalis subsp. lactis Bb12*, and the prebiotic was fructo-oligosaccharides) were observed for patients at potential risk of colorectal cancer, and results were reduction in the risk of colorectal carcinoma [37].

5.4.2. Irritable Bowel Syndrome

Irritable bowel syndrome is a commonly found disorder associated with harm to the large intestine. It is a chronic illness that arises over a long time due to our negligence towards our health, such as developing lousy eating habits, mainly fried food, junk food, carbonated drinks consumption, and excessive intake of spices. The sedentary lifestyle and chronic stress in an individual also enhance the risk factor for this disorder. The illness is associated with abdominal pain, gas, cramping, constipation, or diarrhoea. The administration of synbiotics has shown significant improvement in patients' health conditions with irritable bowel syndrome [38]. In a study among 76 patients with IBS, the supplementation of synbiotics (probiotic used was *L. acidophilus La-5®, B. animalis ssp. lactis BB-12®*, dietary fibers (Beneo)) for nearly four weeks led to an improvement in total IBS-QoL score by 18%. In addition, significant improvements in IBS symptoms like bloating severity, satisfaction with bowel movements were observed. Though several improvements were seen, the area is still in the infancy stage and needs more attention for improvements [39].

6. Delivery Systems for Synbiotics

Synbiotics are functional foods or nutraceuticals having potential health benefits. Their health benefits are undoubtedly numerous, but finding the delivery systems for their administration is an area of concern. Microcapsules can be used as a delivery system for this purpose. The encapsulation of probiotics alone was standard earlier. Now, for synbiotics also a similar process called co-encapsulation is done. In this system, the co(combined) encapsulation of the probiotic bacterium occurs with the prebiotic food component, which will aid this health-promoting bacterium to flourish its growth in the consumer's gut and the increased period of stay inside the digestive system. The co-encapsulation can be achieved by specific techniques like electro-hydrodynamic atomization, i.e., EDHA. Example: The double-layer co-encapsulation of *Lactobacillus plantarum* and *Bifidobacterium lactic* with insulin or resistant starch can be done in the microcapsules of alginate or chitosan using EDHA [40].

Another example is using the microcapsules with combinations of fenugreek gum, alginate, and locust bean gum to co-encapsulate the probiotic strains *Pediococcus pentosaceus* KID7, *Lactobacillus plantarum* KII2, *Lactobacillus fermentum* KLAB6, and *Lactobacillus helveticus* KII13 along with required prebiotic [41]. The use of delivery systems has several associated benefits. The co-encapsulation can enhance or improve the viability of bacteria in microcapsules, and gastrointestinal transit time can also be increased [41] using this system. The bioavailability and bio-accessibility are also enhanced.

Other delivery systems like nano-emulsions (O/W) and microspheres are also available. For example, the aqueous phase of protein incorporated prebiotics emulsion for the nano-emulsion system was used for probiotic *Enterococcus faceium*. Gum Arabica was used as an emulsifier for this. The aqueous phase included whey protein concentrate (WPC), Gum Arabica (GA), and inulin as prebiotic for this study [42]. For the microsphere delivery system, alginate and arabinoxylan can be co-gelated to produce microsphere for a synbiotic combination of probiotic *Lactobacillus plantarum* along with prebiotic arabinoxylan oligosaccharides. Using prebiotic for microsphere will increase the encapsulation efficacy [43]. Each delivery system has its benefits. The majorly and popularly used delivery system observed is co-encapsulation in microcapsules.

CONCLUSION

The never-ending side effects of drugs make them the least picked yet, the only option for someone suffering from any disease or disorder. Their alternatives have always been searched for irrespective of their extensive use for the treatment of any disease. The requirements of pharmaceuticals carrying the potential to cure diseases and prevent them; without causing any harm or side effects to the consumer have always been in the light. The stimulative impact of prebiotics for probiotics laid the foundation of the synbiotic concept, which requires formula used as a food supplement due to its association with numerous health benefits. The potential benefits do not only limit to the digestive tract but spread across the whole body. Probiotics have shown their needed health effects for years, but adding something that can boost their efficiency and potential led to better probiotics and the host they inhabit. Laboratory studies were carried out to develop the most suitable synbiotics by combining desired probiotics with suitable prebiotic. The extensive experimental studies and clinical evidence have shown the potential of synbiotics in the prevention and treatment of certain health concerns. However, more clinical trials need to be done to establish the precise therapeutic application of synbiotics in different groups of populations.

REFERENCES

[1] Zhou, L. and Foster, J. A. (2015). Psychobiotics and the gut–brain axis: in the pursuit of happiness. *Neuropsychiatric Disease and Treatment*, *11*: 715.

[2] Delgado, G. T. C., Tamashiro, W. M. D. S. C., Junior, M. R. M., Moreno, Y. M. F. and Pastore, G. M. (2011). The putative effects of prebiotics as immunomodulatory agents. *Food Research International*, *44*(10): 3167-3173.

[3] Lee, Y. K. and Salminen, S. (2009). *Handbook of probiotics and prebiotics*. John Wiley & Sons.

[4] Delgado, G. T. C., Tamashiro, W. M. D. S. C., Junior, M. R. M., Moreno, Y. M. F. and Pastore, G. M. (2011). The putative effects of prebiotics as immunomodulatory agents. *Food Research International*, *44*(10): 3167-3173.

[5] Parnell, J. A. and Reimer, R. A. (2009). Weight loss during oligofructose supplementation is associated with decreased ghrelin and increased peptide YY in overweight and obese adults. *The American journal of Clinical Nutrition*, *89*(6): 1751-1759.

[6] Luo, J., Van Yperselle, M., Rizkalla, S. W., Rossi, F., Bornet, F. R. and Slama, G. (2000). Chronic consumption of short-chain

fructooligosaccharides does not affect basal hepatic glucose production or insulin resistance in type 2 diabetics. *The Journal of Nutrition*, *130*(6): 1572-1577.

[7] Lu, Z. X., Walker, K. Z., Muir, J. G. and O'Dea, K. (2004). Arabinoxylan fibre improves metabolic control in people with Type II diabetes. *European Journal of Clinical Nutrition*, *58*(4): 621-628.

[8] Daubioul, C. A., Horsmans, Y., Lambert, P., Danse, E. and Delzenne, N. M. (2005). Effects of oligofructose on glucose and lipid metabolism in patients with non-alcoholic steatohepatitis: results of a pilot study. *European Journal of Clinical Nutrition*, *59*(5): 723-726.

[9] Costalos, C., Kapiki, A., Apostolou, M. and Papathoma, E. (2008). The effect of a prebiotic supplemented formula on growth and stool microbiology of term infants. *Early Human Development*, *84*(1): 45-49.

[10] Welters, C. F., Heineman, E., Thunnissen, F. B., van den Bogaard, A. E., Soeters, P. B. and Baeten, C. G. (2002). Effect of dietary inulin supplementation on inflammation of pouch mucosa in patients with an ileal pouch-anal anastomosis. *Diseases of the Colon & Rectum*, *45*(5): 621-627.

[11] Arslanoglu, S., Moro, G. E. and Boehm, G. (2007). Early supplementation of prebiotic oligosaccharides protects formula-fed infants against infections during the first 6 months of life. *The Journal of Nutrition*, *137*(11): 2420-2424.

[12] Azcarate-Peril, M. A., Ritter, A. J., Savaiano, D., Monteagudo-Mera, A., Anderson, C., Magness, S. T. and Klaenhammer, T. R. (2017). Impact of short-chain galactooligosaccharides on the gut microbiome of lactose-intolerant individuals. *Proceedings of the National Academy of Sciences*, *114*(3): E367-E375.

[13] Munjal, U., Glei, M., Pool-Zobel, B. L. and Scharlau, D. (2009). Fermentation products of inulin-type fructans reduce proliferation and induce apoptosis in human colon tumour cells of different stages of carcinogenesis. *British Journal of Nutrition*, *102*(5): 663-671.

[14] Agerholm-Larsen, L., Raben, A., Haulrik, N., Hansen, A. S., Manders, M. and Astrup, A. (2000). Effect of 8 week intake of probiotic milk products on risk factors for cardiovascular diseases. *European Journal of Clinical Nutrition*, *54*(4): 288-297.

[15] Ivey, K. L., Hodgson, J. M., Kerr, D. A., Thompson, P. L., Stojceski, B. and Prince, R. L. (2015). The effect of yoghurt and its probiotics on blood pressure and serum lipid profile; a randomised controlled trial. *Nutrition, Metabolism and Cardiovascular Diseases*, *25*(1): 46-51.

[16] Kadooka, Y., Sato, M., Ogawa, A., Miyoshi, M., Uenishi, H., Ogawa, H. and Tsuchida, T. (2013). Effect of Lactobacillus gasseri SBT2055 in fermented milk on abdominal adiposity in adults in a randomised controlled trial. *British Journal of Nutrition*, *110*(9): 1696-1703.

[17] Rajkumar, H., Mahmood, N., Kumar, M., Varikuti, S. R., Challa, H. R. and Myakala, S. P. (2014). Effect of probiotic (VSL# 3) and omega-3 on lipid profile, insulin sensitivity, inflammatory markers, and gut colonization in overweight adults: a randomized, controlled trial. *Mediators of Inflammation*, doi:10.1155/2014/348959.

[18] Larsen, N., Vogensen, F. K., Gøbel, R. J., Michaelsen, K. F., Forssten, S. D., Lahtinen, S. J. and Jakobsen, M. (2013). Effect of *Lactobacillus salivarius* Ls-33 on fecal microbiota in obese adolescents. *Clinical Nutrition*, *32*(6): 935-940.

[19] Tripolt, N. J., Leber, B., Blattl, D., Eder, M., Wonisch, W., Scharnagl, H. and Sourij, H. (2013). Effect of supplementation with *Lactobacillus casei* Shirota on insulin sensitivity, β-cell function, and markers of endothelial function and inflammation in subjects with metabolic syndrome—A pilot study. *Journal of Dairy Science*, *96*(1): 89-95.

[20] Barreto, F. M., Simão, A. N. C., Morimoto, H. K., Lozovoy, M. A. B., Dichi, I. and da Silva Miglioranza, L. H. (2014). Beneficial effects of *Lactobacillus plantarum* on glycemia and homocysteine levels in postmenopausal women with metabolic syndrome. *Nutrition*, *30* (7-8): 939-942.

[21] Hariri, M., Salehi, R., Feizi, A., Mirlohi, M., Ghiasvand, R. and Habibi, N. (2015). A randomized, double-blind, placebo-controlled, clinical trial on probiotic soy milk and soy milk: effects on epigenetics and oxidative stress in patients with type II diabetes. *Genes & nutrition*, *10*(6): 1-8.

[22] Mohamadshahi, M., Veissi, M., Haidari, F., Javid, A. Z., Mohammadi, F. and Shirbeigi, E. (2014). Effects of probiotic yogurt consumption on lipid profile in type 2 diabetic patients: A randomized controlled clinical trial. *Journal of research in medical sciences: the official journal of Isfahan University of Medical Sciences*, *19*(6): 531.

[23] Tonucci, L. B., Dos Santos, K. M. O., de Oliveira, L. L., Ribeiro, S. M. R. and Martino, H. S. D. (2017). Clinical application of probiotics in type 2 diabetes mellitus: A randomized, double-blind, placebo-controlled study. *Clinical Nutrition*, *36*(1): 85-92.

[24] Ejtahed, H. S., Mohtadi-Nia, J., Homayouni-Rad, A., Niafar, M., Asghari-Jafarabadi, M., Mofid, V. and Akbarian-Moghari, A. (2011). Effect of probiotic yogurt containing *Lactobacillus acidophilus* and *Bifidobacterium lactis* on lipid profile in individuals with type 2 diabetes mellitus. *Journal of Dairy Science*, *94*(7): 3288-3294.

[25] Aller, R., De Luis, D. A., Izaola, O., Conde, R., Gonzalez Sagrado, M., Primo, D. and Gonzalez, J. (2011). Effect of a probiotic on liver aminotransferases in non-alcoholic fatty liver disease patients: a double blind randomized clinical trial. *Eur Rev Med Pharmacol Sci*, *15*(9): 1090-5.

[26] Alisi, A., Bedogni, G., Baviera, G., Giorgio, V., Porro, E., Paris, C. and Nobili, V. (2014). Randomised clinical trial: the beneficial effects of VSL# 3 in obese children with non-alcoholic steatohepatitis. *Alimentary Pharmacology & Therapeutics*, *39*(11): 1276-1285.

[27] Vajro, P., Mandato, C., Licenziati, M. R., Franzese, A., Vitale, D. F., Lenta, S. and Meli, R. (2011). Effects of *Lactobacillus rhamnosus*

strain GG in pediatric obesity-related liver disease. *Journal of Pediatric Gastroenterology and Nutrition*, *52*(6): 740-743.

[28] Niv, E., Naftali, T., Hallak, R., & Vaisman, N. (2005). The efficacy of *Lactobacillus reuteri* ATCC 55730 in the treatment of patients with irritable bowel syndrome—a double blind, placebo-controlled, randomized study. *Clinical Nutrition*, *24*(6): 925-931.

[29] Wickens, K., Black, P. N., Stanley, T. V., Mitchell, E., Fitzharris, P., Tannock, G. W. and Probiotic Study Group. (2008). A differential effect of 2 probiotics in the prevention of eczema and atopy: a double-blind, randomized, placebo-controlled trial. *Journal of Allergy and Clinical Immunology*, *122*(4): 788-794.

[30] Niers, L., Martín, R., Rijkers, G., Sengers, F., Timmerman, H., Van Uden, N, and Hoekstra, M. (2009). The effects of selected probiotic strains on the development of eczema (the PandA study). *Allergy*, *64*(9): 1349-1358.

[31] Markowiak, P. and Śliżewska, K. (2017). Effects of probiotics, prebiotics, and synbiotics on human health. *Nutrients*, *9* (9): 1021.

[32] Manigandan, T., Mangaiyarkarasi, S. P., Hemalatha, R., Hemalatha, V. T. and Murali, N. P. (2012). Probiotics, prebiotics and synbiotics-a review. *Biomedical & Pharmacology Journal*, *5*(2): 295.

[33] Zhou, L. and Foster, J. A. (2015). Psychobiotics and the gut–brain axis: in the pursuit of happiness. *Neuropsychiatric Disease and Treatment*, *11*: 715-723 doi: 10.2147/NDT.S61997.

[34] Eslamparast, T., Poustchi, H., Zamani, F., Sharafkhah, M., Malekzadeh, R. and Hekmatdoost, A. (2014). Synbiotic supplementation in non-alcoholic fatty liver disease: a randomized, double-blind, placebo-controlled pilot study. *The American Journal of Clinical Nutrition*, *99*(3): 535-542.

[35] Misra, S. and Mohanty, D. (2019). Psychobiotics: A new approach for treating mental illness?. *Critical Reviews in Food Science and Nutrition*, *59*(8): 1230-1236.

[36] Scott, A. J., Merrifield, C. A., Younes, J. A. and Pekelharing, E. P. (2018). Pre-, pro-and synbiotics in cancer prevention and

treatment—a review of basic and clinical research. *Ecancermedical Science*, *12*: 869. doi: 10.3332/ecancer.2018.869.

[37] Van Loo, J., Clune, Y., Bennett, M. and Collins, J. K. (2005). The SYNCAN project: goals, set-up, first results and settings of the human intervention study. *British Journal of Nutrition*, *93*(S1): S91-S98.

[38] Ghouri, Y. A., Richards, D. M., Rahimi, E. F., Krill, J. T., Jelinek, K. A. and DuPont, A. W. (2014). Systematic review of randomized controlled trials of probiotics, prebiotics, and synbiotics in inflammatory bowel disease. *Clinical and Experimental Gastroenterology*, *7*: 473-487. doi: 10.2147/CEG.S27530.

[39] Šmid, A., Strniša, L., Bajc, K., Vujić-Podlipec, D., Matijašić, B. B. and Rogelj, I. (2016). Randomized clinical trial: The effect of fermented milk with the probiotic cultures *Lactobacillus acidophilus* La-5® and Bifidobacterium BB-12® and Beneo dietary fibres on health-related quality of life and the symptoms of irritable bowel syndrome in adults. *Journal of Functional Foods*, *24*: 549-557.

[40] Zaeim, D., Sarabi-Jamab, M., Ghorani, B. and Kadkhodaee, R. (2019). Double layer co-encapsulation of probiotics and prebiotics by electro-hydrodynamic atomization. *LWT*, *110*: 102-109.

[41] Damodharan, K., Palaniyandi, S. A., Yang, S. H. and Suh, J. W. (2017). Co-encapsulation of lactic acid bacteria and prebiotic with alginate-fenugreek gum-locust bean gum matrix: Viability of encapsulated bacteria under simulated gastrointestinal condition and during storage time. *Biotechnology and Bioprocess Engineering*, *22*(3): 265-271.

[42] Krithika, B. and Preetha, R. (2019). Formulation of protein based inulin incorporated synbiotic nanoemulsion for enhanced stability of probiotic. *Materials Research Express*, *6*(11): 114003.

[43] Wu, Y. and Zhang, G. (2018). Synbiotic encapsulation of probiotic *Latobacillus plantarum* by alginate-arabinoxylan composite microspheres. *LWT*, *93*: 135-141. doi: 10.1016/j.lwt.2018.03.034.

In: Nutraceuticals ISBN: 978-1-68507-488-3
Editors: A. Kumari and G. Singh

Chapter 11

BIOACTIVE PEPTIDES AND PROTEINS

***Neha Verma*[1], *Shiv Kumar Giri*[2], *Gulab Singh*[2] *and Anil Kumar*[3,*]**

[1]Department of Botany,
Baba Mastnath University, Rohtak, Haryana, India
[2]Department of Biotechnology,
Maharaja Agrasen University, Solan, Himachal Pradesh, India
[3]Centre for Medical Biotechnology,
Maharshi Dayanand University, Rohtak, Haryana, India

ABSTRACT

In recent years, the scientific community has established the connection between a good diet and health. Therefore, people are becoming more aware about the benefits of healthy food and its importance in maintaining well-being. Over the years, with advancements in technology, numerous biologically active peptides (BP) have been identified from various food sources. They are naturally present in milk, eggs, soy, fish, and meat proteins, but can also be synthesized with prior knowledge of the peptide sequence. Although both peptides and proteins differ in their molecular weight; but, they share a

* Corresponding Author's E-mail: anil.cmbt@mdurohtak.ac.in.

common standard building block, i.e., amino acids. The latter are connected by a peptide bond that results from covalent interactions. Once released into the gut and bloodstream, the activity of the peptides and proteins is determined by their sequence and composition of amino acids. The peptides are derived from proteins as small fragments of unequal length by the digestion of various gut enzymes. Out of the complex peptide mixture, a few become active and exert their beneficial effects by binding themselves to gut receptors. Once BP reaches the bloodstream, thereby, they target the cardiovascular, endocrine, nervous, digestive, and immune systems and improve their efficacy. The various studies have demonstrated the antimicrobial, antioxidant, cytomodulatory, and anticancer potential of bioactive peptides. In this chapter, we have discussed various bioactive peptides and their importance in the betterment of human health.

1. INTRODUCTION

The emergence of various lifestyle-related disorders is related to inadequate dietary intake. Therefore, protein and fiber-rich foods are gaining the attention of the scientific community. Protein-rich foods offer nutritional values and enhance the physiological functions of the body [1]. To have potential health benefits from proteins, these need to be hydrolyzed into small fragments known as bioactive peptides (BP). Latter are composed of amino acid residues linked to one another with the help of amide or peptide bonds. Currently, a database named 'Biopep' has been developed that contains more than 4000 BP entries. Inside the parent protein, BP usually remains inactive and can be converted into an active form either by enzymatic hydrolysis or by food processing methods such as cooking, fermentation, and ripening [2]. In recent years, the overall production of BP has increased many folds because of its numerous health benefits. Bioactive peptides specifically interact with specific gut receptors and biomolecules to facilitate the physiological functions of the body [3]. Moreover, depending on their mode of action, they could display hormone or drug-like activities. They have been categorized as antihypertensive, anti-cancerous, antidiabetic, antimicrobial, immunomodulatory, antioxidative, opioid, and mineral binder.

2. Sources of Bioactive Peptides

These can be obtained from both plants (mushrooms, sorghum, pumpkin, soy, rice, wheat, and maize) and animals (bovine blood, meat, eggs) sources. Various fish species (salmon, tuna, sardine, and herring) and dairy products (bovine milk) also provide BP with potential health benefits.

2.1. Vegetal Sources

Bioactive peptide can be produced by *in vitro* gastrointestinal (GI) digestion of soy milk and soybean seeds. Digestion of soy hydrolysates and the soy-fermented foods (natto and tempeh) by a variety of proteases such as trypsin, kidney membrane proteases, pronase, and plasma proteases generates oligopeptides with antithrombotic and angiotensin-converting enzyme (ACE) inhibitory activity [4]. Cereal grains such as wheat, millet, barley, corn, rice, and oats are routinely being used as part of the diet. The cereal proteins contain numerous BP with potential biological effects such as anticancer, antioxidant, antithrombotic, and opioid activities [5]. Roasted (malt, coffee & cocoa) and fermented (beer & aged sake) foods contain cyclic dipeptides such as 2, 5-diketopiperazines (DKPs). Latter are also present in hydrolysates of whey protein and have been reported to have antioxidant properties [6].

The waste material generated from fruit and vegetable processing can also be utilized as a source of BP. The enzymatic digestion of protein-rich fruits such as plums with thermolysin, protease P, and alcalase generates BP with ACE inhibitory and antioxidant properties [7]. A significant amount of waste is produced by the oil industry; for example; the extraction process of olive oil generates a lot of non-biodegradable waste, and attempts have been made to recover BP from such waste material. In one of the studies, enzymatic digestion of waste proteins with alcalase produced a hydrolysate with anti-hypersensitive and antioxidant activity [8].

2.2. Animal Sources

Animal protein is considered as a rich source of BP and is thought to confer various health benefits. The bovine blood obtained from the slaughterhouse contains an abundant amount of BP, and when digested *in vitro*, generates 75 types of unique peptides with profound physiological function [9]. The hydrolysis of blood proteins such as serum albumin with trypsin generates BP with antioxidant, antihypertensive activity and also aids in blood glucose regulation [10]. The peptides derived from meat and fish exhibit antimicrobial, antihypertensive, and antioxidative activities *in vitro* as well as *in vivo*, too. For example, peptides obtained from nebulin (Arg-Pro-Arg) and titin (Ala-Pro-Val-Ala and Pro-Thr-Pro-Val-Pro) in pork digest have been reported to show antihypertensive activity [11]. Eggs are considered a rich source of valuable proteins and are recommended as part of a healthy diet. Eggs have been known to contain numerous types of BP with therapeutic potentials, besides fulfilling the nutritional requirement. The protein powder derived from the egg white is routinely being used in the food industry as it offers extended shelf life and is known to produce numerous BP on hydrolysis with proteases [12]. The BP (Arg-Val-Pro-Ser-Leu) with antihypertensive properties has been derived from egg white protein [13].

2.3. Dairy-Based Products

Dairy-based products, primarily bovine milk, have numerous BP, which act as a critical source of nitrogen and make up for a truly balanced diet. The processing or GI digestion of milk proteins such as immunoglobulins and lactoferrin (Lf) releases numerous active peptides [14]. Milk fermentation with lactic acid bacteria has been reported to generate BP with an antihypertensive effect. Two peptides (Tyr-Pro and Lys-Val-Leu-Pro-Val-Pro-Gln) purified from fermented milk exhibit ACE inhibitory, immunomodulatory, antibacterial, antithrombotic, and anticancer activities [15]. The BP derived from donkey milk protein hydrolysates possess various nutritional properties along with biological activities. One of the studies demonstrated that donkey milk contains five endogenous BP with immunomodulatory and ACE inhibitory effects [16].

2.4. Marine Organisms Derived Sources

The peptides derived from marine organisms have been used effectively for treating various diseases. The potential health effects of such BP include appetite suppression, hypocholesterolemic, antihypertensive, antidiabetic, antimicrobial, anticancer, immunostimulatory, anticoagulant, and antioxidative activities [17]. The BP isolated from crustaceans is known to regulate various biological functions such as metabolism, reproduction, and metamorphosis. Marine algae are considered a powerful source of proteins, and BP isolated from it offers multiple health benefits, including immunomodulatory, antibacterial, and ACE inhibitory activities [18].

3. Production of Bioactive Peptides

They can be obtained from various food proteins by GI digestion, enzymatic hydrolysis, and microbial fermentation. However, with prior knowledge of the peptide sequence, it is also possible to synthesize the BP chemically or using recombinant DNA technology.

3.1. Enzymatic Hydrolysis

Numerous peptides have been obtained from various animal and plant sources by enzymatic treatment. It involves the enzymatic digestion of proteins with proteases such as trypsin, pepsin, alcalase, savinase, elastase, chymotrypsin, and thermolysin in combination or individually [19].

The hydrolysis process involves sample reconstitution at optimum pH, followed by heating at 95°C that releases the proteolytic enzymes and produces hydrolysates. Latter was subjected to centrifugation and several purification techniques to obtain BP [20]. It has been observed that digestion of Neutrase® and thornback ray skin gelatin hydrolysate with protease generates BP with antioxidant and antihypertensive potential [21].

3.2. Gastrointestinal Digestion

Digestion of food-derived proteins inside the GI tract releases several health beneficiaries BP [22]. Once food reaches the stomach, the food proteins get denatured by secretion of hydrochloric acid (HCl) from parietal cells. Low pH favors the conversion of inactive proteases into active ones, and as a result, BP of different lengths are released from proteins by GI enzymes.

3.3. Fermentation

The process of fermentation makes use of a proteolytic enzymatic system of microorganisms. These enzymes hydrolyze the proteins into their building blocks i.e., amino acids and peptides, to fulfill the nitrogen requirement of microbes. For example, fermentation of milk proteins with the help of *L. helveticus* and *S. cerevisiae* is known to produce antihypertensive peptides such as Ile-Pro-Pro (IPP) and Val-Pro-Pro-Pro (VPPP) [23]. Furthermore, *L. helveticus*, *L.-plantarum*, and *S. thermophile* proteolytic systems have also been used to release BP from fermented milk proteins [2].

3.4. Chemical Synthesis

In recent years, peptidomimetics have helped a lot to produce synthetic BP with pharmaceutical properties [24]. Furthermore, advanced technological interventions and in silico approaches could be used for the identification of novel BP. The chemically synthesized tetrapeptide such as Lys-Arg-Glu-Ser (LAGS) has been reported to reduce peroxidation and atherosclerosis [25]. Similarly, the synthesized peptide having the active Pro-His-His (PHH) fragment exhibits remarkable properties of lipoprotein oxidation inhibition [26].

3.5. Genetic Engineering

The synthesis of short-chain peptides using genetic engineering has gained much interest in the last few years. The process involves cloning a peptide coding region in a prokaryotic expression vector that allows mass production of BP with biological activities. The ACE-inhibitory peptides such as His-His-Leu, His-Val-Leu-Pro-Val-Leu, and Gly-His-Ile-Ala-Thr-Phe-Gln-Glu-Arg have been expressed successfully using *E. coli* as host [27].

4. Purification and Characterization of Bioactive Peptides

Various purification and characterization techniques are being used for BP production from crude peptide extracts. The purification of peptides makes use of various separation techniques such as ultrafiltration (UF) and reversed-phase high-performance liquid chromatography (RP-HPLC) [28]. At the same time, characterization is done using spectrometry-based methods such as matrix-assisted laser desorption/ionization mass spectrometry (MALDI-MS)), electrospray ionization mass spectrometry (ESI–MS), and fast atom bombardment mass spectrometry (FAB-MS) [20]. The purity of BP is assessed by methods other than the one used for the purification process.

5. Bioactive Peptides and Health Benefits

These can act as endogenous signals that interact with certain biomolecules and cell receptors and thereby exert various biological effects (Table 1) as discussed below-

5.1. Antihypertensive Peptides

Bioactive peptides (BP) are thought to play an important role in the regulation of ACE. The primary role of ACE is to maintain stable blood pressure by 1) inhibition of angiotensin II synthesis; 2) inhibition of bradykinin degradation; and 3) stimulation of water and Na^+ retention [29]. Peptides with ACE inhibititory activity have been derived from hydrolysates of milk proteins such as α-S1-casein, β-casein, and kappa-casein. Animals (pigs, cattle, and poultry) as well as plants (soybeans, peas, barley, oats, and canola) sources contain numerous BP with antihypertensive potential [30]. A diverse type of amino acid residues in BP contribute significantly towards their ACE inhibitory activity. Latter is thought to be associated with the presence of positively charged (lysine and arginine) and hydrophobic (proline) amino acids at N- and C-terminal, respectively [31]. The BP having branched-chain amino acids (valine and isoleucine) have also been reported to show antihypertensive activity. Likewise, cation ($Zn^{2+)}$) chelators such as serein, histidine, and glutamate residues also could act as potent ACE inhibitors [1].

5.2. Cholesterol-Lowering Peptides

Cholesterol is required for the synthesis of steroid hormones, vitamin D, and bile acids. However, excessive levels of cholesterol (hypercholesterolemia) may lead to plaque formation in arteries and result in arteriosclerosis and ultimately to hypertension [1]. Cholesterol plaques may also result in cardiovascular diseases as they block the arteries and reduce the oxygen supply to the heart. In recent years, BP has been identified with antihypercholesterolemic effects. The peptides obtained from cumin seeds reduce cholesterol by binding to bile acids and by inhibiting micelle formation and lipase activity [32]. Moreover, soybean-derived peptides can lower cholesterol levels by improving low-density lipoproteins (LDL) absorption and by inhibiting β-hydroxy-β-methylglutaryl (HMG)-CoA activity [33]. Similar effects have been observed in rice bran, and cowpea protein hydrolysates, where BP have reduced cholesterol micellar solubilization [34].

5.3. Antioxidants Peptides

Dysfunctional antioxidant system and reactive oxygen species (ROS) such as superoxide anion (O2·) and hydroxyl radical (OH·) generate oxidative stress, and subsequently result in tissue injury. Various risk elements could elevate levels of ROS and it might act as a predisposing factor for various diseases [35]. In the last few years, peptides with antioxidant activities have been derived from animal and plant origin food sources. The hexapeptide (QFQPGL) and tetrapeptides (WNIP & GWNI) derived from milk protein (α-S1-Casein), and egg white protein (ovotransferrin) respectively, have been reported to show antioxidant activity [36].

The antioxidant potential of BP is primarily attributed to its amino acid sequence composition and hydrophobicity. Latter plays a crucial role in promoting antioxidant activity, as it enhances the BP solubilization in lipids and expedites the free radicals' accessibility. The presence of aliphatic (valine and leucine) and aromatic (proline) residues in the peptide sequence significantly contributes to their antioxidant potential [37]. Moreover, the presence of lysine in the BP sequence aids in their antioxidant potential by enhancing their ability to chelate cations (Fe^{2+} and $Cu2^{+}$ ions). It has also been observed that the presence of phenylalanine and histidine residues in BP efficiently quench and scavenge the ROS by acting as proton donors [38].

5.4. Antidiabetic Peptides

Diabetes mellitus (DM) is characterized by elevated blood glucose levels due to inadequate or no insulin secretion from pancreatic cells. It can be categorized as type I (insulin-dependent) and II (insulin-independent). Type 1 results due to insufficient secretion of insulin from dysfunctional β-cells of the pancreas, while; type 2 results from an imbalance between insulin secretion and absorption of blood glucose [29]. The BP derived from various food proteins hydrolysates can act as antidiabetic agents and may aid in balancing of glucose and insulin levels. The BP extracted from fermented soybean proteins has been reported to stimulate insulin-dependent glucose uptake [39]. Similarly, peptides (LSVSVL, ATNPLF, and AKSPLF) obtained from protein hydrolysates of black beans have shown the potential to inhibit the sodium-dependent glucose transporter-1 (SGLT-1) and glucose transporter-2 (GLUT-2), which result in lowering of blood glucose levels. Silver carp protein-derived peptides ((LPIIDI and APGPAGP) have shown the ability to inhibit the activity of dipeptidyl peptidase IV (DPP-IV) [40]. Latter is thought to be involved in the regulation of the incretin hormones [(Glucagon-Like Peptide-1 (GLP-1) and Glucose-Dependent Insulinotropic Polypeptide (GIP)]. The duo is known to normalize blood glucose levels by stimulating insulin secretion from pancreatic cells [41]. Moreover, it has been observed that inhibition of α-amylase and α-glucosidase activity could also be employed as an alternative strategy to regulate the glucose level in diabetic patients [42].

5.5. Antimicrobial Peptides

The peptides with antimicrobial potential are of immense importance, as they also aid in an enhanced immune response against a wide range of microorganisms. The proteins hydrolysates from various sources have been tested extensively for their antimicrobial activity. The BP (AMPSSSEESII) derived from milk protein (β-S2-Casein) can inhibit the growth of various microbes such as *E. coli*, *L. innocua*, *M. luteus*, and *S.*

enteritidis. The antimicrobial property of BP is attributed to the presence of positively charged (arginine and lysine) amino acid residues that facilitate their binding to the negatively charged membrane of microbes [43]. The bovine proteins (lactoferrins, α-lactalbumin, and lactoglobulin) derived peptides have also been found to inhibit the growth of a variety of Gram-positive and negative bacteria [44]. Furthermore, the BP (SIFIQRFTT) obtained from fish protein inhibits the growth of *L. innocua* and *E. coli* [45]. The peptides obtained from blood proteins have also been employed for their antimicrobial potential against *S. aureus C. albicans* and *E. coli* [46]. The presence of hydrophilic and hydrophobic amino acid residues in BP significantly contributes to their antimicrobial potential. Moreover, the mechanism of antimicrobial properties of BP is mainly attributed to their ability to make pores in the microbial cell membrane [47].

5.6. Anticancer Peptides

The modern therapy currently being employed for cancer treatment has many adverse effects as it might result in various types of toxicological complications. Therefore, in recent years, attempts have been made to isolate BP with the anticancer property. Food protein-derived peptides selectively act on different cancer stages and may aid in cancer prevention. The intrinsic properties of BP might be due to their cationic nature, which facilitates their interaction with negatively charged cancer cells [48]. Moreover, the other mechanisms by which BP shows its anticancer potential include angiogenesis prevention, modulation of gene expression, cell cycle arrest, and promotion of apoptosis [49]. The peptides (casein phosphopeptides) derived from milk proteins inhibit intestinal tumor growth by activation of calcium channels that result in apoptosis [50]. Furthermore, the peptide obtained from soybean and chickpea protein hydrolysates downregulates the Bcl-2, PARP, and caspase nine genes and upregulates the expression of p21, p27, and p53 genes in breast cancer cell lines [51-52].

5.7. Immunomodulatory and Anti-Inflammatory Peptides

The BP may also function as anti-inflammatory agents, as they can modulate the immune response. The peptides isolated from soybean and rice protein hydrolysates are known to trigger the activation of non-specific immune responses by stimulating reactive oxygen species. Two synthetic peptides, Tyr-Gly (YG) and Tyr-Gly-Gly (YGG) are thought to stimulate the proliferation of B-cells. The peptides derived from the whey protein concentrate (WPC) and glycomacropeptide (GMP) stimulate IgG and interleukins-1 (IL-1) production and proliferation of phagocytic activities [53]. The peptide obtained from milk proteins such as lactoferrin can stimulate interleukin-8 (IL-8) production from human leukocytes [54]. The egg white protein (EWP) such as ovotransferrin or conalbumin could also boost the immune system by its antimicrobial and anti-inflammatory action [55].

5.8. Opioid Activity of Peptides

Opioids represent a group of compounds that are used as medicine for relaxing body pain. Most of the peptides derived from protein hydrolysates of milk and whey show opioid activity. The peptides obtained from α and β- S1-casein are known as α & β-casomorphins, respectively. The mechanism of casomorphins action may include analgesia production, antidiarrheal effect, controlled secretion of somatostatin and insulin, absorption of amino acids and electrolytes, and also prolong the GI transit time [56, 57].

5.9. Mineral-Binding Peptides

Many food proteins-derived BP functions as mineral binder. These maintain levels of calmodulin (CaM), which controls calcium-dependent cell division and proliferation, and neurotransmission.

Table 1. Sources and function of bioactive peptide

Source	Peptide sequence	Function	Reference
Soybean seeds and soy milk protein	YVVFK, RNLQGENEEEDSGA	Antihypertensive	[60, 42]
Milk protein (β-casein & Lactoferrin)	PFPGPIPN, LHLPLP, QKEPMIGV, KYIPIQ, SLPQNIPPL, HVLPVP, DPYKLRP, PYKLRP, YKLRP, GILRP		[61-63]
Whey from bovine milk	AQSAP, IPAVF, APLRV, AHKAL		[1]
Jellyfish gonads	SY		[64]
β-Lactoglobulin	HIRL, IIAEK	Cholesterol lowering	[65, 66]
Soybean protein and White lupin seed	LPYP, IAVPGEVA, IAVPTGVA HSDADYVLVVLNGR, HGREEEEEEEEEDER, YPSSTKDQQSY		[67]
Black bean and cowpea	YAAAT		[68]
Lentil protein	LLSGTQNQPSFLSGF, NSLTLPILRYL, TLEPNSVFLPVLLH	Antioxidative	[69]
Rice bran protein	YSK		[70]
Casein from bovine milk	PGPIPN, PFPGPIPN, YPFPGPIP, VYPFPGPIPN, MPFPKYPVEP, EPVLGPVRGPFP, QEPVLGPVRGPFP, TPVVVPPFLQPE, TQTPVVVPPFLQPE		[20]
Chicken egg white	AEERYP, DEDTQAMP		[71]
Buffalo ricotta cheese & milk casein	YVEELKPTPEGDL, VLPVPQK		[72, 73]
Whey protein	IPA, IPAVF, VAGTWY	Antidiabetic (T2DM)	[74, 75]
Camel milk protein	ILDKEGIDY, ILDKVGIQY, ILQLA, LLQLE, LAHKPL, ILDKEGIDY VPV, YPI, VPF, VPV, KDLWDDFKGL, MPSKPPLL		[76, 77]
Germinated soybean	NNDDRDS, LSSTEAQQS, NAENNQRN, QQQQQGGSQSQ, EEPQQPQQ, IKSQSES		[42]

Table 1. (Continued)

Source	Peptide sequence	Function	Reference
Milk Protein	LKPTPEGDL, LPYPY, IPIQY, WR, YP, LP, IPI, VPL, IPA, IPAVF		[78, 79]
Meat protein	PPL		[80]
Human milk casein	ELLLNPTHQIYPVTQPLAPV, AMPSSSEESII	Antimicrobial	[81]
Soybean	PGTAVFK		[83]
Forage fish and by product	VNFKLLSHSLLVTLASHL, FPIGMGHGSRPA		[84, 85]
Lactoferrin	RRWQWR, PFWRIRIRR, FKCRRWQWRMKKLGAPSITCV RRAF	Anticancer	[86]
Dendrobium catenatum	RHPFDGPLLPPGD RCGVNAFLPKSYLVHFGWKLLF HFD KPEEVGGAGDRWTC		[87]
Tuna cooking juice	KPEGMDPPLSEPEDRRDGAAGP K, KLPPLLLAKLLMSGKLLAEPCT GR		[88]
Soybean protein	RKQLQGVN, GLTSK, GEGSGA, MPACGSS, LSGNK, MTEEY		[89]
Chickpea	RQSHFANAQP		[90]
Whey protein	Tyr-Gly, Tyr-Gly-Gly	Inflammatory and Immunomodulat ory	[91]
Lactalbumin	GLL		[92]
Casein	VPP, IPP		[93]
Ovotransferrin	IRW, IQW		[94, 95]
Soybean	VPY		[96, 97]

Abbreviations: G: glycine, A: alanine, P: proline, V: valine, L: leucine, I: isoleucine, M: methionine, F: phenylalanine, Y: tyrosine, K: lysine, R: arginine, H: histidine, S: serine, T: threonine, C: cysteine, N: asparagine, Q: glutamine, D: aspartate, E: glutamate, W: tryptophan,

Abnormal CaM levels might aid in the development of cancer and other severe diseases. The CaM-binding peptides have been isolated from hydrolysates of bovine milk proteins such as casein, and they can bind to minerals like Ca, Mg, Fe, Ba, Cr, Ni, Co, and Se [58, 59].

CONCLUSION

Food-derived peptides have the inherent potential to act as therapeutics as they can interact with tissues, cells, enzymes, reactive oxygen species, and specific biomolecules. Such interactions are helpful in the management and treatment of many lifestyle-related disorders, including hypertension, diabetes mellitus, obesity, and cancer. Also, several BP possess antimicrobial potential and that further aids in the enhanced immune response. Although significant progress has been made to isolate and purify BP from several natural sources, there are several issues which still need to be addressed, such as retention of their activity after consumption. Further, studies are needed to be carried out to determine the role of BP in unexplored genetic and metabolic disorders.

REFERENCES

[1] Daliri, Eric Banan-Mwine, Deog H. Oh, and Byong H. Lee. "Bioactive peptides." *Foods* 6, no. 5 (2017): 32.

[2] Yahya, Mohammed A., Omar A. Alhaj, and Abdullrahman S. Al-Khalifah. "Antihypertensive effect of fermented skim camel (*Camelus dromedarius*) milk on spontaneously hypertensive rats." *Nutricion hospitalaria* 34, no. 2 (2017): 416-421.

[3] Hayes, Maria. *Food Proteins and bioactive peptides: New and novel sources, characterisation strategies and applications*. (2018), 38.

[4] Gibbs, Bernard F; Alexandre Zougman, Robert Masse, and Catherine Mulligan. "Production and characterization of bioactive peptides from soy hydrolysate and soy-fermented food." *Food research international* 37, no. 2 (2004): 123-131.

[5] Malaguti, Marco, Giovanni Dinelli, Emanuela Leoncini, Valeria Bregola, Sara Bosi, Arrigo FG Cicero, and Silvana Hrelia. "Bioactive peptides in cereals and legumes: agronomical, biochemical and clinical aspects." *International journal of molecular sciences* 15, no. 11 (2014): 21120-21135.

[6] Kumar, S. Nishanth, J. V. Siji, Bala Nambisan, and C. Mohandas. "Activity and synergistic antimicrobial activity between diketopiperazines against bacteria in vitro." *Applied biochemistry and biotechnology* 168, no. 8 (2012): 2285-2296.

[7] González-García, Estefanía, M. Luisa Marina, and M. Concepción García. "Plum (*Prunus Domestica* L.) by-product as a new and cheap source of bioactive peptides: Extraction method and peptides characterization." *Journal of functional foods* 11 (2014): 428-437.

[8] Esteve, C., M. L. Marina, and M. C. García. "Novel strategy for the revalorization of olive (*Olea europaea*) residues based on the extraction of bioactive peptides." *Food Chemistry* 167 (2015): 272-280.

[9] Caron, Juliette, Gabrielle Chataigné, Jean-Pascal Gimeno, Nathalie Duhal, Jean-François Goossens, Pascal Dhulster, Benoit Cudennec, Rozenn Ravallec, and Christophe Flahaut. "Food peptidomics of in vitro gastrointestinal digestions of partially purified bovine hemoglobin: low-resolution versus high-resolution LC-MS/MS analyses." *Electrophoresis* 37, no. 13 (2016): 1814-1822.

[10] Arrutia, Fátima, Ángela Puente, Francisco A. Riera, Carlos Menéndez, and Ulises A. González. "Influence of heat pre-treatment on BSA tryptic hydrolysis and peptide release." *Food Chemistry* 202 (2016): 40-48.

[11] Ryder, Kate, Alaa El-Din Bekhit, Michelle McConnell, and Alan Carne. "Towards generation of bioactive peptides from meat industry waste proteins: generation of peptides using commercial microbial proteases." *Food chemistry* 208 (2016): 42-50.

[12] Yu, Zhipeng, Yongguang Yin, Wenzhu Zhao, Fei Wang, Yiding Yu, Boqun Liu, Jingbo Liu, and Feng Chen. "Characterization of ACE-inhibitory peptide associated with antioxidant and anticoagulation properties." *Journal of food science* 76, no. 8 (2011): C1149-C1155.

[13] Lin, Songyi, Yang Guo, Qi You, Yongguang Yin, and Jingbo Liu. "Preparation of antioxidant peptide from egg white protein and improvement of its activities assisted by high-intensity pulsed electric field." *Journal of the Science of Food and Agriculture* 92, no. 7 (2012): 1554-1561.

[14] Meisel, Hans, and Richard J. FitzGerald. "Biofunctional peptides from milk proteins: mineral binding and cytomodulatory effects." *Current pharmaceutical design* 9, no. 16 (2003): 1289-1296.

[15] Jäkälä, Pauliina, and Heikki Vapaatalo. "Antihypertensive peptides from milk proteins." *Pharmaceuticals* 3, no. 1 (2010): 251-272.

[16] Chiozzi, Riccardo Zenezini, Anna Laura Capriotti, Chiara Cavaliere, Giorgia La Barbera, Susy Piovesana, Roberto Samperi, and Aldo Laganà. "Purification and identification of endogenous antioxidant and ACE-inhibitory peptides from donkey milk by multidimensional liquid chromatography and nanoHPLC-high resolution mass spectrometry." *Analytical and bioanalytical chemistry* 408, no. 20 (2016): 5657-5666.

[17] Harnedy, Pádraigín A., and Richard J. FitzGerald. "Bioactive peptides from marine processing waste and shellfish: A review." *Journal of functional foods* 4, no. 1 (2012): 6-24.

[18] Harnedy, Pádraigín A., and Richard J. FitzGerald. "Bioactive proteins, peptides, and amino acids from macroalgae 1." *Journal of Phycology* 47, no. 2 (2011): 218-232.

[19] Ugwu, Chizoba Paul, Muawiya Musa Abarshi, Sanusi Bello Mada, Babangida Sanusi, and Humphrey Chukwuemeka Nzelibe. "Camel and horse milk casein hydrolysates exhibit angiotensin converting enzyme inhibitory and antioxidative effects in vitro and in silico." *International Journal of Peptide Research and Therapeutics* 25, no. 4 (2019): 1595-1604.

[20] Bamdad, Fatemeh, Seulki Hazel Shin, Joo-Won Suh, Chamila Nimalaratne, and Hoon Sunwoo. "Anti-inflammatory and antioxidant properties of casein hydrolysate produced using high hydrostatic pressure combined with proteolytic enzymes." *Molecules* 22, no. 4 (2017): 609.

[21] Lassoued, Imen, Leticia Mora, Ahmed Barkia, M-Concepción Aristoy, Moncef Nasri, and Fidel Toldrá. "Bioactive peptides identified in thornback ray skin's gelatin hydrolysates by proteases from *Bacillus subtilis* and *Bacillus amyloliquefaciens*." *Journal of Proteomics* 128 (2015): 8-17.

[22] Mohanty, Debapriya, Rajashree Jena, Prasanta Kumar Choudhury, Ritesh Pattnaik, Swati Mohapatra, and Manish Ranjan Saini. "Milk

derived antimicrobial bioactive peptides: a review." *International Journal of Food Properties* 19, no. 4 (2016): 837-846.

[23] Nakamura, Yasunori, Naoyuki Yamamoto, Kumi Sakai, and Toshiaki Takano. "Antihypertensive effect of sour milk and peptides isolated from it that are inhibitors to angiotensin I-converting enzyme." *Journal of dairy science* 78, no. 6 (1995): 1253-1257.

[24] Kishore, Raghuvansh. "β-Ala containing peptides: potentials in design and construction of bioactive peptides and protein secondary structure mimics." *Current Protein and Peptide Science* 5, no. 6 (2004): 435-455.

[25] Navab, Mohamad, G. M. Anantharamaiah, Srinivasa T. Reddy, Brian J. Van Lenten, Geeta Datta, David Garber, and Alan M. Fogelman. "Human apolipoprotein AI and AI mimetic peptides: potential for atherosclerosis reversal." *Current opinion in lipidology* 15, no. 6 (2004): 645-649.

[26] Nguyen, Su Duy, Tae-Sook Jeong, Mee Ree Kim, and Dai-Eun Sok. "Broad-spectrum antioxidant peptides derived from His residue-containing sequences present in human paraoxonase 1." *Free radical research* 40, no. 4 (2006): 349-358.

[27] Jeong, Do-Won, Dong-Seok Shin, Chang-Won Ahn, In-Sang Song, and Hyong-Joo Lee. "Expression of antihypertensive peptide, His-His-Leu, as tandem repeats in *Escherichia coli*." *Journal of microbiology and biotechnology* 17, no. 6 (2007): 952-959.

[28] Chaudhari, Dipak Dilip, Rajbir Singh, Rashmi Hogarehalli Mallappa, Namita Rokana, Jai Kumar Kaushik, Rajesh Bajaj, Virender Kumar Batish, and Sunita Grover. "Evaluation of casein & whey protein hydrolysates as well as milk fermentates from *Lactobacillus helveticus* for expression of gut hormones." *The Indian journal of medical research* 146, no. 3 (2017): 409.

[29] Cánovas, Juana Mulero, Pilar Zafrilla Rentero, Adela Martínez-Cachá Martínez, Mariano Leal Hernández, and José Abellán Alemán. "Péptidos bioactivos." *Clínica e Investigación en Arteriosclerosis* 23, no. 5 (2011): 219-227. ["Bioactive Peptides." *Clinic and Research in Arteriosclerosis*]

[30] Schanbacher, F. L., R. S. Talhouk, F. A. Murray, L. I. Gherman, and L. B. Willett. "Milk-borne bioactive peptides." *International Dairy Journal* 8, no. 5-6 (1998): 393-403.

[31] Murray, B. A., and R. J. FitzGerald. "Angiotensin converting enzyme inhibitory peptides derived from food proteins: biochemistry, bioactivity and production." *Current pharmaceutical design* 13, no. 8 (2007): 773-791.

[32] Siow, Hwee-Leng, Sy-Bing Choi, and Chee-Yuen Gan. "Structure–activity studies of protease activating, lipase inhibiting, bile acid binding and cholesterol-lowering effects of pre-screened cumin seed bioactive peptides." *Journal of functional foods* 27 (2016): 600-611.

[33] Lammi, Carmen, Chiara Zanoni, and Anna Arnoldi. "IAVPGEVA, IAVPTGVA, and LPYP, three peptides from soy glycinin, modulate cholesterol metabolism in HepG2 cells through the activation of the LDLR-SREBP2 pathway." *Journal of Functional Foods* 14 (2015): 469-478.

[34] Marques, Marcelo Rodrigues, Rosana Aparecida Manólio Soares Freitas, Amanda Caroline Corrêa Carlos, Érica Sayuri Siguemoto, Gustavo Guadagnucci Fontanari, and José Alfredo Gomes Arêas. "Peptides from cowpea present antioxidant activity, inhibit cholesterol synthesis and its solubilisation into micelles." *Food Chemistry* 168 (2015): 288-293.

[35] Panth, Nisha, Keshav Raj Paudel, and Kalpana Parajuli. "Reactive oxygen species: a key hallmark of cardiovascular disease." *Advances in medicine* 2016 (2016).

[36] Shen, Shengwen, Baljot Chahal, Kaustav Majumder, Sun-Jong You, and Jianping Wu. "Identification of novel antioxidative peptides derived from a thermolytic hydrolysate of ovotransferrin by LC-MS/MS." *Journal of agricultural and food chemistry* 58, no. 13 (2010): 7664-7672.

[37] Qian, Zhong-Ji, Won-Kyo Jung, and Se-Kwon Kim. "Free radical scavenging activity of a novel antioxidative peptide purified from hydrolysate of bullfrog skin, *Rana catesbeiana* Shaw." *Bioresource Technology* 99, no. 6 (2008): 1690-1698.

[38] Zou, Tang-Bin, Tai-Ping He, Hua-Bin Li, Huan-Wen Tang, and En-Qin Xia. "The structure-activity relationship of the antioxidant peptides from natural proteins." *Molecules* 21, no. 1 (2016): 72.

[39] Kwon, Dae Young, Sang Mee Hong, Il Sung Ahn, Min Jung Kim, Hye Jeong Yang, and Sunmin Park. "Isoflavonoids and peptides from meju, long-term fermented soybeans, increase insulin sensitivity and exert insulinotropic effects in vitro." *Nutrition* 27, no. 2 (2011): 244-252.

[40] Zhang, Ying, Ran Chen, Xiling Chen, Zhu Zeng, Huiqin Ma, and Shangwu Chen. "Dipeptidyl peptidase IV-inhibitory peptides derived from silver carp (*Hypophthalmichthys molitrix* Val.) proteins." *Journal of agricultural and food chemistry* 64, no. 4 (2016): 831-839.

[41] Deacon, Carolyn F. "Peptide degradation and the role of DPP-4 inhibitors in the treatment of type 2 diabetes." *Peptides* 100 (2018): 150-157.

[42] González-Montoya, Marcela, Blanca Hernández-Ledesma, Rosalva Mora-Escobedo, and Cristina Martínez-Villaluenga. "Bioactive peptides from germinated soybean with anti-diabetic potential by inhibition of dipeptidyl peptidase-IV, α-amylase, and α-glucosidase enzymes." *International journal of molecular sciences* 19, no. 10 (2018): 2883.

[43] Guterstam, Peter, Fatemeh Madani, Hisaaki Hirose, Toshihide Takeuchi, Shiroh Futaki, Samir EL Andaloussi, Astrid Gräslund, and Ülo Langel. "Elucidating cell-penetrating peptide mechanisms of action for membrane interaction, cellular uptake, and translocation utilizing the hydrophobic counter-anion pyrenebutyrate." *Biochimica et biophysica acta (BBA)-biomembranes* 1788, no. 12 (2009): 2509-2517.

[44] Pellegrini, Antonio, Carmen Dettling, Ursula Thomas, and Peter Hunziker. "Isolation and characterization of four bactericidal domains in the bovine β-lactoglobulin." *Biochimica et Biophysica Acta (BBA)-General Subjects* 1526, no. 2 (2001): 131-140.

[45] Guinane, Caitriona M., Robert M. Kent, Sarah Norberg, Paula M. O'Connor, Paul D. Cotter, Colin Hill, Gerald F. Fitzgerald, Catherine Stanton, and R. Paul Ross. "Generation of the antimicrobial peptide caseicin A from casein by hydrolysis with thermolysin enzymes." *International Dairy Journal* 49 (2015): 1-7.

[46] Aissaoui, Neyssene, Jean-Marc Chobert, Thomas Haertlé, M. Nejib Marzouki, and Ferid Abidi. "Purification and biochemical characterization of a neutral serine protease from *Trichoderma harzianum*. Use in antibacterial peptide production from a fish by-product hydrolysate." *Applied biochemistry and biotechnology* 182, no. 2 (2017): 831-845.

[47] Taniguchi, Masayuki, Akihito Ochiai, Hiroshi Kondo, Shun Fukuda, Yohei Ishiyama, Eiichi Saitoh, Tetsuo Kato, and Takaaki Tanaka. "Pyrrhocoricin, a proline-rich antimicrobial peptide derived from insect, inhibits the translation process in the cell-free *Escherichia coli* protein synthesis system." *Journal of bioscience and bioengineering* 121, no. 5 (2016): 591-598.

[48] Hoskin, David W., and Ayyalusamy Ramamoorthy. "Studies on anticancer activities of antimicrobial peptides." *Biochimica et Biophysica Acta (BBA)-Biomembranes* 1778, no. 2 (2008): 357-375.

[49] De Mejia, Elvira Gonzalez, and Vermont P. Dia. "The role of nutraceutical proteins and peptides in apoptosis, angiogenesis, and metastasis of cancer cells." *Cancer and Metastasis Reviews* 29, no. 3 (2010): 511-528.

[50] Perego, Silvia, Stefania Cosentino, Amelia Fiorilli, Guido Tettamanti, and Anita Ferraretto. "Casein phosphopeptides modulate proliferation and apoptosis in HT-29 cell line through their interaction with voltage-operated L-type calcium channels." *The Journal of Nutritional Biochemistry* 23, no. 7 (2012): 808-816.

[51] Hung, Chuan-Chuan, Yu-Hsuan Yang, Pei-Feng Kuo, and Kuo-Chiang Hsu. "Protein hydrolysates from tuna cooking juice inhibit cell growth and induce apoptosis of human breast cancer cell line MCF-7." *Journal of Functional Foods* 11 (2014): 563-570.

[52] Xue, Zhaohui, Haichao Wen, Lijuan Zhai, Yanqing Yu, Yanni Li, Wancong Yu, Aiqing Cheng, Cen Wang, and Xiaohong Kou. "Antioxidant activity and anti-proliferative effect of a bioactive peptide from chickpea (*Cicer arietinum* L.)." *Food Research International* 77 (2015): 75-81.

[53] Li, Esther WY, and Yoshinori Mine. "Immunoenhancing effects of bovine glycomacropeptide and its derivatives on the proliferative response and phagocytic activities of human macrophagelike cells, U937." *Journal of agricultural and food chemistry* 52, no. 9 (2004): 2704-2708.

[54] Shinoda, Ichizo, Mitsunori Takase, Yasuo Fukuwatari, Seiichi Shimamura, Manfred Köller, and Wolfgang König. "Effects of lactoferrin and lactoferricin® on the release of interleukin 8 from human polymorphonuclear leukocytes." *Bioscience, biotechnology, and biochemistry* 60, no. 3 (1996): 521-523.

[55] Wu, Jianping, Kaustav Majumder, and Kristen Gibbons. "Bioactive proteins and peptides from egg proteins." *Bioactive proteins and peptides as functional foods and nutraceuticals* 29 (2010): 247.

[56] Shah, Nagendra P. "Effects of milk-derived bioactives: an overview." *British Journal of Nutrition* 84, no. S1 (2000): 3-10.

[57] Drewnowski, Adam. "Food preferences and the opioid peptide system." *Trends in Food Science & Technology* 3 (1992): 97-99.

[58] FitzGerald, RJ. “Potential uses of caseinophosphopeptides.” *International Dairy Journal* 8, no. 5-6 (1998): 451-457.

[59] Scholz-Ahrens, Katharina E., and J. Schrezenmeir. "Effects of bioactive substances in milk on mineral and trace element metabolism with special reference to casein phosphopeptides." *British Journal of Nutrition* 84, no. S1 (2000): 147-153.

[60] Capriotti, Anna Laura, Giuseppe Caruso, Chiara Cavaliere, Roberto Samperi, Salvatore Ventura, Riccardo Zenezini Chiozzi, and Aldo Laganà. "Identification of potential bioactive peptides generated by simulated gastrointestinal digestion of soybean seeds and soy milk proteins." *Journal of Food Composition and Analysis* 44 (2015): 205-213.

[61] Lin, Songyi, Rong Liang, Peiyu Xue, Shuyu Zhang, Zeyu Liu, and Xiuping Dong. "Antioxidant activity improvement of identified pine nut peptides by pulsed electric field (PEF) and the mechanism exploration." *LWT* 75 (2017): 366-372.

[62] Rodríguez-Figueroa, J. C., A. F. González-Córdova, M. J. Torres-Llanez, H. S. Garcia, and B. Vallejo-Cordoba. "Novel angiotensin I-converting enzyme inhibitory peptides produced in fermented milk by specific wild *Lactococcus lactis* strains." *Journal of Dairy Science* 95, no. 10 (2012): 5536-5543.

[63] Liu, Dong, Haiyan Sun, Lijun Zhang, Shimin Li, and Zhenyu Qin. "High-level expression of milk-derived antihypertensive peptide in *Escherichia coli* and its bioactivity." *Journal of agricultural and food chemistry* 55, no. 13 (2007): 5109-5112.

[64] Zhang, Gaonan, Shujian Zheng, Yuqi Feng, Guo Shen, Shanbai Xiong, and Hongying Du. *Changes in Nutrient Profiling and Antioxidant Activities of Different Fish Soups, Before and After Simulated Gastrointestinal Digestion*. (2018).

[65] Nagaoka, Satoshi, Yu Futamura, Keiji Miwa, Takako Awano, Kouhei Yamauchi, Yoshihiro Kanamaru, Kojima Tadashi, and Tamotsu Kuwata. "Identification of novel hypocholesterolemic peptides derived from bovine milk β-lactoglobulin." *Biochemical and Biophysical Research Communications* 281, no. 1 (2001): 11-17.

[66] Yamauchi, Rena, Kousaku Ohinata, and Masaaki Yoshikawa. "β-Lactotensin and neurotensin rapidly reduce serum cholesterol via NT2 receptor." *Peptides* 24, no. 12 (2003): 1955-1961.

[67] Lammi, Carmen, Chiara Zanoni, and Anna Arnoldi. "IAVPGEVA, IAVPTGVA, and LPYP, three peptides from soy glycinin, modulate cholesterol metabolism in HepG2 cells through the activation of the LDLR-SREBP2 pathway." *Journal of Functional Foods* 14 (2015): 469-478.

[68] Hernandez, Luis M. Real, and Elvira Gonzalez de Mejia. "Bean peptides have higher in silico binding affinities than ezetimibe for

the N-terminal domain of cholesterol receptor Niemann-Pick C1 Like-1." *Peptides* 90 (2017): 83-89.

[69] García-Mora, Patricia, Mercedes Martín-Martínez, María Angeles Bonache, Rosario González-Múniz, Elena Peñas, Juana Frias, and Cristina Martinez-Villaluenga. "Identification, functional gastrointestinal stability and molecular docking studies of lentil peptides with dual antioxidant and angiotensin I converting enzyme inhibitory activities." *Food chemistry* 221 (2017): 464-472.

[70] Wang, Zhujun, and Xuewu Zhang. "Isolation and identification of anti-proliferative peptides from *Spirulina platensis* using three-step hydrolysis." *Journal of the Science of Food and Agriculture* 97, no. 3 (2017): 918-922.

[71] Nimalaratne, Chamila, Nandika Bandara, and Jianping Wu. "Purification and characterization of antioxidant peptides from enzymatically hydrolyzed chicken egg white." *Food chemistry* 188 (2015): 467-472.

[72] Basilicata, Manuela Giovanna, Giacomo Pepe, Simona Adesso, Carmine Ostacolo, Marina Sala, Eduardo Sommella, Maria Carmina Scala, Antonella Messore, Stefania Marzocco, and Pietro Campiglia. "Antioxidant properties of buffalo-milk dairy products: a β-Lg peptide released after gastrointestinal digestion of buffalo ricotta cheese reduces oxidative stress in intestinal epithelial cells." *International journal of molecular sciences* 19, no. 7 (2018): 1955.

[73] Shanmugam, V. P., S. Kapila, T. Kemgang Sonfack, and R. Kapila. "Antioxidative peptide derived from enzymatic digestion of buffalo casein." *International Dairy Journal* 42 (2015): 1-5.

[74] Silveira, Silvana T., Daniel Martínez-Maqueda, Isidra Recio, and Blanca Hernández-Ledesma. "Dipeptidyl peptidase-IV inhibitory peptides generated by tryptic hydrolysis of a whey protein concentrate rich in β-lactoglobulin." *Food chemistry* 141, no. 2 (2013): 1072-1077.

[75] Uenishi, Hiroshi, Toshihide Kabuki, Yasuyuki Seto, Atsushi Serizawa, and Hadjime Nakajima. "Isolation and identification of casein-derived dipeptidyl-peptidase 4 (DPP-4)-inhibitory peptide

LPQNIPPL from gouda-type cheese and its effect on plasma glucose in rats." *International Dairy Journal* 22, no. 1 (2012): 24-30.

[76] Mudgil, Priti, Hina Kamal, Gan Chee Yuen, and Sajid Maqsood. "Characterization and identification of novel antidiabetic and anti-obesity peptides from camel milk protein hydrolysates." *Food chemistry* 259 (2018): 46-54.

[77] Nongonierma, Alice B., Cloé Cadamuro, Aurélien Le Gouic, Priti Mudgil, Sajid Maqsood, and Richard J. FitzGerald. "Dipeptidyl peptidase IV (DPP-IV) inhibitory properties of a camel whey protein enriched hydrolysate preparation." *Food chemistry* 279 (2019): 70-79.

[78] Nongonierma, Alice B., and Richard J. FitzGerald. "Susceptibility of milk protein-derived peptides to dipeptidyl peptidase IV (DPP-IV) hydrolysis." *Food chemistry* 145 (2014): 845-852.

[79] Lacroix, Isabelle ME, Xiu-Min Chen, David D. Kitts, and Eunice CY Li-Chan. "Investigation into the bioavailability of milk protein-derived peptides with dipeptidyl-peptidase IV inhibitory activity using Caco-2 cell monolayers." *Food & function* 8, no. 2 (2017): 701-709.

[80] Lafarga, Tomas, and Maria Hayes. "Bioactive protein hydrolysates in the functional food ingredient industry: Overcoming current challenges." *Food Reviews International* 33, no. 3 (2017): 217-246.

[81] Zhang, Fan, Xianwei Cui, Yanrong Fu, Jun Zhang, Yahui Zhou, Yazhou Sun, Xing Wang, Yun Li, Qianqi Liu, and Ting Chen. "Antimicrobial activity and mechanism of the human milk-sourced peptide Casein201." *Biochemical and biophysical research communications* 485, no. 3 (2017): 698-704.

[82] Adje, Estelle Yaba, Rafik Balti, Didier Lecouturier, Mostafa Kouach, Pascal Dhulster, Didier Guillochon, and Naïma Nedjar-Arroume. "Controlled enzymatic hydrolysis: a new strategy for the discovery of antimicrobial peptides." *Probiotics and antimicrobial proteins* 5, no. 3 (2013): 176-186.

[83] McClean, Stephen, Louise B. Beggs, and Robert W. Welch. "Antimicrobial activity of antihypertensive food-derived peptides

and selected alanine analogues." *Food chemistry* 146 (2014): 443-447.

[84] Tang, Wenting, Hui Zhang, Li Wang, Haifeng Qian, and Xiguang Qi. "Targeted separation of antibacterial peptide from protein hydrolysate of anchovy cooking wastewater by equilibrium dialysis." *Food chemistry* 168 (2015): 115-123.

[85] Aissaoui, Neyssene, Jean-Marc Chobert, Thomas Haertlé, M. Nejib Marzouki, and Ferid Abidi. "Purification and biochemical characterization of a neutral serine protease from *Trichoderma harzianum*. Use in antibacterial peptide production from a fish by-product hydrolysate." *Applied biochemistry and biotechnology* 182, no. 2 (2017): 831-845.

[86] Arias, Mauricio, Ashley L. Hilchie, Evan F. Haney, Jan GM Bolscher, M. Eric Hyndman, Robert EW Hancock, and Hans J. Vogel. "Anticancer activities of bovine and human lactoferricin-derived peptides." *Biochemistry and Cell Biology* 95, no. 1 (2017): 91-98.

[87] Zheng, Qiuping, Daoshou Qiu, Xiaojin Liu, Lei Zhang, Shike Cai, and Xuewu Zhang. "Antiproliferative effect of Dendrobium catenatum Lindley polypeptides against human liver, gastric and breast cancer cell lines." *Food & function* 6, no. 5 (2015): 1489-1495.

[88] Hung, Chuan-Chuan, Yu-Hsuan Yang, Pei-Feng Kuo, and Kuo-Chiang Hsu. "Protein hydrolysates from tuna cooking juice inhibit cell growth and induce apoptosis of human breast cancer cell line MCF-7." *Journal of Functional Foods* 11 (2014): 563-570.

[89] Fernández-Tomé, Samuel, Javier Sanchón, Isidra Recio, and Blanca Hernández-Ledesma. "Transepithelial transport of lunasin and derived peptides: Inhibitory effects on the gastrointestinal cancer cells viability." *Journal of Food Composition and Analysis* 68 (2018): 101-110.

[90] Xue, Zhaohui, Haichao Wen, Lijuan Zhai, Yanqing Yu, Yanni Li, Wancong Yu, Aiqing Cheng, Cen Wang, and Xiaohong Kou. "Antioxidant activity and anti-proliferative effect of a bioactive

peptide from chickpea (*Cicer arietinum* L.)." *Food Research International* 77 (2015): 75-81.

[91] Kayser, Holger, and Hans Meisel. "Stimulation of human peripheral blood lymphocytes by bioactive peptides derived from bovine milk proteins." *FEBS letters* 383, no. 1-2 (1996): 18-20.

[92] Berthou, Jean, D. Migliore-Samour, Alain Lifchitz, Jean Delettré, F. Floc'h, and Pierre Jollès. "Immunostimulating properties and three-dimensional structure of two tripeptides from human and cow caseins." *FEBS letters* 218, no. 1 (1987): 55-58.

[93] Chatterton, Dereck EW, Duc Ninh Nguyen, Stine Brandt Bering, and Per Torp Sangild. "Anti-inflammatory mechanisms of bioactive milk proteins in the intestine of newborns." *The international journal of biochemistry & cell biology* 45, no. 8 (2013): 1730-1747.

[94] Majumder, Kaustav, Subhadeep Chakrabarti, Sandra T. Davidge, and Jianping Wu. "Structure and activity study of egg protein ovotransferrin derived peptides (IRW and IQW) on endothelial inflammatory response and oxidative stress." *Journal of agricultural and food chemistry* 61, no. 9 (2013): 2120-2129.

[95] Huang, Wuyang, Subhadeep Chakrabarti, Kaustav Majumder, Yanyan Jiang, Sandra T. Davidge, and Jianping Wu. "Egg-derived peptide IRW inhibits TNF-α-induced inflammatory response and oxidative stress in endothelial cells." *Journal of agricultural and food chemistry* 58, no. 20 (2010): 10840-10846.

[96] Kovacs-Nolan, Jennifer, Hua Zhang, Masahisa Ibuki, Toshihiro Nakamori, Keiko Yoshiura, Patricia V. Turner, Toshiro Matsui, and Yoshinori Mine. "The PepT1-transportable soy tripeptide VPY reduces intestinal inflammation." *Biochimica et Biophysica Acta (BBA)-General Subjects* 1820, no. 11 (2012): 1753-1763.

[97] Young, Denise, Masahisa Ibuki, Toshihiro Nakamori, Ming Fan, and Yoshinori Mine. "Soy-derived di-and tripeptides alleviate colon and ileum inflammation in pigs with dextran sodium sulfate-induced colitis." *The Journal of nutrition* 142, no. 2 (2012): 363-368.

In: Nutraceuticals
Editors: A. Kumari and G. Singh
ISBN: 978-1-68507-488-3

Chapter 12

INDIAN REGULATIONS ON NUTRACEUTICALS

Meenakshi K. Chauhan* and Monika Targhotra
Department of Pharmaceutics, Delhi Institute of Pharmaceutical Sciences and Research (DPSR-University), New Delhi, India

ABSTRACT

Nutraceuticals described like any food or component of nutrition that delivers beneficial effects like disease prevention or care, now evolved as a requirement among buyers in both developed and emerging nations. Nutraceutical is a new industry in the pharmaceutical world and the growth of the business in the years ahead is massive. The Food Safety and Standard Authority of India [FSSAI] has been the only central authority governing the production of nutraceuticals, delivery, and marketing in India. The availability of a wide variety of foods and dietary products has intensified emphasis on the global health-related food industry. The section summarizes the present situation, potential growth of nutraceuticals in India, the regulatory structure for supplements, and the entire statutory registration process under FSSAI law.

* Corresponding Author's E-mail: meenakshikanwar@yahoo.com.

INTRODUCTION

Nutraceutical is a concept that identifies things that are not nutrients and are commonly accepted (for example, proteins, vitamins, botanicals), yet the biochemical impact on the human body is positive [1, 2]. Stephen Defelice has coined the expression in 1995 and defined it as: food or a part of food contributing medicinal benefits, like disease treatment/prevention. Nutraceuticals are primarily taken in the form of a unit dose, like capsules, pills, or fluids, and are usually recognized as food supplements. There has been substantial growth in the dietary supplements, nutritional, and nutritious food industry globally over the past decades [3]. The fitness and wellness industry is currently projected to be about US$ 1.6 billion and has reached US$ 7.5 to 10 billion in 2015, rising around 30% average growth rate. Historically, various rules and regulations in India have established varied norms, including food additives, toxins, color of food, additives, and branding. India recently introduced the Food Safety and Standard Act 2006, a deeply streamlined dietary regulation to serve only as a central point of focus for the enforcement of food items like food additives, nutritional supplements, and fortified foods. To conform with the international requirements of Europe and the United States, the Food Safety and Standard Act still needs to be significantly effective in terms of facilities and practical sustainability. It is predicted that the act will also have a significant effect on the Indian food products and nutraceutical sector, including the 1994 Dietary Supplements Health Education Act (DSHEA) worldwide.

The determination of the ideal conditions with the use of nutraceuticals must be compatible with data on efficacy and also details on biocompatibility and bioavailability to be able to introduce itself as an efficient tool for the treatment and prevention of certain chronic diseases in subjects not suitable for traditional pharmacological therapy.

With this factor, and because of their organic sources, there is an increasing market for dietary supplements, shading the distinction between pharmaceutical products and food, which also helps farmers expand their cultivation and encourage innovation and technology. Additionally, the key obstacles faced by nutraceuticals are numerous country-specific laws, protection, and safety claim supporting evidence. The primary barrier has been the lack of a standard national policy for nutraceuticals that would consider its existing and emerging function as clinical tool in certain disease states depending on the efficacy, established safety profile, proven scientific effectiveness for both minimizing the possibility of the onset of disease and significantly improving health [4].

Market Growth in India

FMCG (fast-moving consumer goods) producers and the drug industry seem to be the main stakeholders in the Indian nutraceutical sector. Around 64% of the total Indian market for nutritional supplements accounts for nutritional and vitamin supplements. The Indian market for nutraceutical products of $2 billion is subdivided into functional foods and beverages and dietary supplements.

i. Functional food and beverages: This category protect approximately 60% of the global market. Fortified foods are alluded to as the adding of micro-nutrients to the packaged food, like all reinforced beverages, juices, food products, and diet drinks, according to the WHO.
ii. Dietary supplements: This category represents the remaining 40% of the Indian nutraceutical sector. Mineral and vitamin supplements, nutritional supplements, and herbal supplements are included in this class.

The awareness of traditional nutraceutical additives by the local customers is minimal, and nutraceutical producers have to join the cause and inform people of Indians regarding the commodities. The worldwide demand for nutraceuticals sees a substantial increase in recent years. Drinks and food supplements are projected to experience a vividly faster growth rate in India relative to dietary supplements in the next 5 years. The

Indian market for dietary supplements, which increased from in 2008 around $1 billion to 2013 around $1.8 billion, and by 2014 crossed $2 billion also is expected to exceed $4 billion by 2018.

The Indian demand for nutraceuticals is forecast to expand from $4 billion in 2015 to $10 billion in 2022. The estimated demand for practical food and drinks is around $1.5 billion. Approximately 22% CAGR demand for functional foods and beverages is projected to hit $4.11 billion by the year 2022. The dietary supplement market is forecast to hit 19% of CAGR by the year 2022 at $5.90 bn. The Indian demand for nutraceuticals reached only about 10% at all levels of India. The prevalence rate is high at 22.51% for urban areas, where it is just 6.32% in remote areas of India. India accounts for only 2% of the world nutraceuticals industry and is well far behind in terms of total expenditure on nutraceuticals at only US$ 2.5 relative to the world average of around US$ 21. The Indian market for nutraceuticals is especially pronounced throughout South India, led via East India, with West Bengal, Tamil Nadu, and Andhra Pradesh leading the three top states. With the recent invasion of lifestyle disorders in rural areas, the urban orientated Indian demand for nutraceuticals is steadily making inroads in south villages. Almost one-third of the sector is occupied by rural India [5].

Laws Regulating Nutraceutical - Food Safety and Standard Act 2006

It combines the core elements of the 1954 Food Adulteration Prevention Act and is protected by international standards, recommendations, and the Codex Alimentarius Committee [6]. The core elements of the law as a whole are:

a) This act serves as a carrier for eight laws regulating the food industry and establishes Sector Regulatory Body for Food Safety and Standards Authority (FSSA) and other associated bodies. Requirements or products, toxins, pesticide contamination, environmental risks, tags, and many others may be used in these specifications.

b) In the food processing industry, it is required that everybody can receive a license or certification issued by local entities.
c) The bill shall be enforced by federal food safety commissioners and municipal food safety commissioners' officials.
d) The clause ensures for a system with a structural fine in which the sentence relies on the seriousness of the breach.
e) Framing and controlling requirements for nutraceuticals are the responsibility of the Food Safety and Standards Authority of India (FSSAI) as defined in the law on Food Safety 2006.

Advantages of the Act's implementation

This act includes certain advantages as mentioned below:

- Coexistence of eight rules, i.e., harmonized steps
- Integration worldwide of regulations
- Science-based criteria
- Description and uniformity in new emerging food sectors
- Help to overcome misconduct

Complications of Law enforcement

Regulation broadly defines the intervention of government in industry. And in the context of food, regulations mainly regulate the consistency of the material. Each legal framework has had its benefits and drawbacks. However, the advantages of enforcing the Food Safety and Standards Act have overshadowed the obstacles that occur due to enforcing this act. Except for across the US, in which the DSHEA is always in force to control these commodities, the government is working on drafting a bill in India to regulate the production, distribution, and distribution of food supplements, herbal supplements as well as other nutraceuticals [7].

Regulation- Food Regulations History in India

India is the leading vegetable and fruit market globally, but a limited proportion of consumable farm goods are managed at about 2%, compared to 80% as in the United States [7]. Weak infrastructure and logistical and rigid food safety regulations are obstacles to development in the food market. The Arbitrary Committee on Food and Agriculture Sectors was created in 1998 by the Executive Council on Enterprise and Market by the Prime Minister, which introduced integrated legislation under the Single Food Regulatory Authority. In the year 2002, the key goal of a national non-profit organization was to ensure that any food processing industry could provide its products with research assistance to preserve customers and encourage and maintain a legislative climate favourable to the sector. In 2003, an advisory group study from the Ministry of Health suggested a need to build current dietary laws and additional frameworks for governing dietary supplements and nutritional food. It is suggested that necessary safety monitoring for these items should be done. The Bureau of Indian Standards (BIS) and the National Standards Committee, comprising representatives from various core food industry stakeholders, have established ethical guidelines in India. Eventually, passed into law, the Indian Food Safety Standard Bill 2005 pledged a significant influence on the industry of Indian food processing. The Indian Food Safety and Standard Act with key aims was enforced in 2006 to - implement a single food-related law to allow for the technological advancement of the food manufacturing industry [8].

Requirements for License and Registration

- Registration must be obtained by any food owner in the nation and enforcement in compliance with the FSSAI protocol (licensing food company and registration) as per Regulation 2011.
- A producer may not start the business until he is enrolled or consists of a valid license.

- Previous licenses/approvals must be translated before August 5, 2012 (presently prolonged by a few months) into FSSAI licenses/enrolments.
- The request to issue the license must be sent to the licensing authority as per Schedule 2 in Form B; the license shall be granted within 60 days of the date of approval of the ID number of the request.
- Upon issue of the request ID number, the licensing body may inform the food safety agent for the examination of the property in compliance with such regulations as specified by FSSAI.
- The regulatory authority must grant a license in compliance with Schedule 2 in Format C.
- The certification or license issued in compliance with these regulations must be precise and legal, subsisting for a total of 1-5 years, except as stated [8].

Regulatory Entry Requirements in India

Nutraceuticals could be manufactured and marketed in form of capsules, tablets, and syrups. This should achieve the USP, IP, and BP quality specifications and standards. The composition must, wherever needed, be based on good therapeutic or dietary concepts backed by established scientific evidence. To be processed into capsules, tablets, or syrup, it should not be just a mixture of minerals and vitamins; this is not a product until minerals and vitamins are applied to a food component depending on the product type. Steroids, psychotropic additives, or hormones are not allowed to be included in the preparation. Any requirement must not be surpassed by the number of nutrients recommended for daily consumption (RDA) by the Indian Council of Medical Research (ICMR). When those standards are not defined, then standards will be included in the Codex Alimentarius Commission. Usage of licensed additives and colouring agents, natural, similar, or artificial flavours as allowed in Schedule VF of such rules, as permitted in the Food

Safety and Standards (Food Product Standards and Food Additives) Regulations, 2011 [5].

To access the Indian nutraceutical industry, the focus should be on the fields that involve product assessment, real product review, sourcing, product evaluation, permits, and India-specific health and claims for production.

Evaluation of Products

The 2011 Food Safety and Standards Rules demonstrate the framework and processes which the central government intends to establish for regulatory compliance. The system includes a structural hierarchy from, food security officers to a number of officers and demarcations who may be participating in the substance testing process at all times, like product safety officers, food inspectors, etc.

Various measures in the study of the product involve [8]:

a) Construction and authentication of samples of records by the concerned body
b) Collecting samples
c) Dispatch of a sample by the authorized body
d) Food testing and analysis
e) If the review is not done within a defined amount of time, the assigned officer will take an additional plan of action
f) Proceedings in Adjudication

Licenses

Around 4-5 licenses to have the commodity licensed in India may be needed, based on the exact state of the commodity, like:

- The business needs to market bulk medications or complete formulations
- The business imports bulk ingredient products.
- The company has a license for packing,
- A manufacturing license is required.

- This requires a marketing license.
- The number of records to be supplied by the food importer including registration dossiers to the government agency.

Health and Label Claims

The critical element to be concentrated on while accessing the Indian market is to establish health and label statements relevant to Indian government regulations. Foreign as well as national customers have a range of concerns regarding the specifications of Indian packaging and labelling, consignment packing, the need for sample material and registration declaration, the structure of the shipment and its approach, the quality of the mark, the layout claim and the product claim [8].

Registration Process in India

For the registration of license following steps need to be taken-

a) Submit under Form A to the State Licensing Authority for the registration of the site.
b) The request shall be dealt with for a limited time of seven days by the body.
c) An inspection shall take place in 30 days and then the site acceptance is done and if not suitable, denial of the proposal shall operate concurrently.
d) Manufacture must request for a manufacturing license in form B and then, within 60 days, a specific application number has been issued.
e) Queries shall be communicated within 60 days to the application on the incomplete application.
f) In 60 days, after getting the incomplete data inspection will be conducted on the premises.
g) After 30 days' authorization or denying of a license is done.

Conclusion

In summary, the introduction of the Food Safety and Standard Act 2006 had been an important step, but further needs to be done to remove the duplication of existing rules and regulations. There were several rules and legislation regulating food safety and standards after FSSA. Later, it was important to preserve in 2006 the unified and scientific growth of the food manufacturing industry, all current legislation was combined to create a new regulation. The Food Safety and Standards Regulations of 2011 confirmed in the Indian Gazette entered into power on 5 August 2011 to govern the storage, delivery, and allocation of dietary supplements, dietary food products, and supplements in India.

References

[1] Kalra, E. K. Nutraceutical: Definition and Introduction (2003). *AAPS Pharmaceutical Scientists*, 5:27-28. https://dx.doi.org/10.1208/ps050325.

[2] Hardy, G. Nutraceutical and Functional foods: Introduction and Meaning. *(2001) Nutrition,* 16(7-8):688-89. http://doi.org/10.1016/s0899-9007(00)00332-4.

[3] Brower, V. Nutraceuticals: poised for healthy slice of healthcare market? (1998) *Nature Biotechnology*, 16(8):728-73. http://doi.org/10.1038/nbt0898-728.

[4] Borchers, A. T., Keen, C. L., Gerswin, M. E. The basis of structure/function claims of nutraceuticals. *(2016) Clinical Reviews Allergy and Immunology,* 51(3):370–82. http://doi.org/10.1007/s12016-016-8536-9.

[5] Indian Nutraceuticals Industry: Current Scenario and Future Trends (2017) *Assocham India*. http://www.mrssindia.in/uploads/reports/pdf/assocham-knowledge-report-on-nutracuticals-released-at-3rd-national-symposium-6.pdf. [cited 15 Nov 2020].

[6] *Food Safety and Standard Act* 2006. https://fssai.gov.in/cms/ food-safety-and-standards-act-2006.php.

[7] Food Safety and Standards Bill, 2005: *Legislative Brief.* http://www.indiatogether.org/foodsafe-laws. [cited 10 Nov 2020].

[8] Keservani, R. K., Sharma, A. K., Ahmad, F., Baig, M. Nutraceutical and Functional Food Regulations in India. (2014) Nutraceutical and Functional Food Regulations in the United States and around the World (Second Edition), *Food Science and Technology,* Ch 19, pp 327-42. http://doi.org/10.1016/B978-0-12-405870-5.00019-0.

In: Nutraceuticals ISBN: 978-1-68507-488-3
Editors: A. Kumari and G. Singh

Chapter 13

ROLE OF SPICES AND HERBS IN FUNCTIONAL FOODS

Muthupalani Muthukumar[1]*, Basappa M. Naveena, Rituparna Banerjee and Sanganabhatla Divya
ICAR- National Research Centre on Meat,
Hyderabad, Telangana, India

ABSTRACT

Spices and herbs have been utilized for a very long time in traditional medicine. They possess several health-beneficial properties including antimicrobial, antioxidant, anti-inflammatory, antidiabetic, anticarcinogenic, carminative, spasmolytic, hypolipidemic, hepato-protective, etc. Though herbs and spices have been used since antiquity to provide unique sensory properties to foods and beverages, their potential to be utilized as natural preservatives in processed foods has gained importance in the recent past. Spices and herbs exert their antioxidant activity through free radical scavenging, decomposing peroxides, and quenching singlet and triplet oxygen. Several studies also confirmed their antimicrobial properties against antibiotic-resistant

* Corresponding Author's E-mail: muthukumar55@rediffmail.com.

bacteria responsible for various infectious diseases in humans. The presence of several bioactive compounds viz., alkaloids, flavonoids, phenolic acids, sulfur-containing compounds, tannins, terpenes, and anthocyanins increased their potential to be used as a natural alternative to synthetic preservatives. Therefore, fortifying foods with spices and herbs could improve the nutritional, therapeutic, and functional attributes of food products. Several *in-vivo* and *in-vitro* experiments have proved the efficacy of herbs and spices to be used as functional foods or nutraceuticals. This chapter summarizes an overview of the classification, chemical composition, and functional properties of herbs and spices, their potential as functional foods and impacts on human health, issues in the use of spices and herbs, and suggestions for future research.

1. Introduction

Food is viewed as a means to improve health and well-being. Functional foods are developed to positively impact human health besides providing the essential nutritional requirements for good health. Increased health consciousness, the growing cost of health care, longevity, and longing for a better quality of life have fueled the surge in demand for functional foods [1]. Numerous bioactive compounds inherently biosynthesized by plants can promote health and combat the risk of degenerative diseases in humans. The undesirable oxidative damage induced by reactive oxygen species (ROS) could be fought either through supplementation of exogenous antioxidants or enhancing the defense of endogenous antioxidants present in our body. Though the health-enhancing properties of plants have been known for thousands of years, the frequent ineffectiveness of conventional medicine has contributed to the renaissance of phytotherapy [2].

Herbs and spices used in culinary preparation often help in enhancing certain physiological functions and preventing risk factors associated with a variety of diseases in the consumer, in addition to imparting specific aroma, taste, piquancy, and colour to food [3]. Generally, the fresh green leafy part of a cooking plant is referred as a culinary herb. Other parts of the plant like tree bark (cinnamon), bud (cloves), root (ginger), fruit/berry (pepper), seeds (cardamom), and even the flower stigma (saffron), often dried as a spice. Turmeric, garlic, ginger, clove, black pepper, cinnamon, oregano, basil, thyme, and rosemary are candidates extensively employed in food preparation.

Concerns about the possible health problems and toxicity associated with the use of synthetic food additives are growing over time. To overcome this, natural plant-based compounds are being employed in the food industry. Using natural phytochemicals as preservatives could be an alternative approach to produce food products with high nutritional content, clean label, and synthetic chemical additives. Phytochemicals are classified as "Generally recognised as safe" (GRAS) food additives for human consumption by the Food and Drug Administration. Most of them have a wide range of health-promoting properties and offer more significant advantages over their artificial counterparts [4]. In addition to enhancing the organoleptic quality of foods, these compounds can slowdown the autoxidation of unsaturated triacylglycerols and the growth of pathogenic microbes [5]. With increasing consumer demand, the spice trade has been growing globally. The global value of the spice market was about US $11.5 billion in 2019 and is likely to increase to about US $14.5 billion by the year 2025.

2. Chemical Composition

Spices and herbs used in food preparations are not only rich in nutrients, especially minerals and fiber, but they also have substantial quantities of bioactive phytochemicals of various kinds. These phytochemicals constitute a broader group of bioactive compounds viz., alkaloids, flavonoids, carotenoids, lignans, phenolics, saponins, sterols, glucosinolates, sulfur compounds, etc. Table 1 depicts the major chemical constituents present in commonly used spices and herbs.

Table 1. Chemical composition of culinary spices and herbs

Spices and herbs	Important chemical constituents
Basil	Anthocyanins, apigenin, α-terpinene, β-sitosterol, carvacrol, catechin, cinnamate, citronellol, 1,8-cineole eugenol, p-coumaric acid, farnesol, geraniol, limonene, menthol, methyl kaempferol, quercetin, rosmarinic acid, rutin, safrole, tannin and ursolic acid.
Bay leaf	1,8-cineole and cinnamtannin.
Black pepper	α-pinene, β-pinene, camphene, limonene, guineesine, isoquercetin, pinene, piperidine, piperine, terpenes, sarmentine and sarmentine.
Cardamon	Caffeic acid, 1,8-cineole, kaempferol, limonene, luteolin, myrcene, pelargonidin, terpinolene and quercetin.
Cinnamon	Benzaldehyde, catechins, cinnamic aldehyde, cinnamate, eugenol, 2-hydroxycinnamaldehyde limonene, linalool, methyleugenol, myristicin, pinene, proanthocyanidins, safrole, tannins and terpineol.
Clove	Acetyleugenol, flavonoids, gallicacid, isoeugenol, phenolic acids, pinene, rugenol, sesquiterpene and vanillin
Coriander	Borneol, caffeic, cumene, n-coumaric acids, ferulic, geraniol, kaempferol, linalool, pinene, pyrogallol quercetin, rutin, terpineol, terpinene, tocopherols, and vanillicacids.
Fenugreek	Aromatic aldehydes, sesquiterpenes and terpenes.
Garlic	Allicin, diallyl sulfide, allyl isothiocyanate, S-allyl cysteine, diallyl disulfide and diallyl trisulfide.
Ginger	Borneol, camphene, curcumin, gingerol, geraniol, geranial, linalool, paradol, shogaols, turmeric, zingerol, zingiberon, zingiberene, zingerone and zerumbone.
Mustard	Allyl isothiocyanate, carotene, isorhamnetin, isorhamnetin-7-O-glucoside and kaempferol glucoside.
Nutmeg	Argenteane, caffeic acid, catechins, lignans, myricetin and orgentin.
Oregano	Apigenin, carvacrol, P-coumaric acid, diosmetin, eriodictyol, luteolin, myricetin, protocatechuic acid, quercetin, rosmarinic acid and thymol.
Red pepper	Ascorbic acid, capsanthin, capsaicin, carotene, lutein, quercetin and tocopherol.

Spices and herbs	Important chemical constituents
Rosemary	Apigenin, α-pinene, β-carotene, caffeic acids, carnosol, carnosic acid, 1,8-cineole, diosmetin, geraniol, limonene, luteolin, naringin, pinene, rosmarinic acid, rosmanol, ursolic acid and vanillic acid.
Sage	Apigenin, α-pinene, β-sitosterol, β-carotene, β-pinene, carnosol, caffeic acids, carnosol, carnosine, catechins, citral,1,8-cineole, farnesol, ferulic acid, gallic acid, geraniol, limonene, luteolin, pinene, perillyl alcohol, rosmarinic acid, saponin, ursolic acid and vanillic acid.
Thyme	Apigenin, α-pinene, β-carotene, caffeic acid, carnosic acid, carvacrol, 1,8-cineole, cismaritin, diosmetin, gallic acid, hispidulin, hesperidin, kaempferol, limonene, luteolin, naringenin, quercetin, rosmarinic acid, thymol and ursolic acid.
Turmeric	Curcumins, curcuminoids, β-turmerin, essential oils, eugenol, carotene, ascorbic acid, caffeic, p-coumaric, protocatechuic, syringic and vanillic acid.

2.1. Polyphenolic Compounds

Phenolic compounds are composed of a hydroxyl group (—OH) bound to an aromatic hydrocarbon group. More significant variation exists in molecular weight as well as the structure among phenols. Most polyphenols available as glycosides have diverse sugar units and acylated sugars at different polyphenol skeletons. The ability to scavenge a wide range of reactive oxygen spices (ROS) and chelating metal ions is primarily determined by the presence of hydroxyl groups in phenolic compounds. Polyphenols exert their antioxidant capacity either through retarding enzymes involved in the formation of ROS or scavenging them. They also exhibit their diverse antimicrobial mechanism through the synergistic activity via reducing the efflux pumps' function and working as efflux pumps inhibitors strand. Phenolic compounds are also associated with some of the vital enzymes like β Ketoacyl acyl carrier protein synthase (KAS) II and III, which are accountable for forming precursors of bacterial cell membranes. Further, phenolics also engaged with FabG, FabI, and FabZ, which are involved in the fatty acid biosynthesis elongation cycle. Curcumin is one such potential compound that has diverse mechanisms of action [6].

2.1.1. Phenolic Acids

These are important classes of plant phenolic substances having one carboxylic acid group. Many of the phenolic acids viz., amides, esters, and glycosides exist in bound form, and very few of them are in free form. Phenolic acids are classified into hydroxybenzoic and hydroxycinnamic acid based on their C_1-C_6 and C_3-C_6 backbones. Phenolic acids possess antioxidant, antimicrobial, and anti-inflammatory properties. Hydroxycinnamic acids and hydroxybenzoic acids are derived from cinnamic acid and benzoic acid, respectively. Hydroxycinnamic acids are present in foods primarily as simple esters with quinic acid or glucose. The four most common hydroxycinnamic acids are caffeic, ferulic, p-coumaric, and sinapic acids. Hydroxybenzoic acids have a standard structure of C_6-C_1 and are present in soluble form (conjugated with sugars or organic acids). They are attached with cell wall fractions as lignin. The more predominant hydroxybenzoic acids include p-hydroxybenzoic, protocatechuic, vanillic, and syringic acids. Phenolic acids exert their inhibitory effect on microbial growth by disrupting the cytoplasmic membrane, altering their permeability, inhibiting extracellular microbial enzymes, directly affecting microbial metabolism, and depriving the substrates needed for the growth of microbes [7].

2.1.2. Flavonoids

These phenolic substances have a hydroxyl group synthesized by plants in reaction to microbial infection. Almost all flavonoids are synthesized from phenylalanine, tyrosine, and aromatic amino acids. Flavonoids are C15 compounds with the general structural backbone of C6-C3-C6. They are a broad group of polyphenols with a benzo-γ-pyrone structure. Flavonoids are divided into six main subclasses: flavones, isoflavones, flavanols, flavanones, catechins, and anthocyanidins. The structure determines the chemical properties of flavonoids, degree of hydroxylation, degree of polymerization, other substitutions, and conjugations. The antioxidant effects of functional hydroxyl groups are mediated either by free radicals scavenging or metal ions chelation. Metal ions chelation prevents damage to the target's biomolecules. The therapeutic efficacy of flavonoids against several bacterial and viral infections, degenerative diseases (cardiovascular, cancers), and senile-associated ailments have been reported [8].

2.1.3. Other Polyphenols

2.1.3.1. Curcumin

It is a well-known antimicrobial compound obtained from turmeric. The antibacterial activity of curcumin is due to its ability to damage bacterial cell membranes. The amphipathic and lipophilic chemical structure enables curcumin to enter into membrane bilayer and improve the permeability of the bilayer membrane [9].

2.1.3.2. Lignans

These are fibre associated phytoestrogens compounds found in plants. These polyphenolic compounds are derived from phenylalanine and include pinoresinol, lericiresinol, secoisolariciresinol, syringaresinol, and sesamin. Plant lignans are converted into various metabolites like enterolignans, enterodiol, and enterolactone by a group of intestinal bacteria. Due to their tissue-specific estrogen receptor activation, anti-inflammatory, and apoptotic effects, enterolignans reduce the risk of disease in humans.

2.1.3.3. Tannins

Tannins are more efficient free radical scavengers and delay tissue oxidation more effectively than β-carotene, vitamin C, and vitamin E. Tannins ably neutralise the hydroxyl free radical (•OH) and also noncompetitively inhibit xanthine oxidase, the enzyme involved in the generation of significant amount of free radicals in cellular metabolism. Tannins are used effectively as antiseptic, antibacterial, and antifungal due to their capacity to form complexes with macromolecules, specifically with proteins [10].

2.2. Carotenoids

These compounds are predominantly antioxidants and efficient scavengers of singlet molecular oxygen and peroxyl radicals. In biological systems, several factors affect the antioxidant activity of dietary carotenoids. The antioxidant potential of carotenoids could be enhanced by using them along with other synergistic antioxidants and also using mixtures of carotenoids rather than single compounds [11].

2.3. Organosulfur Compounds

S-alk(en)yl-l-cysteine sulfoxides are usually present in the plants of the *Allium* genus (family *Amaryllidaceae*), such as garlic, onion. In contrast, S-methyl cysteine-l-sulfoxide is found in plants belonging to the *Brassica* genus or cruciferous families (*Brassicaceae*) like cabbage, cauliflower, and mustard. Organosulfur compounds viz., ajoene, allicin, and isothiocyanates can display antibacterial, antifungal and antiviral, activities.

2.3.1. Allicin

Allicin/diallylthiosulfinate, the compound responsible for the pungent smell of garlic (*Allium sativum*), also possesses strong bactericidal activity against both Gram-negative and Gram-positive bacteria. Allicin exerts its antimicrobial activity through inhibiting sulfhydryl-dependent enzymes such as RNA polymerase, alcohol dehydrogenase, and thioredoxin reductase. It targets intracellular enzymes by interacting with their free SH groups and also plays a role in partially inhibiting the DNA and protein synthesis. Allicin could exert its bactericidal activity against several bacteria with LD_{50} values ranging from 3 to >100 μg/mL. However, the bacteriostatic effect requires lesser concentrations (0.15–1.5 μg/mL). Allicin inhibits RNA synthesis by reducing or hindering cell protection mechanisms due to which it is potentially used in bacteriostasis [12].

2.3.2. Ajoene

Ajoene, one of the significant constituents of garlic, is also an organosulfur compound. The E and Z ajoene are the two main stereoisomers. Ajoene has proven broad-spectrum antimicrobial activity against Gram-negative, Gram-positive bacteria, fungi, and protozoa [13].

2.3.3. Isothiocyanates (ITCs)

Isothiocyanates are volatile organosulfur phytochemicals synthesized by the plant upon damage. They are produced by enzymatic cleavage of released glucosinolates from intracellular compartments by membrane-bound myrosinase. Myrosinase helps hydrolysis and intramolecular rearrangement of intermediates, leading to the synthesis of nitriles, thiocyanates, and isothiocyanates. Plants belonging to the Brassica family viz., broccoli, cabbage, cauliflower, and mustard are significant sources of these compounds. Among these compounds, isothiocyanates are considered potent antimicrobial candidates as they are more effective inhibitors of variety of pathogenic bacteria. Allyl isothiocyanate (AITC), both in vapour and liquid form, is effective against a wide range of food spoilage as well as pathogenic microbes. They exert their action by disrupting the cell wall integrity, which results in the leakage of cellular metabolites. Exposure to AITC causes the formation of holes on the bacterial cell membranes and leads to intracellular substances leakage. AITC impedes O_2 uptake of yeasts and uncouples the oxidative phosphorylation by inhibiting cytochrome c oxidase in the electron transport chain.

2.4. Terpenes

Terpenes are hydrocarbons formed through the amalgamation of several isoprene units. These are the modified forms of terpenes where methyl groups are changed or removed or added with oxygen atoms. Carvacrol, linalool, menthol, and thymol are efficient terpenoids with potent antimicrobial activity against pathogenic microbes. Generally, terpenes are more effective against Gram-positive bacteria than Gram-negative ones. The lipophilic properties have a more significant impact on antimicrobial mechanism of terpenes. By enhancing membrane fluidity and permeability, changing the topology of its proteins and creating disruptions across the respiration chain, monoterpenes impact the membrane structures.

2.4.1. Carvacrol

It is a phenolic monoterpenoid primary found in aromatic plants such as oregano and thyme. Carvacrol exerts its antimicrobial action by causing damages to the structural and functional properties of the cell membrane. The damage to the outer membrane of bacteria leads to release of lipopolysaccharides.

2.4.2. Thymol

Like carvacrol, thyamol has a hydroxyl group at various positions of the phenolic ring. Thymol is a potent antifungal compound and its antibacterial activities involve disruption of the inner and outer cell membrane, interaction with membrane proteins and intracellular targets, which leads to membrane potential loss, cellular uptake of ethidium bromide, and leakage of carboxyfluorescein, ATP and potassium ions. Food industries could use thymol, carvacrol, eugenol, and menthol as substitutes for synthetic chemical fungicides. These phytochemicals are beneficial to a greater extent against several food spoiling fungi [15].

2.4.3. Eugenol

Clove essential oil is one of the vital sources of Eugenol. Because of its non-specific permeabilization, eugenol is capable of permeating the cell membrane and interacting with proteins. Eugenol exerts its antimicrobial action on membranes primarily by inhibiting biofilm formation, interrupting communication between cells, eradicating the pre-formed biofilms, and killing the microbes in biofilms [16].

2.5. Alkaloids

Alkaloids are one of the largest groups of secondary plant metabolites. This group includes neuroactive compounds like caffeine and nicotine, life-saving drugs such as emetine used for oral poisoning, as well as the antitumoral molecules viz. vinblastine and vincristine. Alkaloids are known to be toxic against pathogens and predators, making them defense compounds in plants. Quantity, duration of exposure, the sensitivity of the individual, site of action, and developmental stage determine the severity of toxic effects. Further, depending on the ecological or pharmacological context, the toxicity effects could be harmful and beneficial.

3. Health Benefits

Recent developments in biomedical have proved that the reactive oxygen species (ROS) -superoxide (O_2● -), hydrogen peroxide (H_2O_2), and nitric oxide (NO●), which cause oxidative damage in living tissues, are the major contributors to most of the old age-related illness such as CVDs, cancer, cataracts, impaired immune system and degenerative nervous diseases like Alzheimer's and Parkinson's disease. The reaction of free radicals with DNA of the plasmid, dividing the double helix and causing oxidative damage, leads to carcinogenesis, degenerative nerve diseases, and abnormalities in the chromosomes. The innate defences of the consumer need to be strengthened through exogenous antioxidant food supplementation. The polyphenols, in addition to functioning as free radical scavengers, also modulate the cellular signaling processes during inflammation and serve as signaling age [17]. The balance between the antioxidant defence mechanism and the generated reactive oxygen species is required for normal cellular functions. The inclusion of plant antioxidants in the diet can support the human body in neutralising free radicals generated and reducing the damage due to oxidative stress. Many antioxidant phytochemicals counter ROS directly or restore antioxidants through enhancing the regenerative systems. The therapeutic potential of bioactive compounds found in herbs and spices in inhibiting allergies, preventing cancer and heart diseases by suppressing cell growth, viral replication, and abrupting several other pathological conditions has been demonstrated through extensive *in-vitro* as well as *in-vivo* models.

In addition, to improve the organoleptic properties of foods, culinary herbs and spices could also be used as functional ingredients in various food products.

Table 2. Health benefits of commonly used culinary spices and herbs

Spices	Health benefits
Bay leaf	Bay leaf oil possesses several beneficial effects against fungal and bacterial infections
Cardamom	Effective in controlling digestive disorders and helps in coping with diabetes
Chili	Helps in maintaining cholesterol level and also helps burning calories
Cinnamon	Helps in reducing blood glucose level in type 2 diabetic patients by supporting the natural production of insulin. It has anti-blood clot properties. It also lowers triglycerides, LDL, and total cholesterol.
Clove	Excellent remedy in the problems associated with tooth ache and sore gums. Clove is also effective in treating digestive problems, fever, cough, and cold.
Coriander	Seeds are suitable to treat sore throat, hay fever, allergies, and digestion problems
Cumin	Cumin contains high amount of B - complex and is also rich in various minerals like iron, copper, calcium, potassium, and zinc. It also has many antiseptic properties and helps to keep the immune system healthy.
Curry leaves	Beneficial for reducing blood sugar levels
Fenugreek	Helpful in treating diabetes and lowering cholesterol levels in the body
Garlic	Garlic has antibiotic properties. It helps in coping with cough and cold
Ginger	Excellent digestive aid. It helps in the absorption of food and the elimination of various gastric disorders.
Mustard	Excellent source of minerals viz., calcium, iron, manganese, and zinc and also consists of omega-3 fatty acids
Nutmeg	Helpful in treatments of heart disorder and asthma
Pepper	Beneficial for digestive problems, muscle pains, and especially with cold, cough, various infections, etc.
Saffron	Helpful in the treatment of skin and respiratory system disorders
Star anise	Improves digestion and is beneficial for treating rheumatism
Turmeric	Good antiseptic aid, a powerful antioxidant, anti-inflammatory and helps in coping with diabetes easier

Due to the presence of biologically active compounds, many of the plants have been demonstrated to possess broad spectrum of health-promoting properties such as antioxidative, antiallergenic, anti-inflammatory, antidiabetic, spasmolytic, hepatoprotective, antimutagenic, anticarcinogenic, and antimicrobial activities. They also function as enhancers of numerous body functions in humans such as gastrointestinal activity, immune-modulators, boosters, and reducing agents for blood pressure and cholesterol [18].

Consumption of antioxidant-enriched foods protects the human body from oxidation caused by free radicals, superoxide, other oxygen radicals, and oxidation-causing substances. Several studies have established the role of spices in promoting health benefits to the consumer.

Fenugreek is one such spice found to reduce blood cholesterol and glucose levels in both laboratory diabetic animals and humans. As food products serve as unique carrier to supply bioactive compounds and other nutraceuticals for promoting consumer health and wellbeing, designating them as functional foods [19].

In recent years, multiple approaches have been contemplated to overcome the antimicrobial resistance. Combining other molecules/compounds with the failing antibiotics could be one of the strategies to restore the desirable antibacterial activity. Phytochemicals having solid antimicrobial properties are increasingly employed to overcome microbial resistance. Lately, different methodologies have been recommended to defeat the obstruction of anti-toxins. One of the prescribed procedures to accomplish this objective has included the blend of different atoms with weak antimicrobials, which re-establishes the alluring antibacterial movement.

4. Prevention of Formation of Harmful Substances

Spices and herbs, in addition to enhancing the safety and shelf life of food products by effectively retorting lipid oxidation and microbial growth, also play a significant role in decreasing the formation of toxic compounds like heterocyclic amines (HCAs). Cooking meat at a high temperature or for a long duration leads to the formation of these mutagenic compounds from the reaction of creatine or creatinine, amino acids, and sugar. The 2-amino- 3, 8-dimethylimidazo (4,5-f) quinoxaline (MeIQx), and 2-amino-1-methyl-6-phenylimidazo (4,5 b) pyridine (PhIP) are the most predominant heterocyclic amines linked with red

. Maillard reaction between carbohydrates and amino acids produces heterocyclic amines with N-methyl 1-2-aminoimidazole moiety in the meat matrix by producing intermediate Schiff bases and imine form aldol. This imine undergoes heat catalysis to produce dialkyl-pyrazine radical or by condensation of glyoxal, and another imine generates pyridine radical. In the meat, the free radicals produced react with creatinine, and a particular HCA compound is generated depending upon the type of aldehyde added.

The antioxidants present in the spices could reduce HCA production by inactivating free radicals of pyrazinium and pyridinium generated as intermediates during the Maillard reaction. Several studies reported that rosemary, sage, thyme, garlic, onion, red chili, black pepper and turmeric could reduce 50 to 100% concentration of various HCA compounds. The ability of these spices to inhibit HCAs is related to the amount of polyphenolic antioxidant compounds present in them. Black pepper (1% w/w) inhibited (12-100%) the formation of HCA (IQ, MeIQ, MeIQx, 4,8-DiMeIQx, PhIP) when used as a surface marinade for 12 hours before frying high-fat meat balls [20]. The concentration of heterocyclic amines (HCA) was found to be reduced in grilled beef with the addition of spices such as turmeric (4 g/100 g), torch ginger, lemongrass, and curry leaves (10 g/100 g) [21].

5. Role of Spices and Herbs in Food Preservation

High moisture and high-fat food products are more prone to both microbial and oxidative spoilage, and therefore it is desirable to use preservatives with antioxidant and antimicrobial properties. As the public becomes more suspicious of artificial food additives, a common requirement is the availability of bioactive phytochemicals with preservative properties to protect the food from oxidation by enzymes and spoilage by microbes. An ideal replacement for toxic artificial food additives could be the use of potent antioxidant and antimicrobial extracts from various plants. Among the several functional properties of plants, antioxidant and antimicrobial activities have been widely exploited for protection and enhancing the shelf life of foods. Bioactive substances such as polyphenols, flavonoids, alkaloids, quinines, polypeptides, or oxygen substituted derivatives present in spices and herbs exert antioxidant, antibacterial and antifungal activities. The incorporation of bioactive antioxidant compounds extracted from spices and herbs retards the oxidative rancidity in different meat products, dairy products, and oils and improves their shelf life.

Similarly, the growth of pathogenic bacteria and fungi was inhibited by the antimicrobial phenolic compounds extracted from herbs and spices. Examples of these compounds are coumaric acid, oleuropein, tea catechins, ellagic acid, and ferulic acid. Hence, spices rich in bioactive compounds could be considered as an alternative to the synthetic antioxidant and antimicrobial chemicals employed in safe food production.

5.1. Natural Antioxidants from Herbs and Spices

Virtually all living organisms and biological systems, including food, undergo autoxidation processes, lipid peroxidation, and other oxidation forms, similar to the human body—the reaction of fats and oils or lipid contained in foods with atmospheric oxygen results in oxidative rancidity. The reaction products viz., aldehydes, ketones, acids, and alcohols formed from oxidative rancidity impart undesirable off-odors, off-flavours, and undesirable changes in colour and texture in many fat-rich foods [22]. Thus, lipid oxidation renders the foods rancid and makes them unpalatable and unacceptable to consumers. In addition to the loss in nutritional value resulting from the deterioration of fat-soluble vitamins and essential fatty acids, oxidative degradation of fat also leads to the production of harmful substances such as lipid peroxide and malonaldehyde. These toxic compounds are found to induce mutagenesis and carcinogenesis in consumers. Hence, the utilisation of antioxidants has become an imperative necessity to protect fat-rich food products from this type of chemical deterioration.

Dietary antioxidants incorporated into foodstuffs exert their function similar to the endogenous antioxidants of the human body in protecting the foods from oxidative deterioration and conserving their sensory and texture properties, and keeping them safe for consumption. The photo-antioxidant compounds derived from herbs and spices delay or retard oxidative rancidity mainly through their ability to absorb and neutralise free radicals such as hydroxyl, peroxyl, and superoxide radicals, quench singlet and triplet oxygen or decompose peroxides and form complexes with metal ions, that catalyse the oxidation reactions. The shelf life of foods could also be enhanced by employing natural polyhydroxylated aromatic compounds like flavonoids, which retard the development of oxidative rancidity by inhibiting the activities of lipoxygenase and cyclooxygenase enzymes. The color of meat could be stabilised with carnosic acid extracted from rosemary, which has a higher role in controlling lipid oxidation and metmyoglobin oxidation [23]. Further, plant-based antioxidants in addition to having desirable properties viz., being natural, non-genetically modified organisms (GMO) and having clean-label ingredients, have never been associated with promotion of carcinogenesis as that of synthetic antioxidants.

Fishes are more prone to autoxidation among animal foods because of the higher proportion of unsaturated fatty acids and the higher probability of having transition metals in their flesh and skin. *Moringa oleifera* leaves extracts inhibited lipid oxidation in ground pork patties while having no effect on sensory attributes [24].

Similarly, vegetable oils and dried fruits are more susceptible to autoxidation among vegetarian foods because of the formation of alkanes/alcohols or homolytic removal of β carbon-carbon bonds. In case of protein autoxidation, the damage comes from photochemical reactions, metal-ion interactions, and carbonyl groups. Amino acids viz., arginine, cysteine, serine lysine, histidine, proline, methionine, tryptophane, and valine are susceptible to protein autoxidation. Food antioxidants are employed to stabilise oils, fats, lipids, and other food products. Food antioxidants are used to stabilise lipids and other food products by avoiding and arresting their series of oxidative reactions.

Several analytical methods viz., ABTS (2, 2'-azino-bis (3-ethylbenzothiazoline6-sulphonic acid)), cupric reducing antioxidant capacity (CUPRAC), DPPH (2, 2-diphenyl-1-picrylhydrazyl), ferric reducing antioxidant power (FRAP), oxygen radical absorbance capacity (ORAC), total radical-trapping antioxidant parameter (TRAP), Trolox equivalent antioxidant capacity (TEAC) and total phenolics content are employed to establish the antioxidant activity and potency of herbs and spices. Several methods need to be employed to determine the antioxidant capacity of these extracts since the phenolic compounds present in them have different molecular structures and different reaction natures either by reducing Fe^{+3} ions or by acting against hydroxyl group, ABTS, and DPPH radicals. The methods of extraction as well as the kind of solvents employed for extraction possibly cause variation in the total phenolic levels of the same spice or herb. Further, the variety of plants, the location where cultivated, fertilisation and prevailing climate are some of the natural variations influence the antioxidant activities of spices orherbs.

Several studies have proved the presence of significant levels of variety of natural phenolic antioxidants in number of spices and herbs. This includes flavones (chrysin, apigenin, luteolin and diosmetin) in parsley and thyme; flavanols (kaempferol, isorhamnetin, myricetin, quercetin, rutin) in onion; carnosic acid, carnosol, rosmanol, rosmarinic acid in rosemary; carnosol, carnosic acid, lateolin, rosmanul, rosmarinic acid in sage; flavones (eriodictyol, dihydroquercetin, apigenin and dihydrokaempherol); flavonols, flavanones in oregano; kaempherol, quercetin, and rhamnetin in black pepper. Due to multiple hydroxyl groups, flavonoids exert stronger antioxidant potential activities against peroxyl radicals compared to phenolic acids. Several studies have demonstrated a possible synergistic effect between natural extracts and synthetic antioxidants.

Table 3. Total phenolic and flavonoid content of essential herbs and spices

Species	Substances and types of substances	Mode of action
Rosemary	Carnosic acid, carnosol, rosmarinic acid, rosmanol	Scavenge superoxide radicals, lipid antioxidant, and metal chelator
Sage	Carnosol, carnosic acid, rosmanol, rosmarinic acid	Scavenge free radicals
Oregano	Derivatives of phenolic acids, flavonoids, tocopherols	Scavenge free radicals
Thyme	Thymol, carvacrol, p-cumene-2,3-diol, biphenyls, flavonoids	Scavenge free radicals
Ginger	Gingerol related compounds, diarylheptanoids	Scavenge free radicals
Turmeric	Curcumins	Scavenge free radicals
Black pepper	Phenolic amides, flavonoids	Scavenge free radicals
Red pepper	Capsaicin	Scavenge free radicals
Clove	Eugenol, gallates	Free radical scavenger and metal chelator

5.2. Antimicrobial Activity of Major Spices and Herbs

Antimicrobial phytochemicals in plants defend themselves against microorganisms and are also crucial for the proper functioning of the plant. Several categories of phytochemicals viz., phenolic compounds, terpenoids, alkaloids, essential oils, and polypeptides present in spices possess antimicrobial properties. These phytochemicals have been utilised for several centuries to treat various health problems, including infections. Spices could also be employed to reduce the risk of food poisoning, spoilage and improve food safety and products shelf life [25]. The spices and herbs exert their antibacterial activity by interfering with the formation or functioning of essential components of microbes. The hydrophobic essential oils disturb the bacterial cell structure by accumulating in the lipid bilayer of the bacterial cell membrane and mitochondria and increase the permeability of the cell wall, which leads to the loss of the liquid constituents of the cell. Several research studies have documented the minimum inhibitory concentration of spices against food spoilage, pathogenic and harmful bacteria varying from 0.01% to 1%.

Numerous spices and herbs possess an inhibitory effect on at least one or more pathogen. Garlic is one such spice that has broad antibacterial activity against a variety of pathogenic and foodborne microbes, including *Clostridium*, *Escherichia*, *Salmonella*, *Staphylococcus*, *Streptococcus*, *Mycobacterium,* and *Helicobacter* species. Spices such as clove, oregano, rosemary, cumin, thyme and cinnamon have antimicrobial action against a larger group of food spoilage bacteria, harmful fungi, pathogens and even antibioticresistant microorganisms. Carvacrol and thymol are the two major phenols, principally responsible for the antimicrobial activity in thyme. In the case of cinnamon, cinnamaldehyde possesses antimicrobial effects by inhibiting the biosynthesis of the cell wall and interfering with the function of the membrane and also specific enzyme activities. Rosmarinic acid, carnosic acid, carnosol, rosmaridiphenol, epirosmanol, rosmanol and isorosmanol are responsible for the microbial inhibitory effect of spices and herbs. Based on their antimicrobial activities, spices are grouped into strong (clove, mustard and cinnamon), medium (cumin, coriander, bay leaf, sage, thyme, oregano and rosemary), and weak (ginger and black pepper) categories. Food model studies carried out with spices and their extracts have shown strong antimicrobial effects on wide range of foodborne pathogens both in the laboratory medium as well as *in-vitro* studies.

Table 4. Minimum inhibitory concentration (MIC) of certain essential oils (mg/ mL) [26]

Microorganisms	Minimum inhibitory concentration (mg/ml)			
	Basil	Cinnamon	Oregano	Rosemary
S. aureus	2.5	5	5	10.0
L. monocytogenes	1.25	2.5	2.5	2.5
B. cereus	1.25	2.5	2.5	5
Y. enterocolitica	0.075	0.075	0.075	0.075
E. coli	1.25	5.0	5.0	2.5
S. typhimurium	2.5	5.0	5.0	10.0

Table 5. Antimicrobial activity of natural extracts against target pathogen

Name of natural extract	MIC Value	Test organism(s)	Refs
Mint EO from Menthapiperita	1.2%v/v	Staphylococcus aureus and Salmonella enteritidis	[27]
Carvacrol, thymol, TC Cravone	1-3 mM 10 mM	Escherichia coli and Salmonella typhimurium	[28]
Finger root extract	0.2% v/v, 0.4% v/v	Listeria monocytogenes, Bacillus cereus and S. aureus	[29]
Galangal extract	8% v/v	E. coli O157:H7	[29]
Clove, cinnamon, and mustard extracts	1% v/v	E. coli O157:H7, S. aureus and B. cereus	[30]
Thyme	1000 ppm	Vibrio parahaemolyticus	[31]
Thyme	0.05%(v/v)	S. aureus	[32]
Carvacol	1 mM	E. coli O157:H7 ATCC 43895	[33]
Carvacol, P-cymene	7.8-800 µg/ml	Campylobacter spp.	[34]
EO and extracts from clove	0.5-5.5 mg/ml	L.monocytogenes	[35]
Phenolic contents from dried fruits of cinnamon	5 mg/plate	Salmonella typhimurium TA 100	[36]

6. Challenges and Future Perspective in the Usage of Spices and Herbs

Food industry is continuously exploring innovative ways and means to enhance the functional attributes of food products that offer greater value and potential health benefits to users. Food products have been successfully employed as a unique carrier to supply bioactive plant chemicals and additional nutrients to promote consumers health. However, before making any claim, the science based relationship of phytochemicals to health benefits should be established through extensive pharmacological and chemical experiments together with human metabolic studies [37]. Several technological hurdles have to be tackled to develop safe functional food products fortified with herbs and spices. In general, herbs and spices are recognized as safe, at least at concentration normally present in foods. However, for promotion of health and prevention of disease, the amount of spices and their bioactive compounds that need to be employed exceeds those normally found in food. It is imperative to conduct scientific studies to identify any possible safety issues associated with different levels of their usage. The synergistic effect of combining blended spices/herbs or blended natural compounds could be a solution to this problem.

Although several *in-vitro* assays revealed that many phytochemicals have strong antioxidant potentials, only few of these compounds could express their therapeutic value under *in-vivo* conditions due to their interference with several physio-pharmacological activities such as absorption, circulation, metabolism and excretion. The biggest challenge in the development of products with novel natural compounds is the transformation of *in-vitro* experiments to *in-vivo* trials and lastly clinical trials with human subjects. Similarly, the usage of natural products as antimicrobial substance is also more critical, because the microbial inhibition efficacy of the natural phytochemicals mainly depends upon maximum plasma concentration, tissue penetration and their bioavailability. As various external factors like the climate, location, time of harvest, nature of the extracts and mode of extraction have a significant impact on the chemical composition of plants, these aspects need to be taken into account to optimise the application of natural extracts in different food systems. The current pressing needs are refinement of existing procedures and development of innovative methods for optimised extraction and separation of active components from herbs and spices. Stability is one of the biggest challenges for wider usage of natural phytochemicals as dietary antioxidants. Generally, the degradation of antioxidant activity of many phytocompounds (polyphenols, carotenoids, terpenes) starts at the point of extraction and continues thereafter, which leads into a loss of potential. Enzymes, chemicals and the passage of time and processing at elevated temperatures are some of the factors responsible for loss of activity. Technological advancements such as encapsulation, vacuum packaging, modified atmospheric packing with inert gas, using oxygen scavengers, and more antioxidants have been employed to retard the degradation of antioxidant efficacy of extract compounds.

CONCLUSION

Phytochemicals with varied chemical nature and mechanisms of action offer promising therapeutic tools for developing functional products to promote health and well-being in consumers. Besides enhancing the organoleptic properties, especially flavor, aroma, and color, herbs and spices, because of their antioxidant and antimicrobial potential and lower toxicity, could be utilized as an effective food preservative. Natural plant-based compounds could be employed together with conventional food preservation methods to enhance the antioxidant and antimicrobial potential and ensure food products' safety. The interest in exploring the exact mechanisms of actions and physio-pharmacological properties of the bioactive compound and the development of newer technologies for better ways of utilization of spices and herbs will continue to grow in the years to come.

REFERENCES

[1] Kearney, J. (2010). Food consumption trends and drivers. Philosophical Transactions of the Royal Society of London. *Series B, Biological Sciences,* 365(1554): 2793-2807.

[2] Kasote, D. M., Katyare, S. S., Hegde, M. V. and Bae, H. (2015). Significance of antioxidant potential of plants and its relevance to therapeutic applications. *International Journal of Biological Sciences,* 11(8): 982–991.

[3] Srinivasan, K. (2005). Role of spices beyond food flavoring: Nutraceuticals with multiple health effects. *Food Reviews International,* 21(2): 167-188.

[4] Lourenço, S. C., Moldao Martins, M. and Alves, V. D. (2019). Antioxidants of natural plant origins: from sources to food industry applications. *Molecules* (Basel, Switzerland), 24(22): 4132.

[5] Yashin, A., Yashin, Y., Xia, X. and Nemzer, B. (2017). Antioxidant activity of spices and their impact on human health: A review. *Antioxidants* (Basel, Switzerland), 6(3): 70.

[6] Tsao, R. (2010). Chemistry and biochemistry of dietary polyphenols. *Nutrients*, 2(12): 1231–1246.

[7] Kumar, N. and Goel, N. (2019). Phenolic acids: Natural, versatile molecules with promising therapeutic applications. *Biotechnology Reports* (Amsterdam, Netherlands), 24: e00370.

[8] Tungmunnithum, D., Thongboonyou, A., Pholboon, A. and Yangsabai, A. (2018). Flavonoids and other phenolic compounds from medicinal plants for pharmaceutical and medical aspects: An overview. *Medicines* (Basel, Switzerland), 5(3): 93.

[9] Tyagi, P., Singh, M., Kumari, H., Kumari, A. and Mukhopadhyay, K. (2015). Bactericidal activity of curcumin I is associated with damaging of bacterial membrane. *PloS one,* 10(3): e0121313.

[10] Angela, O. A., Ludy, C. P., Margarita, R. R. F. and Maria, P. C. G. (2018). Plant extracts as antioxidant additives for food industry. Antioxidants in Foods and its Applications. *Intech Open*, doi 10.5772/intechopen.75444.

[11] Fiedor, J. and Burda, K. (2014). Potential role of carotenoids as antioxidants in human health and disease. *Nutrients,* 6(2): 466–488.

[12] Nakamoto, M., Kunimura, K., Suzuki, J. I. and Kodera, Y. (2020). Antimicrobial properties of hydrophobic compounds in garlic: Allicin, vinyldithiin, ajoene and diallyl polysulfides. *Experimental and Therapeutic Medicine*, 19(2): 1550–1553. https://doi.org/10.3892/etm.2019.8388.

[13] El-Saber Batiha, G., Magdy Beshbishy, A., G Wasef, L., Elewa, Y., A Al-Sagan, A., Abd El-Hack, M. E., Taha, A. E., M Abd-Elhakim, Y. and Prasad Devkota, H. (2020). Chemical constituents and pharmacological activities of garlic (*Allium sativum* L.): A review. *Nutrients*, 12(3): 872.

[14] Mitsiogianni, M., Koutsidis, G., Mavroudis, N., Trafalis, D. T., Botaitis, S., Franco, R., Zoumpourlis, V., Amery, T., Galanis, A., Pappa, A. and Panayiotidis, M. I. (2019). The role of isothiocyanates as cancer chemo-preventive, chemo-therapeutic and anti-melanoma agents. *Antioxidants* (Basel, Switzerland), 8(4): 106.

[15] Nazzaro, F., Fratianni, F., De Martino, L., Coppola, R. and De Feo, V. (2013). Effect of essential oils on pathogenic bacteria. *Pharmaceuticals* (Basel, Switzerland), 6(12): 1451–1474.

[16] Nuñez, L. and Aquino, M. D. (2012). Microbicide activity of clove essential oil (*Eugenia caryophyllata*). *Brazilian Journal of Microbiology*, 43(4): 1255–1260.

[17] Aggarwal, B. B. and Shishodia S. (2006) Molecular targets of dietary agents for prevention and therapy of cancer. *Biochemical Pharmacology*, 71:1397–421.

[18] Vaishali, R. M., Ramanath, P. V., Pratapchandra, K. H. and Hegde S. (2013). Preliminary phytochemical screening of members of Lamiaceae family: *Leucas linifolia, Coleus aromaticus* and *Pogestemon patchouli*. *International Journal of Pharmaceutical Sciences Review and Research*, 21: 131–137.

[19] Lobo, V., Patil, A., Phatak, A. and Chandra, N. (2010). Free radicals, antioxidants and functional foods: Impact on human health. *Pharmacognosy Reviews*, 4(8): 118–126.

[20] Oz, F. and Kaya, M. (2011). The inhibitory effect of black pepper on formation of heterocyclic aromatic amines in high-fat meatball. *Food Control*, 22: 596-600.

[21] Jinap, S., Iqbal, S., Talib, N. and Hasnol, N. D. (2016). Heterocyclic aromatic amines in deep fried meat: The influence of spices marination and sensory quality. *Journal of Food Science and Technology*, 53(3): 1411–1417.

[22] Johnson, D. and Decker, E. (2015). The Role of oxygen in lipid oxidation reactions: A review. *Annual Review of Food Science and Technology,* 6(1):171-190.

[23] Naveena, B. M., Vaithiyanathan, S., Muthukumar, M., Sen, A. R., Kumar, Y. P., Kiran, M., Shaju, V. A. and Chandran, K. R. (2013). Relationship between the solubility, dosage and antioxidant capacity of carnosic acid in raw and cooked ground buffalo meat patties and chicken patties. *Meat Science*, 95: 195–202.

[24] Muthukumar, M., Naveena, B. M., Vaithiyanathan, S., Sen, A. R. and Sureshkumar, K. (2014). Effect of incorporation of Moringa oleifera leaves extract on quality of ground pork patties. *Journal of Food Science and Technology*, 51(11): 3172–3180.

[25] Chouhan, S., Sharma, K. and Guleria, S. (2017). Antimicrobial activity of some essential oils-present status and future perspectives. *Medicines* (Basel, Switzerland), 4(3): 58.

[26] Sheila, M. D. S., Anildo, C. J. R., Gerson, N. S., Fábio, L. S., Silvani, V., Marisete, K. and Cleide, R. W. V. (2012). Chemical composition and antimicrobial activity of essential oils from selected herbs cultivated in the South of Brazil against food spoilage and foodborne pathogens. *Ciência Rural*, 71(3):471-480.

[27] Tassou, C., Koutsoumanis, K. and Nychas, G. J. E. (2000). Inhibition of Salmonella enteritidis and *Staphylococcus aureus* in nutrient broth by mint essential oil. *Food Research International*, 33 (3): 273–280.

[28] Helander, I. M., Alakomi, H. L., Kyösti, L. K. and Tiina M. S. (1998). Characterization of the action of selected essential oil

components on Gram negative bacteria. *Journal of Agricultural and Food Chemistry*, 46 (9): 3590–3595.

[29] Aktug, S. E. and Karapinar, M. (1986). Sensitivity of some common food-poisoning bacteria to thyme, mint and bay leaves. *International Journal of Food Microbiology*, 3 (6): 349–354.

[30] Sofia, P. K., Prasad, R., Vijay, V. K. and Srivastava, A. K. (2007). Evaluation of antibacterial activity of Indian spices against common foodborne pathogens. *International Journal of Food Science and Technology*, 42 (8): 910–915.

[31] Amiri H (2012) Essential oils composition and antioxidant properties of three *Thymus* species. *Evidence Based Complementary and Alternative Medicine: eCAM*, 2012: 728065.

[32] Kwon, H. A., Kwon, Y. J., Kwon, D. Y. and Lee, J. H. (2008). Evaluation of antibacterial effects of a combination of *Coptidisrhizoma, Mume fructus*, and *Schizandrae fructus* against Salmonella. *International Journal of Food Microbiology*, 127 (1–2):180–183.

[33] Burt, S. A., van der Zee, R., Koets, A. P., de Graaff, A. M., van Knapen, F., Gaastra, W., Haagsman, H. P. and Veldhuizen, E. J. (2007). Carvacrol induces heat shock protein 60 and inhibits synthesis of flagellin in *Escherichia coli* O157:H7. *Applied and Environmental Microbiology*, 73(14): 4484–4490.

[34] Aslim, B. and Yucel, N. (2008). *In vitro* antimicrobial activity of essential oil from endemic *Origanum minutiflorum* on ciprofloxacin-resistant *Campylobacter* spp. *Food Chemistry*, 107 (2), 602–606.

[35] Hoque, M., Bari, M. L., Juneja, V. K. and Kawamoto, S. (2008). Antimicrobial activity of cloves and cinnamon extracts against foodborne pathogens and spoilage bacteria and inactivation of *Listeria monocytogenes* in ground chicken meat with their essential oils. *Report of National Food Research Institute*, 72: 9–21.

[36] Jayaprakasha, G. K., Negi, P. S., Jena, B. S. and Rao, L. J. M. (2007). Antioxidant and antimutagenic activities of *Cinnamomum zeylanicum* fruit extracts. *Journal of Food Composition and Analysis*, 20 (3): 330–336.

[37] Clare, M. and Hasler (2002). Functional foods: benefits, concerns and challenges—A position paper from the American council on science and health. *The Journal of Nutrition*, 132(12): 3772–3781.

In: Nutraceuticals
Editors: A. Kumari and G. Singh
ISBN: 978-1-68507-488-3

Chapter 14

OMEGA 3 FATTY ACID AS A NOVEL NUTRACEUTICAL AND ITS ROLE IN HUMAN HEALTH

Priyanka Kumari Singh*[1]*, Rajni Chopra*[2,*]*,
***Meenakshi Garg*[3] *and Akriti Dhyani*[1]**

[1]Department of Food Technology,
Institute of Home Economics, University of Delhi, New Delhi, India
[2]Department of Food Science and Technology, National Institute of Food Technology Entrepreneurship and Management, Sonipat, Haryana, India
[3]Department of Food Technology, Bhaskaracharya College of Applied Science, University of Delhi, New Delhi, India

ABSTRACT

In recent years, there has been growing interest in the potential benefits of nutraceuticals in the food industry and consumers. The chapter is designed to provide in-depth information on the potential

*Corresponding Author's E-mail: rajnichopra.niftem@gmail.com.

capacity of fatty acids as nutraceuticals. Nutraceuticals are a group of nutritional substances extracted from natural foods that offer nourishing and therapeutic benefits, and is one of the vital nutraceuticals. As the awareness of a healthy diet and lifestyle increases, omega-3 attracts a lot of interest among the food industry and consumers. Out of various clinical trials and epidemiological studies, it has been established that omega-3 possesses multiple nutraceutical properties. It also reduces the risk of several diseases like inflammatory diseases, cancer therapy, cardiovascular symptoms, bipolar disorder, kidney disease, attention deficit disorder, and many more. However, beneficial outcomes are not accomplished by conventional modes of delivery and applications. The omega-3 is prone to oxidation, and by-products may lead to adverse effects which limit its uses. This shortcoming can be overcome using emerging technology such as nanoemulsion, nanoparticles, and liposomes to improve their oxidative stability and the effectiveness of omega-3.

INTRODUCTION

Increased public awareness and health awareness have attracted researchers and consumers towards the nutraceuticals industry. To fulfill the nutritional requirement of the human body, food, and other novel foods should be included in the diet. Therefore, the incorporation of health supplements and nutraceuticals fulfills the nutritional requirement. The word nutraceuticals is derived from two terminologies: "nutrition" and "pharmaceuticals." This refers "to a substance that is procured by extraction or purification of foods and is sold in forms that are commonly not related to food and similar characteristics to pharmaceuticals, having potential physiological advantages" [1]. Kathleen and Stephen [2] defined "Nutraceutical is any non-toxic food component with scientifically proven health benefits including prevention and treatment of disease." Thus, a nutraceutical exhibits physiological benefits or protection against chronic disease.

OMEGA -3 AS NUTRACEUTICALS

People are nowadays becoming more nutrition savvy and moving towards foods having various health benefits. Omega-3 is one of the primary interests of food professionals and industries because of its health benefits. Omega-3 fatty acids are essential polyunsaturated fatty acids mainly consisting of EPA and DHA. EPA and DHA are referred to as "miracle food of the 21st century" [3, 4]. EPA and DHA are bioactive lipids [5]. Mammals cannot synthesize omega-3 because they lack enzymes that help form double bonds at the omega-3 position. Hence, mammals require omega-3 in their diet. EPA and DHA have shown anti-inflammatory effects [6] and are essential for proper fetal growth and development [7]. EPA and DHA also act as precursors of lipid mediators, which play a crucial role in numerous health diseases [8].

Several studies on omega-3 suggested that omega-3 is beneficial in several diseases. The omega-3 produces several lipid mediators beneficial in cardiovascular disease, inflammation, cancer, depression, other mental disorders, and metabolic syndrome [9]. The evidence and epidemiological study recommended that omega-3 in the diet helps in altering blood lipids profile. Oxidative stability and bioavailability of omega-3 is a primary concern for food technologists and the food industry. The different novel delivery systems were used to increase the bioavailability and functional capabilities of omega-3. It also protects the bioactive agent from oxidation. Therefore, industry and researchers are adapting delivery systems such as encapsulation, nanoemulsion, nano liposomal, solid lipid nanoparticle (SLN) [10].

CHEMISTRY AND NOMENCLATURE OF OMEGA -3

"Fatty acids are hydrocarbon chains consist of a carboxyl group (COOH) at one end and a methyl group (CH_3) at the other end"[9]. The fatty acid is of two different types, i.e., saturated FA and unsaturated FA. The saturated fatty acid (SFAs) consists of a single bond, whereas unsaturated fatty acids (USFA's) comprise one or more than one double bond. Further, the unsaturated fatty acid is categorized into two, i.e., monounsaturated fatty acids (MUFAs) (having one double bond) and polyunsaturated fatty acids (PUFAs) (more than one double bond) [11]. PUFAs were further sub-grouped based on the location of the first double bond. The two main compound groups are omega-3 and omega-6 families [12].

TYPES OF OMEGA-3 FATTY ACID

Polyunsaturated fatty acids are further classified as short-chain and long-chain unsaturated fatty acids [13]. The FAO/WHO Expert [14] defines "Unsaturated short-chain fatty acids with 19 or fewer carbon atoms and unsaturated long-chain fatty acids with 20-24 carbon atoms". Thus, there are majorly three different types of omega-3 fatty acids:

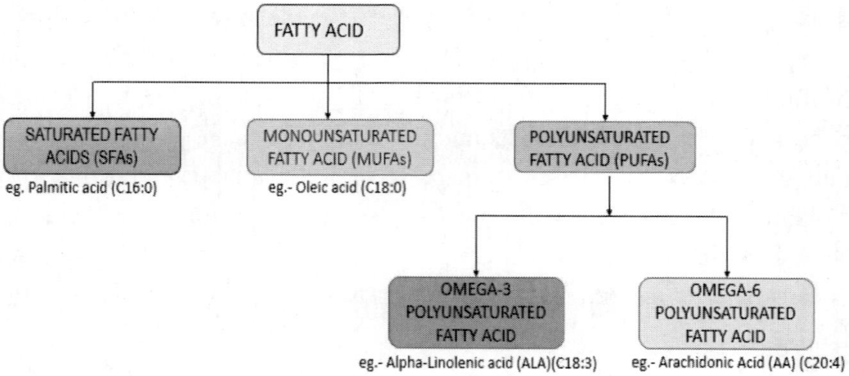

Figure 1. Classification of fatty acids based on the number of the bond.

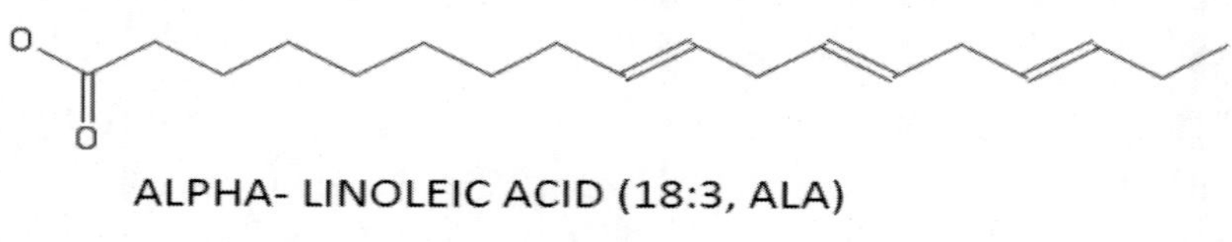

Figure 2. Structure of Alpha-linolenic acid.

Alpha-Linoleic Acid (ALA)

Alpha-linoleic acid [ALA; C18:3 (n-3)] is an essential omega-3 fatty acid obtained from the diet. It consists of an 18-carbon chain and three cis double bonds. Thus, it is also named as-cis-9,12,15-octadecatrienoic acid. The abundant sources of ALA are green leafy vegetables, flaxseed, linseed, perilla seed, and nuts. Cyclooxygenases and lipoxygenases convert arachidonic acid to prostaglandins and leukotrienes.

Eicosapentaenoic Acid (EPA)

EPA [EPA; 20:5(n-3)] is a long-chain omega-3 of 20 carbon chains with 5 *cis* double bond omega-3. Rich sources of EPA are some fish, e.g., salmon, cod, menhaden, and sardine. The other sources of EPA are different types of edible algae. It is also formed from the ALA by the desaturation and elongation process. These are the precursor for prostaglandin-3, Thromboxane -3, prostacyclin I-3, and Leukotriene-5 eicosanoids. These eicosanoids have shown positive effects on some human disorders like inflammation, platelet aggregation.

Docosahexaenoic Acid (DHA)

Docosahexaenoic acid (DHA) is unsaturated, long-chain PUFA having 22 carbon atoms [15]. DHA is the major component of the CNS and retina. However, the human body cannot synthesize it, and the requirement of DHA in the body can be either accomplished from diet or synthesized from ALA [16]. DHA is found in fish and other seafood [17]. DHA plays an essential role in brain development. It is also crucial during prenatal development. Therefore, pregnant women should include DHA in their diet [18]. DHA improves cognitive functions as well as neuroprotective functions [19].

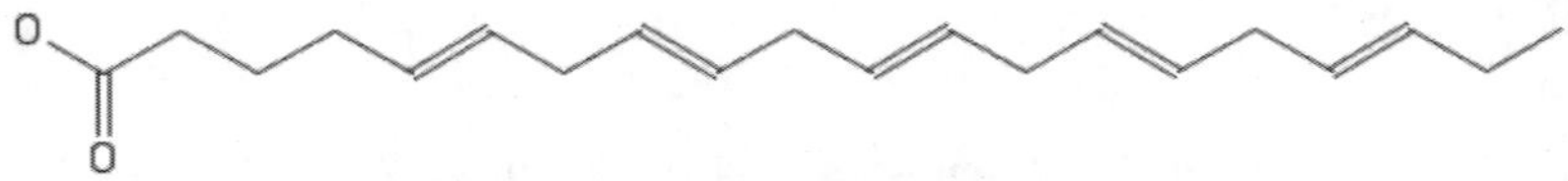

EICOSAPENTAENOIC ACID (20:5, EPA)

Figure 3. Structure of Eicosapentaenoic acid.

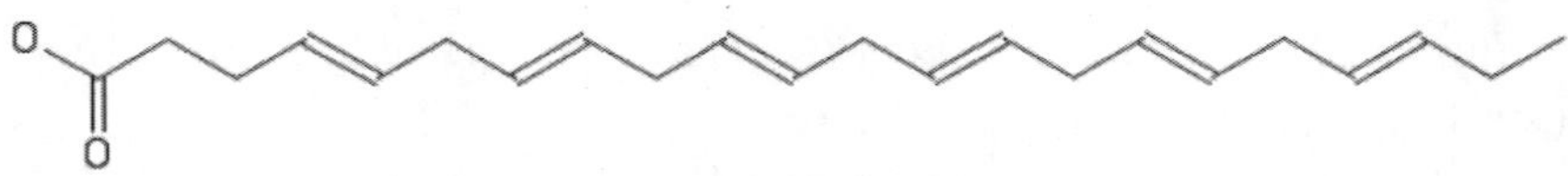

DOCOSAHEXANOIC ACID (22:6, DHA)

Figure 4. Structure of Docosahexaenoic acid.

SOURCES OF OMEGA-3 FATTY ACID

Nowadays, omega-3 exists naturally in some food and is also available as fortified food such as egg, milk, yogurt, juices, infant formulas, etc. [20, 21].

Plant Sources

Alpha-linolenic acid (ALA) is the only plant source found in green leafy vegetables (GLVs) and seeds. Other sources of omega-3 are seeds, seed oils, and some nuts. Soybean oil, rapeseed oil, walnut, corn oil is good sources of α-linolenic acid. Flaxseed and their oil consist of 40-55% FA as alpha-linolenic acid [22].

Vegetable oil such as Echium oil comprised approximately 36.3% of ALA. It also contains a significant amount of LC- PUFA, i.e., stearidonic acid (18:4n-3; SDA) [23]. Perilla seed oil (Perilla frutescens) is also an abundant source of omega-3 containing up to 54% of ALA [24].

Fish and Marine Sources

Fish and seafood are primary sources of long-chain omega-3 fatty acids such as EPA, DHA, DPA which appear to be highly bioactive [25]. Fish such as cod, mackerel, herring, salmon, and tuna are rich sources of omega-3, including EPA and DHA [26]. Different types of fish contain different amounts of these fatty acids depending upon the type of fish and its metabolic characteristics, diet of the fish, the temperature of the water, and season [27].

Single Cell Oil

Due to the beneficial health effects of omega-3, the demand for omega-3 PUFA has been increased in the past years. Fishes are rich sources of omega-3 fatty acids, but high consumption, environmental pollution and mishandling by humans leads to the shortage of existing fish stocks [28, 29]. Single-cell oil from microalgae or yeast is an alternative source of omega-3 fatty acids. These are sustainable alternatives capable of fulfilling the global demand and vegetarian sources of omega-3 [30]. However, these sources have drawbacks as oils from microalgae require high cost and investment. So, there is a need for an economical method for the production and cultivation the microalgae.

METABOLISM OF OMEGA-3

The ALA serves as a parent component in the formation of another form of omega-3. The process of desaturation (double bond formation) and elongation (insertion of two carbon atoms) is required in the metabolic pathway [31]. As the same enzymes (Δ6-desaturase) are used for the conversion, LA and ALA fight for the synthesis of eicosanoids. The Δ6-desaturase enzyme is rate-limiting in the pathway [32]. Most plants, algae, some fungi have Δ12-desaturases to convert oleic acid into LA and Δ15-

desaturases for the conversion of LA to ALA [33]. ALA cannot be synthesized from oleic acid in the human body due to a deficiency of desaturase enzyme [34].

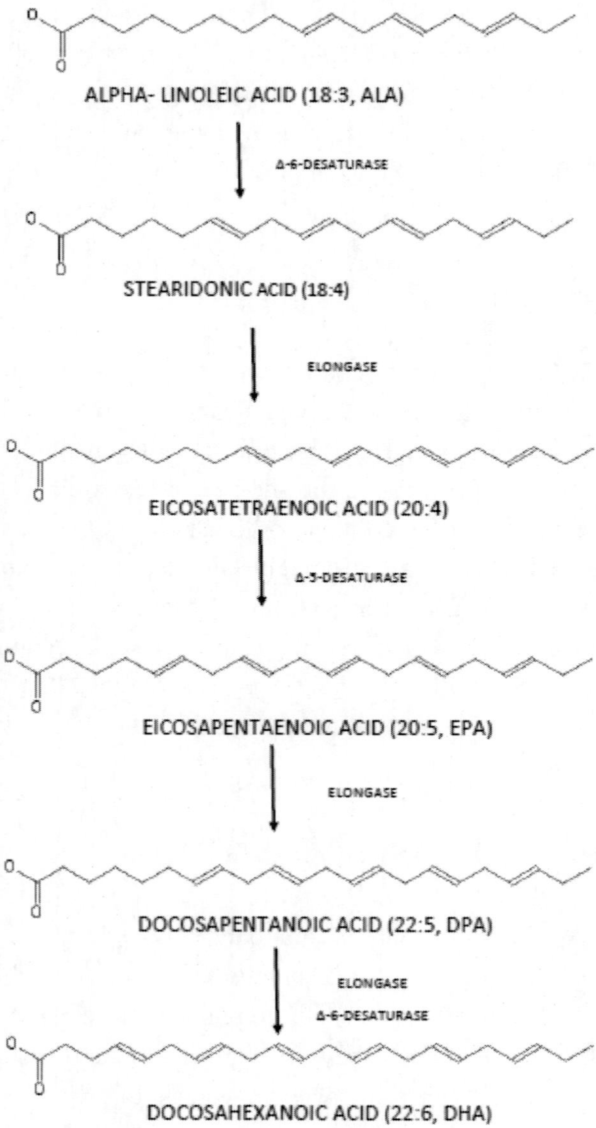

Figure 5. Metabolic pathway of omega-3.

In animals, biosynthesis of LC-PUFAs initiates in the endoplasmic reticulum of liver cells. The first step is the desaturation step, in which the addition of a double bond at the 6th C-C bond was done to get stearidonic acid (SDA). Further stearidonic acid elongated to 20: 4n-3 (Eicosatetraenoic acid). Finally, one more desaturation was carried out by Δ5-desaturase producing eicosapentaenoic acid (20:5n-3, EPA) (34, 35). Further, EPA elongated to produce docosapentaenoic acid (DPA; C22:5, n−3) and tetracosanolpentaenoic acid (C24:5, n−3) in two successive elongation cycles. Then, Δ6-desaturation further desaturates this fatty acid to get tetracosahexaenoic acid (THA; C24:6, n−3). The last step is the translocation of 24:6n-3 from the endoplasmic reticulum to peroxisomes, where it is subjected to β-oxidation in which two carbon were removed to yield DHA [36, 34].

One of the main functions of these EFAs is the biosynthesis of the hormone known as eicosanoids. The member of eicosanoids includes prostaglandins, prostacyclin, thromboxane, leukotrienes, and hydroxyl fatty acids. These mediators help in inflammation, immunity, platelet reactivity, and smooth muscle contraction [37].

HEALTH BENEFITS

Several clinical trials and epidemiological studies suggested that omega-3 is a vital component of the human diet. Its deficiency in the human diet can cause several abnormalities in the human body. Therefore, omega-3 fatty acids have attracted a lot of researchers. The research investigates the effect of omega-3 fatty acids in different areas of health such as premature infant health [38, 39], asthma [40, 41], bipolar and depressive disorders [42, 43] diabetes [44, 45].

Omega-3 and Cardiovascular Health

Cardiovascular diseases (CVD) are used for a class of heart and blood vessels [46, 47]. CVD is one of the significant reasons for worldwide death and is responsible for 33% of worldwide deaths [48]. The primary risk factor of CVD is the consumption of high saturated fat and trans-fat intake and smoking, high sodium intake, genetic predisposition.

The omega-3 fatty acids have been associated with substrate competition for the production of cyclooxygenase (COX) enzymes between omega-3 and arachidonic (AA), which results in the formation of PGs and TX [49]. This substrate competition by omega-3 results in vasodilation and decreased platelet aggregation. The omega-3 and arachidonic acid compete for the Δ-6 desaturase enzyme and the sn-2 position, which could lead to inhibition of AA, which further leads to a decrease in AA levels in the body [50].

Research studies performed using animal models for CVD have signified the potential protective role of omega-3. It was found that omega-3 lowers the possibilities of coronary outcomes by regulating cholesterol levels, lipogenesis, inflammation, thrombosis, and arterial stiffness [34, 51, 52, 53, 54 55, 56]. Research studies have observed that omega-3 fatty acid in diet results in the reduction of VLDL synthesis. Hence, serum levels of triglycerides are reduced, which further improves the degradation of fatty acids, facilitating triglyceride clearance from the plasma [53, 57]. It also helps prevent arrhythmias, reduced heart rate, and systemic vascular resistance [58, 59] in a study of the physiological effect of omega-3 derived from seafood on cardiovascular health. It was reported that omega-3 fatty acids in the diet help lower triglycerides, blood pressure, and heart rate (53). He et al. (2004) reported that one fish/week reduced CVD risk by 16%. It was also found that increasing the servings/week up to 2-4 reduced the risk by 21% [60]. A systematic review and meta-analysis study showed a reciprocal relationship between inclusion of fish consumption and mortality from CVD. They also reported improved blood pressure and blood lipid profile [61]. Therefore, research evidence indicates that the inclusion of omega-3 rich fish in the diet reduces CHD risk.

EPA and DHA also play a vital role in the reduction of clinical risk of CVD disease. EPA and DHA have been shown to have triglyceride-lowering effects, which helps in increasing the genes encoding proteins involved in fatty acid oxidation [62]. In addition, researchers suggested dietary supplementation of DHA and EPA helps lower the risk by positively improving the lipid profile and platelet aggregation [15]. In another study, it was found that DHA in the diet is inversely correlated with the risk of CHD [63]. Evidence from research studies shows that DHA exhibits antiplatelet aggregatory potential, TAG-lowering effect as well as antiarrhythmic effects. In a double-blinded, placebo-controlled trial, subjects with abnormal lipid were supplemented with 1.2g DHA/day for six weeks. The researchers found a significant drop in TAG (20% approx.) and increased HDL levels in the subject [64].

Omega-3 and Cancer

Cancer is the second most contributor to death globally [65]. Cancer is an uncontrolled and unnatural growth of the cells. The most common cancer types are breast cancer, prostate cancer, lung cancer, cervical cancer, and infection-related cancer [65].

Numerous clinical and epidemiology studies, as well as experimental data, have shown the consequences of ω-3 fatty acids in the diet on various types of cancer like breast cancer [66, 67], colorectal cancer [68], gastrointestinal cancer [69], prostate cancer [70, 71]. An increase in dietary intake of omega-3 leads to increased incorporation of omega-3 in cell membranes. Thus, omega-3 helps reduce the proliferation, invasion, and metastasis of cancerous cells [72].

Based on some *in-vitro* studies, PG-2 shows carcinogenic actions. In contrast, omega -3 serves as an antagonist effect on the formation of COX by invading and inhibiting the growth of tumor cell growth by lowering the amount of cyclooxygenase COX-2 and PGE2[34, 73]. In addition, omega-3 fatty acid helps optimize receptor functioning and signaling transduction pathways, leading to the arrest of cancer cell growth [74, 75].

A research study was carried out on women to evaluate the consequences of consuming fish rich in omega-3 on breast cancer. They reported that fatty fish exhibits a protective effect on breast cancer in Asian patients [66].

Omega-3 and Neurological Disorders

Omega-3 rich diet is a promising nutritional approach for neurological disorders. Omega-3 fatty acids have been positively associated with depressive conditions such as bipolar disorder and unipolar depression [26].

Depression and Other Mood Disorders

Depression is a complex multifactorial brain disorder in which an individual feels lethargic, irritable, has frequent mood swings, and loses in pleasure activities [76]. The contributing factors are drug and alcohol abuse or altered production of neurotransmitters which further results in organic changes in the brain. Besides these factors, stress, social isolation, genetic influences also contribute to the risk factors. Pharmaceuticals are not too effective against depression and also have side effects. However, omega-3 PUFA was found to have a positive impact on depression.

The effectiveness of omega-3 supplementation in depression depends on two factors: proportion and dosage. A recent meta-analysis study supplementation of EPA or DHA in diet indicated that omega-3 PUFA with EPA 60% or more effectively improves depression [77]. Several meta-analyses indicated that the EPA has significantly greater efficacy than DHA [78, 79, 80]. Therefore, research studies recommended that the range of 2:1 to 3:1 ratio of EPA to DHA is more efficacious [81, 82]. Omega-3 fatty acids help to transmit the chemical messengers that activate serotonin and dopamine, linked with positive emotion and emotional stability [83, 26]. Serotonin also plays a vital role in depression. Serotonin is a chemical that regulates a wide variety of brain functions. Essential fatty acids such as omega-3 are necessary for serotonin production [26]. Evidence showed that EPA and DHA in the diet might help in increasing the release, which promotes emotional stability [84]. In contrast, a research study by Carabelli et al. (2015) on omega-3 dietary supplementations suggested that incorporating omega-3 can correct this deficiency in the diet.

Alzheimer's Disease (AD)

Alzheimer's disease is the most common brain disease which results in memory loss. The existence of amyloid β (Aβ) in the brain is associated with the development of Alzheimer's disease. Its deposition in the brain leads to inflammation, diminishes neuronal function, and triggers oxidative stress, contributing to the onset of Alzheimer's disease [85, 86, 87].

Evidence from various epidemiological and clinical studies indicated that omega 3 in the diet might impact risk and outcome in Alzheimer's disease.

Supplementation improves memory complaints, mild cognitive impairment, and Alzheimer's disease [34, 88]. Researchers also recommended that increased intake of the DHA is correlated with a lower risk of Alzheimer's disease [89]. They also suggested a diet rich in DHA helps to protect the brain against cell damage and memory loss caused by Alzheimer's disease.The research was conducted on aged mice with a genetic mutation (linked to Alzheimer's disease). Supplementation of a DHA-rich diet on mice leads to reduced pathogens linked to Alzheimer's [90].

A plethora of animal studies were conducted on DHA dietary intervention, and it was recommended that consumption of DHA is beneficial in neuropathy such as amyloid-beta, inflammation [91, 92].

Attention Deficit Hyperactivity Disorder (ADHD)

ADHD is a neurodevelopmental disorder in children. Symptoms related to these disorders are attention deficit, hyperactivity, and impulsivity [93]. Omega-3 supplementation is also correlated with ADHD. A study found that the level of omega-3 fatty acids is low in ADHD patients compared to healthy patients [101]. In addition, several clinical and epidemiological studies have indicated that a lack of DHA and EPA in the diet may increase the risk of ADHD [94, 95].

It was found that children with ADHD, when administered with omega-3, improve the quality of life in terms of increased concentration and sleep quality in ADHD children [96]. A study revealed a negative correlation between omega-3 fatty acid levels in blood and ADHD symptoms. In a meta-analysis study, it was found that most of the RCT reported that there is a positive outcome of supplementation of omega-3, EPA, DHA in terms of learning ability, cognitive behavior [97]. Therefore, a low amount of omega-3 in the diet of patients with ADHD further deteriorates omega-3 because of oxidative stress, resulting in the development and exacerbation of the ADHD symptoms.

Omega-3 and Inflammation

PUFA plays a vital role in regulating inflammation and acts as a precursor for lipid mediators. Omega-6 fatty acid promotes inflammation, whereas omega-3 acts as an anti-inflammatory [98]. Different studies have recognized the role of omega-3 in inflammation. However, several recent investigations showed that increasing the omega-3 in the diet lowers the risk of inflammatory disease [99, 100]. By competing with enzymes, omega-3 prevents the production of AA-derived pro-inflammatory mediators and creates anti-inflammatory mediators such as resolvins, protectins, and maresins [101, 31, 17]. These actively terminate the inflammatory reactions by removing dead cells and increasing phagocytosis [102, 103].

The mechanism underlying the anti-inflammatory action of EPA & DHA includes the change in fatty acid composition in the cell membrane, disruption of lipid rafts, inhibiting the formation of pro-inflammatory cytokines, and modifying lipid mediators synthesis [17]. Studies evaluated the effectiveness of supplementation of the omega-3 FA in rheumatoid arthritis [104, 105], inflammatory bowel diseases (Crohn's disease and ulcerative colitis), and asthma [115,116]. It was indicated that the inclusion of omega-3 rich foods reduces the inflammation in the diseases. In addition, a research study reported that intake of omega-3 helps in the management of asthma. A cross-sectional study of 642 subjects found that seafood information and an increase in serum PUFA level decreased the risk of having non-specific bronchial hyper-responsiveness (NSBH) [106].

Novel Delivery System for Omega-3

Omega-3 has many positive health effects, but these are highly susceptible to oxidative degradation. The oxidative degradation of leads to free radicals and hydroperoxides, unstable intermediary compounds [107]. The oxidation results in a reduction in the nutritional properties, the formation of free radical during oxidation contribute to the development of atherosclerosis [108]. Also, the efficacy of in different health problems is contradictory.A wide variety of individual variance and lack of uniformity in the outcomes have been demonstrated by various studies. Due to differences in chemical composition, the type of dosage used (pills, tablets, gels, etc.) can affect bioavailability and stability [109]. Different delivery methods such as nanoemulsion, encapsulation, and microencapsulation are widely used in various food products. They improve oxidation rate in both initial stages (raw materials) and the final food product [109]. It also helps to cope with processing stresses faced during manufacturing and checks the undesirable interaction during the formulation of nutraceuticals [110].

Encapsulation

To protect PUFAs against oxidative degradation, encapsulation has been one of the widely used methods. Encapsulation is the process of entrapping an active component (nutraceutical) within a secondary material (encapsulant). The encapsulation of the active components results in protection from the outer environment up to it being released by a trigger at a desired site and time [111]. In addition, encapsulation masks the unpleasant taste and odor of omega-3 fatty acids while delivering them in food and supplements [111]. Several techniques such as spray-drying emulsions, extrusion, coacervation, biopolymer gelation, and fluidized bed coating are used to encapsulate omega-3 [112]. The most common method used to encapsulate omega-3 is diffusing the lipid in a polymer solution and then eliminating the water phase by spray drying to produce powder in the dry form [107, 113]. The spray drying temperature was decreased to minimize oxidative degradation, and the air was replaced by nitrogen [114].The human trial has reported that omega-3 oils supplemented in the microencapsulated complex are bioequivalent to oils supplemented in the form of soft gels [115].

Nanoemulsion

Enrichment of food products with omega-3 FA is in great demand for the food industry and technologists. Nanoemulsion helps encapsulate PUFA rich oils. Nanoemulsion is defined as systems that consist of small droplet size particles with materials in nanometer (approximately 10−9 m) [116, 117]. Its droplets' size varies from 20 nm to 200 nm [118] or smaller than 300nm [119]. Nanoemulsion is of three types: Water-in-oil nanoemulsion, oil-in-water nanoemulsion, and bi-continuous nanoemulsion [10]. The small size particles prevent flocculation and produce fine dispersion. Due to high solubilization, it can be used to produce hydrophobic drug delivery [120]. These tiny droplets result in easy transportation in cell membranes which further increases its bioavailability. Tiny droplets increase their bioavailability and result in increased blood plasma and erythrocyte concentrations [121].

In a research study, omega-3 is encapsulation was done using oil-in-water nanoemulsion along with dual-channel microfluidics techniques. It was found that encapsulation reduces the risk of coronary heart disease and inflammation [122]. Study omega-3 fatty acid enrich Solid Lipid Nanoparticles (SLN) was fed to colorectal cancer patients. The SLN prepared improved the physicochemical properties and also lowered the rate of oxidation of omega-3. It also enhanced the delivery of omega-3 and antineoplastic efficacy [123].

Nanoliposomal

Liposomes are flexible, biocompatible, non-toxic lipid-based carriers. It helps to protect the active ingredient against enzymatic degradation [10]. They help in the encapsulation and delivery of both hydrophilic and hydrophobic bioactive components. This technology is widely used in many studies and industrial products to encapsulate and efficiently release food compounds [124]. The incorporation of Krill oil into liposomes was found to affect inflammatory bowel disease (IBW) compared to oral delivery [123]. Natural biomimetic lecithin formulations rich in omega-3 LC-PUFAs extracted from salmon heads and processed to nanoliposomes have been studied. The nanoliposomes reduced stiffness and proliferation and showed an anti-cancer effect on breast cancer [125].

Conclusion

Several clinical and meta-analysis studies indicate that omega-3 acts as potent nutraceuticals. Many research studies show that supplementation of omega-3 rich oils and fish oils is positively beneficial in diseases like CVD, mental illness, atherosclerosis, rheumatoid arthritis, obesity, diabetes, and cancer. The different delivery systems such as nanoemulsions, microemulsions increase the functional performance and efficacy of the omega-3. Therefore, omega-3 consumption should be recommended for all age groups. Further, research on the toxicity of this newly developed delivery system needs to explore.

REFERENCES

[1] Gupta, S. C., Kim, J. H., Prasad, S. & Aggarwal, B. B. (2010). Regulation of survival, proliferation, invasion, angiogenesis, and metastasis of tumor cells through modulation of inflammatory pathways by nutraceuticals. *Cancer and Metastasis Reviews*, *29*(3), 405-434.

[2] Kathleen, C. & Stephen, D. (2013). Nutraceuticals: What Are They and Do They Work? Kentucky Equine Research. *Inc. Versailles, KY*,*7*(4).

[3] Bourre, J. M. (2007). Dietary omega-3 fatty acids for women.*Biomedicine &Pharmacotherapy*,*61*(2-3),105-112.

[4] Amiri-Jami, M. & Griffiths, M. W. (2010). Recombinant production of omega-3 fatty acids in *Escherichia coli* using a gene cluster isolated from *Shewanellabaltica* MAC1. *Journal of Applied Microbiology*,*109*(6), 1897-1905.

[5] Khan, S. A. & Makki, A. (2017). Dietary changes with omega-3 fatty acids improve the blood lipid profile of Wistar albino rats with hypercholesterolemia. *International Journal of Medical Research & Health Sciences*,*6*(3), 34-40.

[6] Tortosa-Caparrós, E., Navas-Carrillo, D., Marín, F. & Orenes-Piñero, E. (2017). Anti-inflammatory effects of omega 3 and omega 6 polyunsaturated fatty acids in cardiovascular disease and metabolic syndrome.*Critical Reviews in Food Science and Nutrition*, *57*(16),3421-3429.

[7] Dunstan, J. A., Mori, T. A., Barden, A., Beilin, L. J., Taylor, A. L., Holt, P. G. & Prescott, S. L. (2003). Maternal fish oil supplementation in pregnancy reduces interleukin-13 levels in cord

blood of infants at high risk of atopy. *Clinical & Experimental Allergy*, *33*(4), 442-448.

[8] Dona, M., Fredman, G., Schwab, J. M., Chiang, N., Arita, M., Goodarzi, A. & Serhan, C. N. (2008). Resolvin E1, an EPA-derived mediator in whole blood, selectively counterregulates leukocytes and platelets. *Blood, The Journal of the American Society of Hematology*, *112*(3), 848-855.

[9] Calder, P. C. & Yaqoob, P. (2009). Omega-3 polyunsaturated fatty acids and human health outcomes. *Biofactors*, *35*(3),266-272.

[10] Helal, N. A., Eassa, H. A., Amer, A. M., Eltokhy, M. A., Edafiogho, I. & Nounou, M. I. (2019). Nutraceuticals' Novel Formulations: The Good, the Bad, the Unknown and Patents Involved. *Recent Patents on Drug Delivery & Formulation*, *13*(2), 105-156.

[11] Patkowska-Sokola, B., Usydus, Z., Szlinder-Richert, J. & Bodkowski, R. (2009). Technology for recovering omega-3 fatty acids from family from fish oils and protecting them against oxidative changes. *PrzemyslChemiczny*, *88*(5),548-552.

[12] Bodkowski, R., Szlinder-Richert, J., Usydus, Z. & Patkowska-Sokola, B. (2011). An attempt of optimization of fish oil crystallization at low temperature. *Przemysl Chemiczny*, *90*(5), 703-706.

[13] Insel, P. M. (2013). *Discovering nutrition*. Jones & Bartlett Publishers.

[14] Joint, F. A. O. & Consultation, W. E. (2008). Fats and fatty acids in human nutrition. *Proceedings of the joint FAO/WHO expert consultation*.

[15] Senanayake, S. P. J. N. & Fichtali, J. (2006). Single-cell oils as sources of nutraceutical and specialty lipids: processing technologies and applications. *Nutraceutical Science and Technology*, *5*, 251.

[16] Sun, G. Y., Simonyi, A., Fritsche, K. L., Chuang, D. Y., Hannink, M., Gu, Z. & Beversdorf, D. Q. (2018). Docosahexaenoic acid (DHA): An essential nutrient and a nutraceutical for brain health and

diseases. *Prostaglandins, Leukotrienes and Essential Fatty Acids*, *136*,3-13.

[17] Calder, P. C. (2017). Omega-3 fatty acids and inflammatory processes: from molecules to man. *Biochemical Society Transactions*, *45*(5), 1105-1115.

[18] Singh, M. (2005). Essential fatty acids, DHA and human brain. *The Indian Journal of Pediatrics*, *72*(3), 239-242.

[19] Echeverría, F., Valenzuela, R., Hernandez-Rodas, M. C. & Valenzuela, A. (2017). Docosahexaenoic acid (DHA), a fundamental fatty acid for the brain: New dietary sources. *Prostaglandins, Leukotrienes and Essential Fatty Acids*, *124*, 1-10.

[20] Ofosu, F. K., Daliri, E. B., Lee, B., Yu, X. & Chuncheon-si, G. D. (2017). Current trends and future perspectives on omega-3 fatty acids. *Res. Rev. J. Biol*, *5*, 11-20.

[21] Puri, M. (2017). Algal biotechnology for pursuing omega-3 fatty acid (bioactive) production. *Microbiology Australia*, *38*(2), 85-88.

[22] Harper, C. R., Edwards, M. C. & Jacobson, T. A. (2006). Flaxseed oil supplementation does not affect plasma lipoprotein concentration or particle size in human subjects. *The Journal of Nutrition*, *136*(11), 2844-2848.

[23] Kuhnt, K., Weiß, S., Kiehntopf, M. & Jahreis, G. (2016). Consumption of echium oil increases EPA and DPA in blood fractions more efficiently compared to linseed oil in humans. *Lipids in Health and Disease*, *15*(1), 32.

[24] Asif, M. (2011). Health effects of omega-3, 6, 9 fatty acids: Perilla frutescens is a good example of plant oils. *Oriental Pharmacy & Experimental Medicine*, *11*(1), 51-59.

[25] Burdge, G. C. & Calder, P. C. (2006). Dietary α-linolenic acid and health-related outcomes: a metabolic perspective. *Nutrition Research Reviews*, *19*(1), 26-52.

[26] Chandola, H. M. & Tanna, I. (2014). Role of omega-3 fatty acids in brain and neurological health with special reference to clinical depression. In *Omega-3 fatty acids in brain and neurological health* (pp. 163-179). Academic Press.

[27] Calder, P. C. (2013). Omega-3 polyunsaturated fatty acids and inflammatory processes: nutrition or pharmacology?. *British journal of Clinica lPpharmacology*, *75*(3), 645-662.

[28] Tocher, D. (2009). Issues surrounding fish as a source of omega-3 long-chain polyunsaturated fatty acids. *Lipid Technology*, *21*(1), 13-16.

[29] Hixson, S. M. (2014). Fish Nutrition and Current Issues in Aquaculture: The Balance in Providing Safe and Nutritious Seafood, in an Environmentally Sustainable Manner. *J Aquac Res Development*, *5*(234), 2.

[30] Winwood, R. J. (2013). Recent developments in the commercial production of DHA and EPA rich oils from micro-algae. *Oil*, *20*(6), D604.

[31] Punia, S., Sandhu, K. S., Siroha, A. K. & Dhull, S. B. (2019). Omega 3-metabolism, absorption, bioavailability and health benefits–A review. *Pharma Nutrition*, *10*, 100162.

[32] Calder, P. C. (2012). Mechanisms of action of (n-3) fatty acids. *The Journal of nutrition*, *142*(3), 592S-599S.

[33] Ruiz-Lopez, N., Haslam, R. P., Napier, J. A. & Sayanova, O. (2014). Successful high-level accumulation of fish oil omega-3 long-chain polyunsaturated fatty acids in a transgenic oilseed crop. *The Plant Journal*, *77*(2), 198-208.

[34] Saini, R. K. & Keum, Y. S. (2018). Omega-3 and omega-6 polyunsaturated fatty acids: Dietary sources, metabolism, and significance—A review. *Life sciences*, *203*,255-267.

[35] Baker, E. J., Miles, E. A., Burdge, G. C., Yaqoob, P. & Calder, P. C. (2016). Metabolism and functional effects of plant-derived omega-3 fatty acids in humans. *Progress in Lipid Research*, *64*, 30-56.

[36] Arnoldussen, I. A. & Kiliaan, A. J. (2014). Impact of DHA on metabolic diseases from womb to tomb. *Marine Drugs*, *12*(12), 6190-6212.

[37] Shahidi, F. (2013). Oilseed Processing and Fat Modification. In *Biochemistry of Foods* (pp. 363-384). Academic Press.

[38] Quin, C., Erland, B. M., Loeppky, J. L. & Gibson, D. L. (2016). Omega-3 polyunsaturated fatty acid supplementation during the pre and post-natal period: A meta-analysis and systematic review of randomized and semi-randomized controlled trials. *Journal of Nutrition & Intermediary Metabolism*, *5*, 34-54.

[39] Tinoco, S. M. B., Sichieri, R., Setta, C. L., Moura, A. S. & do Carmo, M. G. T. (2009). n-3 polyunsaturated fatty acids in milk is associate with weight gain and growth in premature infants. *Lipids in Health and Disease*, *8*(1), 23.

[40] Reisman, J., Schachter, H. M., Dales, R. E., Tran, K., Kourad, K., Barnes, D. &Blackman, J. (2006). Treating asthma with omega-3 fatty acids: where is the evidence? A systematic review. *BMC Complementary and Alternative Medicine*, *6*(1), 26.

[41] Miyata, J. & Arita, M. (2015). Role of omega-3 fatty acids and their metabolites in asthma and allergic diseases. *Allergology International*, *64*(1), 27-34.

[42] Balanzá-Martínez, V., Fries, G. R., Colpo, G. D., Silveira, P. P., Portella, A. K., Tabarés-Seisdedos, R. & Kapczinski. (2011). Therapeutic use of omega-3 fatty acids in bipolar disorder. *Expert Review of Neurotherapeutics*, *11*(7), 1029-1047.

[43] Sarris, J., Mischoulon, D. & Schweitzer, I. (2012). Omega-3 for bipolar disorder: meta-analyses of use in mania and bipolar depression. *Journal of Clinical Psychiatry*, *73*(1), 81.

[44] Hartweg, J., Perera, R., Montori, V. M., Dinneen, S. F., Neil, A. H. & Farmer, A. J. (2008). Omega-3 polyunsaturated fatty acids (PUFA) for type 2 diabetes mellitus. *Cochrane Database of Systematic Reviews*, (1).

[45] Brown, T. J., Brainard, J., Song, F., Wang, X., Abdelhamid, A. & Hooper, L. (2019). Omega-3, omega-6, and total dietary polyunsaturated fat for prevention and treatment of type 2 diabetes mellitus: systematic review and meta-analysis of randomised controlled trials. *BMJ*, *366*, l4697.

[46] Vilahur, G., Badimon, J. J., Bugiardini, R. & Badimon, L. (2014). Perspectives: The burden of cardiovascular risk factors and coronary

heart disease in Europe and worldwide. *European Heart Journal Supplements*, *16*(suppl_A), A7-A11.

[47] Mendis, S., Puska, P., Norrving, B. & World Health Organization. (2011). *Global atlas on cardiovascular disease prevention and control*. World Health Organization.

[48] Joseph, P., Leong, D., McKee, M., Anand, S. S., Schwalm, J. D., Teo, K. & Yusuf, S. (2017). Reducing the global burden of cardiovascular disease, part 1: the epidemiology and risk factors. *Circulation Research*, *121*(6), 677-694.

[49] Shahidi, F. & Miraliakbari, H. (2004). Omega-3 (n-3) fatty acids in health and disease: part 1—cardiovascular disease and cancer. *Journal of Medicinal Food*, *7*(4), 387-401.

[50] Shahidi, F. &Ambigaipalan, P. (2018). Omega-3 polyunsaturated fatty acids and their health benefits. *Annual Review Of Food Science and Technology*, *9*, 345-381.

[51] Harris, L., Fritsche, H., Mennel, R., Norton, L., Ravdin, P., Taube, S. & BastJr, R. C. (2007). American Society of Clinical Oncology 2007 update of recommendations for the use of tumor markers in breast cancer. *Journal of Clinical Oncology*, *25*(33), 5287-5312.

[52] Macchia, A., Grancelli, H., Varini, S., Nul, D., Laffaye, N., Mariani, J. & Tognoni, G. (2013). Omega-3 fatty acids for the prevention of recurrent symptomatic atrial fibrillation: results of the FORWARD (Randomized Trial to Assess Efficacy of PUFA for the Maintenance of Sinus Rhythm in Persistent Atrial Fibrillation) trial. *Journal of the American College of Cardiology*, *61*(4), 463-468.

[53] Mozaffarian, D., Hao, T., Rimm, E. B., Willett, W. C. & Hu, F. B. (2011). Changes in diet and lifestyle and long-term weight gain in women and men. *New England Journal of Medicine*, *364*(25), 2392-2404.

[54] O'Connell, T. D., Block, R. C., Huang, S. P. & Shearer, G. C. (2017). ω3-Polyunsaturated fatty acids for heart failure: Effects of dose on efficacy and novel signaling through free fatty acid receptor 4. *Journal of molecular and cellular cardiology*, *103*, 74-92.

[55] Siscovick, D. S., Barringer, T. A., Fretts, A. M., Wu, J. H., Lichtenstein, A. H., Costello, R. B. & Appel, L. J. (2017). Omega-3 polyunsaturated fatty acid (fish oil) supplementation and the prevention of clinical cardiovascular disease: a science advisory from the American Heart Association. *Circulation*, *135*(15), e867-e884.

[56] Colussi, G., Catena, C., Novello, M., Bertin, N. & Sechi, L. A. (2017). Impact of omega-3 polyunsaturated fatty acids on vascular function and blood pressure: Relevance for cardiovascular outcomes. *Nutrition, Metabolism and Cardiovascular Diseases*, *27*(3), 191-200.

[57] Jacobson, T. A., Glickstein, S. B., Rowe, J. D. & Soni, P. N. (2012). Effects of eicosapentaenoic acid and docosahexaenoic acid on low-density lipoprotein cholesterol and other lipids: a review. *Journal of clinical lipidology*, *6*(1), 5-18.

[58] Arnesen, H. & Seljeflot, I. (2010). Studies on very long chain marine n-3 fatty acids in patients with atherosclerotic heart disease with special focus on mechanisms, dosage and formulas of supplementation. *Cellular and Molecular Biology (Noisy-Le-Grand, France)*, *56*(1), 18-27.

[59] Den Ruijter, H. M., Verkerk, A. O. &Coronel, R. (2010). Incorporated fish oil fatty acids prevent action potential shortening induced by circulating fish oil fatty acids. *Frontiers in Physiology*, *1*, 149.

[60] He, K., Song, Y., Daviglus, M. L., Liu, K., Van Horn, L., Dyer, A. R. &Greenland, P. (2004). Fish consumption and incidence of stroke: a meta-analysis of cohort studies. *Stroke*, *35*(7), 1538-1542.

[61] Chowdhury, R., Stevens, S., Gorman, D., Pan, A., Warnakula, S., Chowdhury, S. & Franco, O. H. (2012). Association between fish consumption, long-chain Omega 3 fatty acids, and risk of cerebrovascular disease: systematic review and meta-analysis. *BMJ*, *345*, e6698.

[62] Hunter, P. M. & Hegele, R. A. (2017). Functional foods and dietary supplements for the management of dyslipidaemia. *Nature Reviews Endocrinology*, *13*(5), 278-288.

[63] Gil, A. & Gil, F. (2015). Fish, a Mediterranean source of n-3 PUFA: benefits do not justify limiting consumption. *British Journal of Nutrition*, *113*(S2), S58-S67.

[64] Gupta, R. C., Srivastava, A. & Lall, R. (Eds.). (2019). *Nutraceuticals in Veterinary Medicine*. Springer.

[65] Wiseman, M. J. (2019). Nutrition and cancer: prevention and survival.*British Journal of Nutrition*,*122*(5), 481-487.

[66] Nindrea, R. D., Aryandono, T., Lazuardi, L. & Dwiprahasto, I. (2019). Protective effect of omega-3 fatty acids in fish consumption against breast cancer in Asian patients: a meta-analysis.*Asian Pacific journal of cancer prevention: APJCP*,*20*(2), 327.

[67] Pizato, N., Luzete, B. C., Kiffer, L. F. M. V., Corrêa, L. H., de Oliveira Santos, I., Assumpção, J. A. F. & Magalhães, K. G. (2018). Omega-3 docosahexaenoic acid induces pyroptosis cell death in triple-negative breast cancer cells. *Scientific reports*,*8*(1), 1-12.

[68] Volpato, M. & Hull, M. A. (2018). Omega-3 polyunsaturated fatty acids as adjuvant therapy of colorectal cancer. *Cancer and Metastasis Reviews*, *37*(2-3), 545-555.

[69] Eltweri, A. M., Thomas, A. L., Metcalfe, M., Calder, P. C., Dennison, A. R. & Bowrey, D. J. (2017). Potential applications of fish oils rich in omega-3 polyunsaturated fatty acids in the management of gastrointestinal cancer. *Clinical Nutrition*, *36*(1), 65-78.

[70] Liang, Z., Xie, B., Li, J., Wang, X., Wang, S., Meng, S. & Xie, L. (2016). Hypertension and risk of prostate cancer: a systematic review and meta-analysis. *Scientific Reports*, *6*, 31358.

[71] Aucoin, M., Cooley, K., Knee, C., Fritz, H., Balneaves, L. G., Breau, R. & Seely, D. (2017). Fish-derived omega-3 fatty acids and prostate cancer: a systematic review. *Integrative Cancer Therapies*, *16*(1), 32-62.

[72] Berquin, I. M., Edwards, I. J. & Chen, Y. Q. (2008). Multi-targeted therapy of cancer by omega-3 fatty acids. *Cancer Letters*, *269*(2), 363-377.

[73] Kobayashi, M. &Yamamoto, M. (2006). Nrf2-Keap1 regulation of cellular defense mechanisms against electrophiles and reactive oxygen species. *Advances in enzyme regulation*, *46*(1), 113-140.

[74] Serini, S. & Calviello, G. (2017). Modulation of Ras/ERK and phosphoinositide signaling by long-chain n-3 PUFA in breast cancer and their potential complementary role in combination with targeted drugs. *Nutrients*, *9*(3), 185.

[75] Yang, X., Han, H., De Carvalho, D. D., Lay, F. D., Jones, P. A. & Liang, G. (2014). Gene body methylation can alter gene expression and is a therapeutic target in cancer.*Cancer Cell*, *26*(4), 577-590.

[76] Gago-Dominguez, Manuela, J. Esteban Castelao, Can-Lan Sun, David Van Den Berg, Woon-PuayKoh, Hin-Peng Lee.& Mimi C. Yu. (2004). "Marine n-3 fatty acid intake, glutathione S-transferase polymorphisms and breast cancer risk in postmenopausal Chinese women in Singapore." *Carcinogenesis*, *25*, no. 11, 2143-2147.

[77] Liao, Y., Xie, B., Zhang, H., He, Q., Guo, L., Subramaniapillai, M., ...&Mclntyer, R. S. (2019). Efficacy of omega-3 PUFAs in depression: a meta-analysis.*Translational psychiatry*, *9*(1), 1-9.

[78] Grosso, G., Galvano, F., Marventano, S., Malaguarnera, M., Bucolo, C., Drago, F. & Caraci, F. (2014). Omega-3 fatty acids and depression: Scientific evidence and biological mechanisms. *Oxidative Medicine and Cellular Longevity*, *2014*, 313570.

[79] Sublette, M. E., Ellis, S. P., Geant, A. L. & Mann, J. J. (2011). Meta-analysis: effects of eicosapentaenoic acid in clinical trials in depression. *The Journal of Clinical Psychiatry*, *72*(12), 1577.

[80] Hallahan, B., Ryan, T., Hibbeln, J. R., Murray, I. T., Glynn, S., Ramsden, C. E. & Davis, J. M. (2016). Efficacy of omega-3 highly unsaturated fatty acids in the treatment of depression. *The British Journal of Psychiatry*, *209*(3), 192-201.

[81] Peet, M. & Horrobin, D. F. (2002). A dose-ranging study of the effects of ethyl-eicosapentaenoate in patients with ongoing

depression despite apparently adequate treatment with standard drugs. *Archives of General Psychiatry*, *59*(10), 913-919.

[82] Mischoulon, D., Papakostas, G. I., Dording, C. M., Farabaugh, A. H., Sonawalla, S. B., Agoston, M. & Nierenberg, A. A. (2009). A double-blind randomized controlled trial of ethyl-eicosapentaenoic (EPA-E) for major depressive disorder. *The Journal of Clinical Psychiatry*, *70*(12),1636.

[83] Yehuda, S., Rabinovitz, S., Carasso, R. L. & Mostofsky, D. I. (2002). The role of polyunsaturated fatty acids in restoring the aging neuronal membrane. *Neurobiology of Aging*, *23*(5), 843-853.

[84] Carabelli, B., Delattre, A. M., Pudell, C., Mori, M. A., Suchecki, D., Machado, R. B. & Lima, M. M. (2015). The Antidepressant-Like Effect of Fish Oil: Possible Role of Ventral Hippocampal 5-HT 1A Post-synaptic Receptor. *Molecular neurobiology*, *52*(1), 206-215.

[85] Heras-Sandoval, D., Pedraza-Chaverri, J. & Pérez-Rojas, J. M. (2016). Role of docosahexaenoic acid in the modulation of glial cells in Alzheimer's disease. *Journal of neuroinflammation*, *13*(1), 1-13.

[86] Mucke, L. & Selkoe, D. J. (2012). Neurotoxicity of amyloid β-protein: synaptic and network dysfunction. *Cold Spring Harbor perspectives in medicine*, *2*(7), a006338.

[87] Yonezawa, K., Kusumoto, Y., Kanchi, N., Kinoshita, H., Kanegae, S., Yamaguchi, N. & Ozawa, H. (2020). Recent trends in mental illness and omega-3 fatty acids. *Journal of Neural Transmission*, *127*(11).

[88] Cederholm, T., Barazzoni, R. O. C. C. O., Austin, P., Ballmer, P., Biolo, G. I. A. N. N. I., Bischoff, S. C. & Jensen, G. L. (2017). ESPEN guidelines on definitions and terminology of clinical nutrition. *Clinical nutrition*, *36*(1),49-64.

[89] Calon, F., Lim, G. P., Yang, F., Morihara, T., Teter, B., Ubeda, O. & Frautschy, S. A. (2004). Docosahexaenoic acid protects from dendritic pathology in an Alzheimer's disease mouse model. *Neuron*, *43*(5), 633-645.

[90] Lim, G. P., Calon, F., Morihara, T., Yang, F., Teter, B., Ubeda, O. &Cole, G. M. (2005). A diet enriched with the omega-3 fatty acid

docosahexaenoic acid reduces amyloid burden in an aged Alzheimer mouse model. *Journal of Neuroscience*, *25*(12), 3032-3040.

[91] Torres-Guzman, A. M., Morado-Urbina, C. E., Alvarado-Vazquez, P. A., Acosta-Gonzalez, R. I., Chávez-Piña, A. E., Montiel-Ruiz, R. M. & Jimenez-Andrade, J. M. (2014). Chronic oral or intraarticular administration of docosahexaenoic acid reduces nociception and knee edema and improves functional outcomes in a mouse model of Complete Freund's Adjuvant–induced knee arthritis. *Arthritis Research & Therapy*, *16*(2),R64.

[92] Oksman, M., Iivonen, H., Hogyes, E., Amtul, Z., Penke, B., Leenders, I. & Tanila, H. (2006). Impact of different saturated fatty acid, polyunsaturated fatty acid and cholesterol containing diets on beta-amyloid accumulation in APP/PS1 transgenic mice. *Neurobiology of Disease*, *23*(3), 563-572.

[93] Devassy, J. G., Leng, S., Gabbs, M., Monirujjaman, M. & Aukema, H. M. (2016). Omega-3 polyunsaturated fatty acids and oxylipins in neuroinflammation and management of Alzheimer disease. *Advances in Nutrition*, *7*(5), 905-916.

[94] Hawkey, E. & Nigg, J. T. (2014). Omega− 3 fatty acid and ADHD: Blood level analysis and meta-analytic extension of supplementation trials. *Clinical Psychology Review*, *34*(6), 496-505.

[95] Farooqui, A. A. & Horrocks, L. A. (2001). Plasmalogens, phospholipase A 2, and docosahexaenoic acid turnover in brain tissue. *Journal of Molecular Neuroscience*, *16*(2-3), 263-272.

[96] Ross, B. M., McKenzie, I., Glen, I. & Bennett, C. P. W. (2003). Increased levels of ethane, a non-invasive marker of n-3 fatty acid oxidation, in breath of children with attention deficit hyperactivity disorder. *Nutritional Neuroscience*, *6*(5), 277-281.

[97] Yehuda, S., Rabinovitz-Shenkar, S. & Carasso, R. L. (2011). Effects of essential fatty acids in iron-deficient and sleep-disturbed attention deficit hyperactivity disorder (ADHD) children. *European Journal of Clinical Nutrition*, *65*(10), 1167-1169.

[98] Crippa, A., Agostoni, C., Mauri, M., Molteni, M. & Nobile, M. (2018). Polyunsaturated fatty acids are associated with behavior but

not with cognition in children with and without ADHD: an Italian study. *Journal of Attention Disorders*, *22*(10), 971-983.

[99] Kang, J. X. & Weylandt, K. H. (2008). Modulation of inflammatory cytokines by omega-3 fatty acids. In *Lipids in Health and Disease* (pp. 133-143). Springer, Dordrecht.

[100] Calder, P. C. (2015). Marine omega-3 fatty acids and inflammatory processes: Effects, mechanisms and clinical relevance. *Biochimicaet BiophysicaActa (BBA)-Molecular and Cell Biology of Lipids*, *1851*(4), 469-484.

[101] Calder, P. C. (2008). Polyunsaturated fatty acids, inflammatory processes and inflammatory bowel diseases. *Molecular Nutrition & Food Research*, *52*(8), 885-897.

[102] Engler, H., Brendt, P., Wischermann, J., Wegner, A., Röhling, R., Schoemberg, T. & Schedlowski, M. (2017). Selective increase of cerebrospinal fluid IL-6 during experimental systemic inflammation in humans: association with depressive symptoms. *Molecular psychiatry*, *22*(10), 1448-1454.

[103] Serhan, C. N. &Levy, B. D. (2018). Resolvins in inflammation: emergence of the pro-resolving superfamily of mediators. *The Journal of Clinical Investigation*, *128*(7), 2657-2669.

[104] Bäck, M. & Hansson, G. K. (2019). Omega-3 fatty acids, cardiovascular risk, and the resolution of inflammation. *The FASEB Journal*, *33*(2), 1536-1539.

[105] Gan, R. W., Bemis, E. A., Demoruelle, M. K., Striebich, C. C., Brake, S., Feser, M. L. & Norris, J. M. (2017). The association between omega-3 fatty acid biomarkers and inflammatory arthritis in an anti-citrullinated protein antibody-positive population. *Rheumatology*, *56*(12), 2229-2236.

[106] Saidane, O., Semerano, L. & Sellam, J. (2018). Could omega-3 fatty acids prevent rheumatoid arthritis?. *Joint Bone Spine*, *85*(June 2018).

[107] Stoodley, I., Garg, M., Scott, H., Macdonald-Wicks, L., Berthon, B. & Wood, L. (2020). Higher omega-3 index is associated with better

asthma control and lower medication dose: A cross-sectional study. *Nutrients*, *12*(1), 74.

[108] Adams, S., Lopata, A. L., Smuts, C. M., Baatjies, R. & Jeebhay, M. F. (2019). Relationship between serum omega-3 fatty acid and asthma endpoints. *International journal of environmental research and public health*, *16*(1), 43.

[109] Moomand, K. & Lim, L. T. (2014). Oxidative stability of encapsulated fish oil in electrospunzeinfibres. *Food Research International*, *62*, 523-532.

[110] Espín, J. C., García-Conesa, M. T. & Tomás-Barberán, F. A. (2007). Nutraceuticals: facts and fiction. *Phytochemistry*, *68* (22-24), 2986-3008.

[111] Taneja, A. & Singh, H. (2012). Challenges for the delivery of long-chain n-3 fatty acids in functional foods. *Annual Review of Food Science and Technology*, *3*, 105-123.

[112] Augustin, M. A. & Sanguansri, L. (2015). Challenges and solutions to incorporation of nutraceuticals in foods. *Annual Review of Food Science And Technology*, *6*, 463-477. Augustin,

[113] Drusch, S. & Mannino, S. (2009). Patent-based review on industrial approaches for the microencapsulation of oils rich in polyunsaturated fatty acids. *Trends in Food Science & Technology*, *20*(6-7), 237-244.

[114] Baik, M. Y., Suhendro, E. L., Nawar, W. W., McClements, D. J., Decker, E. A. & Chinachoti, P. (2004). Effects of antioxidants and humidity on the oxidative stability of microencapsulated fish oil. *Journal of the American Oil Chemists' Society*, *81*(4), 355-360.

[115] Drusch, S., Serfert, Y., Scampicchio, M., Schmidt-Hansberg, B. & Schwarz, K. (2007). Impact of physicochemical characteristics on the oxidative stability of fish oil microencapsulated by spray-drying. *Journal of Agricultural and Food Chemistry*, *55*(26), 11044-11051

[116] Barrow, C. J., Nolan, C. & Holub, B. J. (2009). Bioequivalence of encapsulated and microencapsulated fish-oil supplementation. *Journal of Functional Foods*, *1*(1), 38-43.

[117] Gutiérrez, J. M., González, C., Maestro, A., Solè, I. M. P. C., Pey, C. M. & Nolla, J. (2008). Nano-emulsions: New applications and optimization of their preparation. *Current Opinion in Colloid & Interface Science*, *13*(4), 245-251.

[118] McClements, D. J. & Rao, J. (2011). Food-grade nanoemulsions: formulation, fabrication, properties, performance, biological fate, and potential toxicity. *Critical Reviews in Food Science and Nutrition*, *51*(4), 285-330.

[119] Jaafari, M. R., Ghafarian, A., Farrokh-Gisour, A., Samiei, A., Kheiri, M. T., Mahboudi, F. A &McMaster, W. R. (2006). Immune response and protection assay of recombinant major surface glycoprotein of Leishmania (rgp63) reconstituted with liposomes in BALB/c mice. *Vaccine*, *24*(29-30), 5708-5717.

[120] Anton, N. & Vandamme, T. F. (2009). The universality of low-energy nano-emulsification. *International Journal of Pharmaceutics*, *377*(1-2), 142-147.

[121] Crommelin, D. J., Park, K. & Florence, A. (2010). Pharmaceutical nanotechnology: unmet needs in drug delivery. *Journal of Controlled Release.*, *141*(3), 263-264.

[122] Yu, H. & Huang, Q. (2013). Investigation of the cytotoxicity of food-grade nanoemulsions in Caco-2 cell monolayers and HepG2 cells. *Food Chemistry*, *141*(1), 29-33.

[123] Uluata, S., McClements, D. J. & Decker, E. A. (2015). Physical stability, autoxidation, and photosensitized oxidation of ω-3 oils in nanoemulsions prepared with natural and synthetic surfactants. *Journal of Agricultural and Food Chemistry*, *63*(42), 9333-9340.

[124] Kim, J. H., Hong, S. S., Lee, M., Lee, E. H., Rhee, I., Chang, S. Y. & Lim, S. J. (2019). Krill Oil-Incorporated Liposomes As An Effective Nano vehicle To Ameliorate The Inflammatory Responses Of DSS-Induced Colitis. *International Journal of Nanomedicine*, *14*, 8305.

[125] Hadian, Z. (2016). A review of nanoliposomal delivery system for stabilization of bioactive omega-3 fatty acids. *Electronic Physician*, *8*(1), 1776.

[126] Li, X., Diao, W., Xue, H., Wu, F., Wang, W., Jiang, B. & Yu, W. (2020). Improved efficacy of doxorubicin delivery by a novel dual-ligand-modified liposome in hepatocellular carcinoma. *Cancer Letters*, *489*, 163-173.

In: Nutraceuticals
Editors: A. Kumari and G. Singh
ISBN: 978-1-68507-488-3

Chapter 15

ROLE OF NUTRACEUTICALS IN FOOD PACKAGING

***Priyanka Prajapati*[1], *Meenakshi Garg*[1,*] *and Rajni Chopra*[2]**

[1]Department of Food Technology,
Bhaskaracharya College of Applied Sciences,
University of Delhi, Delhi, India
[3]Department of Food Science and Technology,
National Institute of Food Technology Entrepreneurship
and Management, Sonipat, Haryana, India

ABSTRACT

The increased consumer's interest in healthy and nutritious food has driven the demand for food containing biologically active molecules, especially antimicrobials, antioxidants, phenols, and essential oils. Food packaging seems to provide a promising delivering system for incorporating these nutraceuticals compounds into food. Packaging plays four passive functions of containment, protection, convenience, and

[1*] Corresponding Author's E-mail: meenakshi.garg@bcas.du.ac.in.

communication. The packaging material should also be pertinent to better shelf life and maintaining nutritional quality and safety until the food package is open. Nutraceuticals or bioactive compounds can be used as a vehicle to deliver antioxidants, enzymes, flavors, anti-browning agents, antibacterial, and other active compounds to extend the shelf life of food products even after opening. Dynamic packaging technology is based on the concept of the incorporation of specific components into packaging systems that release or absorb substances from or into the packed food or the surrounding environment of the food product. Hence, it gives rise to a new conceptual approach to developing a new packaging material containing nutraceuticals compounds called "Bioactive" packaging systems. A bioactive packaging system protects the food product through the sustained release of functional compounds encapsulated in the food package. Nanotechnology is one such method of encapsulating the nutraceuticals compound into food packaging material. Reducing the particle size of a bioactive compound using micro/ nanotechnologies may improve the bioactive compound's biological availability, delivery properties, and solubility. Emulsification, liposomes, electrospinning, and coatings are widely used techniques to develop nano or microparticulate systems. This chapter will focus on adding different nutraceutical compounds into packaging materials, solving the problem of direct application of synthetic preservatives to the food product. Thus, protecting food from the external environment with minimal use of artificial preservatives can be achieved by applying bioactive packaging.

INTRODUCTION

Over the past few years, the public interest and consumer demand have shifted towards nutraceutical compounds and functional foods, powered by progressive scientific research to identify potential applications of nutraceutical substances in the food chain system. In 1989 the term "nutraceutical" was coined from "nutrition" and "pharmaceutical." According to Dr. Stephen DeFelice, nutraceutical can be defined as "a food (or part of a food) that provides medical or health benefits, including the prevention and treatment of a disease" [1]. Nutraceuticals are nutritional supplements that confer a health benefit to the consumer other than nutrition. Some popular nutraceuticals compounds include tocopherol, green tea, calcium, probiotic culture, glucosamine, echinacea, omega-3, folic acid, and plant essential oils. The majority of the functional compounds may provide multiple health benefits to the consumer [2]. For example, the high amount of antioxidant compounds (catechins and (-)-epigallocatechin-3-gallate) present in green tea leaves could be beneficial in preventing several types of cardiovascular disease, respiratory disease, and liver disease [3]. Garlic essential oil has the potential to act as antimicrobial, cardioprotective, antioxidant, antitumor, and antidiabetic due to the presence of diallyl mono-, di- and trisulphide compounds [4]. Due to the beneficial property of these nutraceuticals compounds, they find a scope in food packaging to deliver high-quality and nutritious foods to consumers.

With the growing demand to provide quality food products, the packaging industry tries to fabricate packaging material loaded with active molecules, which are also highly responsive to any surrounding condition changes. The essential fundamental operation of the packaging material is to serve as a container for food, prevent any physical damage, and protect it from unwanted chemical and biological changes during storage and transportation. Among many other packaging functionalities, one function is to deliver fresh, wholesome food to the consumer. This leads to the setting of a new domain in packaging technology, known as bioactive packaging, in which packaging material or coating contains active compounds. The active compounds encapsulated in packaging material are phytochemicals, vitamins, nanofibers, and prebiotics [5]. Bioactive packaging is a novel approach that may benefit the consumer beyond the nutrients that a food originally inherits. Active packaging system founds the basis of bioactive packaging [6]. A typical active packaging provides the traditionally passive role of food protection and interacts with the food product or headspace to extend food stability. Besides fulfilling all these functions, bioactive packaging contains functional compounds that may provide additional health benefits to the consumer. Edible coating and films have found application in bioactive packaging as a carrier of nutraceuticals compounds.

ANTIMICROBIAL PACKAGING

The increased consumer interest in easy-to-consume food products makes the food manufacturers focus more on food safety and quality [7]. In this regard, food researchers are developing novel techniques to preserve food by using natural compounds. Applying these natural compounds can solve food spoilage and safety issues with no harmful effects on consumer health. For example, antimicrobials are directly added to food products to protect food from spoilage in food industries.

Artificial antimicrobials used in food are chemical compounds that are intentionally added to food in the permissible amount to inhibit the growth of pathogenic and spoilage bacteria. Common synthetic antimicrobials used in food products are nitrites, tartaric acid, potassium metabisulfite, citric acid, sodium benzoate, nisin, chlorides, and sulfates. Generally, synthetic antimicrobials belong to the family of organic acids and their salts. Regulatory agencies approve these compounds and their amounts that are added to food, and food manufacturers must use these artificial antimicrobials in the permissible limit. However, using some artificial antimicrobials also menaces consumer health as they may bind with some nutrients and make them unavailable or degrade them. For example, sulfites are used as an antimicrobial agent in some beverages, and cordials can bind with thiamine (Vitamin B1) and make them unavailable to be absorbed by the body [8]. This reflects the need to find an alternative method to introduce antimicrobial in food products to enhance their shelf life. Furthermore, almost 99% of the food products and beverages have to be packed in packaging material; thus, adding antimicrobials on the packaging material can be a promising technology to preserve food from antimicrobials yet maintain its natural characteristics. The critical advantage of this method would be the protection of packaged food products from the direct addition of synthetic antimicrobials in the food composition.

Generally, three modes of action have been shown by antimicrobials agents used in the packaging system.

- Release: Antimicrobial agents migrate from the packaging material to the package's headspace or diffuse into the food and inhibit the growth of spoilage microorganisms.
- Absorption: These antimicrobials remove or bind with the essential factors required by the microorganism for their growth.
- Immobilization. Immobilization type of antimicrobials is more effective for liquid food products than solid food because they only inhibit the microorganism that comes in their direct contact. Hence, whole liquid food is more likely to contact the antimicrobial compound immobilized within the matrix of packaging material [9].

Several different forms are developed for antimicrobial packaging systems, which include:

1. Sachets/pads containing volatile antimicrobial agents are added to the food package.
2. Direct addition of volatile and non-volatile antimicrobial agents into the polymer used for the food product packaging.
3. Coating of the polymer by the antimicrobial agent.
4. Edible coating of the food by the polymer containing an antimicrobial agent.
5. Immobilization of antimicrobials compounds to polymers enclosing food products by ion or covalent bonding.
6. Use of polymers that is inherently antimicrobial in nature [10].

The main objective of the antimicrobial packaging system is to reduce, retard or completely inhibit the growth of spoilage microorganisms. This is achieved only when antimicrobial compounds directly contact the microbial cell and cause cell lysis by disrupting their mechanism. In addition, volatile antimicrobial compounds have killed microorganisms in the headspace of the food package (indirect contact).

Table 1. Packaging material having antibacterial compounds of plant origin

S.No.	Packaging material	Nutraceutical compound	Bacterial strain	References
1	Polylactic acid films	Cinnamaldehyde	*E. coli* W1485 *B. Cerus* ATCC 14799	Makwana, S. et al. 2014
2	Sodium caseinate film	*Lactobacillus sakei*	*Listieria monocytogens*	H. Gialamas et al. 2010
3	Chitosan film	Rosemary essential oil	*Listeria monocytogenes* (PTCC 1163, *Streptococcus agalactiae* (PTCC 1768), *Escherichia coli* (PTCC 1533)	Abdollahi, M. et al. (2012)
4	Zein films	Zataria multiflora Boiss. essential oil	*Listeria monocytogenes, Escherichia coli*	Kashiri, M. et al. 2017
5	Cape hake by-products proteins	Garlic essential oil	*Shewanella putrefaciens*	Teixeira, B. et al. 2014
6	Chitosan film	Lemongrass essential oil	Fungus	Ali, A. et al. (2015)
7	Gelatin film	Lemongrass essential oil	Lactic acid bacteria (LAB), psychrophilic bacteria, H2S-producing bacteria and Enterobacteriaceae	Ahmad, M, et al. (2012)
8	Linear low-density polyethylene	Clove essential oil	*Salmonella Typhimurium* and *Listeria monocytogenes*	Mulla, M. et al. 2017
9	Chitosan film	Thyme essential oil	yeast cells	Quesada, J. et al. (2016)
10	Carboxymethyl cellulose (CMC)-polyvinyl alcohol (PVA) based films	Cinnamon essential oil	*Penicillium. digitatum*	Fasihi, H. et al. (2016)
11	Polylactic acid film	*Origanum vulgare L. virens* oil	*Staphylococcus aureus, Yersinia enterocolitica, Listeria monocytogenes, Enterococcus faecalis and Staphylococcus carnosus.*	Llana-Ruiz-Cabello, M, et al. (2016)

Table 2. Sachets or absorbents pads containing antimicrobial compounds to protect food

S.No.	Essential oil	Target microorganism	Food matrix	References
1	oregano essential oil (OEO)	*Escherichia coli*, *Salmonella* Enteritidis, *Penicillium* sp. Yeast Molds	Bread	Passarinho, A. T. P. et al. (2014)
2	Cinnamon essential oil OEO Lemongrass essential oil	*Alternaria alternata* *Fusarium semitectum*, *Lasiodiplodia theobromae* *Rhizopus stolonifera* Mesophilic bacteria	Papaya	Espitia, P. J. P. et al. (2012)
3	OEO	Psychrotrophs, Pseudomonads, Enterobacteriaceae, yeasts, and lactic acid bacteria	Chicken drumsticks	Oral, N. et al. (2009)
4	Allyl isothiocyanate (main component mustard essential oil)	*Aspergillus flavus*	Peanuts	C. G. Otoni et al. (2014)
5	Rosemary and Thyme essential oil	*Listeria monocytogenes*	Mozzarella cheese	Jung H. Han. et al. (2014)

This fact implies that the food product must have come in contact with antimicrobial agents, either in the vapor phase (indirect contact) or by direct contact between the antimicrobial agent and food [9].

Some essential oils obtained from food spices are volatile in nature. The volatility of essential oils is due to their aromatic compounds of low molecular weight. Essential oils contain a good amount of phenolic compounds, up to 85%. These phenolic compounds like (eugenol, thymol, and carvacrol show antimicrobial activity by disrupting the cell membrane or genetic material and damaging the bacterial cell's enzymatic function [11]. Essential oils extracted from garlic, cinnamon, lemongrass, and thyme have been most commonly used in packaging films for their antimicrobial activity. In addition, the essential oils obtained from plants have some therapeutic value, like some of these oils are potential antioxidants [12].

Most researches are focused on the sustained release of antimicrobial compounds from packaging material into the food product. Sachets and absorbent pads have expanded their application to be used as antimicrobial packaging. They are incorporated with an antimicrobial agent and allow sustained release of these antimicrobials to kill foodborne pathogens.

The main principle behind applying carrier/emitting sachets is the subsequent release of active compounds that are adsorbed on the (sachet material) permeable to the active compound. The sachet material used for the manufacture of sachets must withstand handling and transportation to avoid failure and leakage of the active compound [8].

ANTIOXIDANT PACKAGING

Oxidation of food and microbial growth are the leading causes affecting food products' safety and quality. Lipid oxidation is the main criterion of spoilage of a great variety of fat-rich food, such as nuts, fatty fish, red meats, whole milk powders, oils, and salad dressing.

It leads to the formation of toxic aldehydes and epoxy compounds that affect both sensory attributes and nutritional quality of food. Existing approaches to control lipid oxidation in foods is the direct addition of an artificial 'antioxidant's compounds (BHT, BHA) or packaging under modified atmospheres in which oxygen in the food package headspace is limited. However, these synthetic antioxidants (polyphenol, organophosphate, and thioester compounds) cause potential adverse health effects on human health in the long run. A recent approach to these methods is the fabrication of packaging material with antioxidant compounds whose main advantage is that it can provide control release of antioxidants with antioxidant compounds whose main advantage is to control antioxidants' release during storage.

Two methods describe the working of antioxidant packages:

- Release of antioxidant compounds into the headspace of a food package or diffusion in the food product.
- The scavenging of prooxidant compounds such as oxygen species, free radical species that initiate chain reaction or metal ions from the headspace of a food package or the food itself.

Selection of suitable antioxidant and packaging material to incorporate that antioxidant is crucial. The packaging material and antioxidant compound should be compatible with sustaining the antioxidant's release to inhibit or retard oxidation. The homogenous distribution of antioxidants throughout the packaging material and the partition of coefficient in different phases should favor its release into the food particle or package that contains food. The solubility of antioxidants in the food product determines its effectiveness after releasing from packaging film. Thus, the type of antioxidant compounds should be selected while taking into consideration its solubility in food.

Table 3. Natural antioxidant incorporated into packaging films

S.No.	Packaging material	Antioxidant compound	Antioxidant activity	References
1	Cassava starch films	Polyphenols-rich rosemary extracts	4.4 - 13.6 mg of gallic acid equivalents per gram.	Piñeros-Hernandez, D., et al. (2017)
2	Polypropylene	Green tea extract	13.7 mg of gallic acid equivalent per gram	C. López de Dicastillo et al. (2013)
4.	Chitosan film	Mango leaf extract	3.47 mg of gallic acid equivalent per gram	Rambabu, K et al. (2018)
5	Chitosan film	Olive pomace flour	1.18-2.73 µmol Trolox equivalents/g dried film)	De Moraes Crizel, T et al. (2018)
6	PVA	Apple pomace	7.9 mg mg of gallic acid equivalent per gram	K.K. Gaikwad et al. 2016

Two types of mechanisms are employed for developing antioxidant packaging systems:

- Independent devices: An independent device includes sachets, pads, or labels containing the antioxidant agent separately from the food product. It is added within the carrier material and placed individually beside food.
- Antioxidant packaging materials: In this packaging system, the polymer contains antioxidant agents that help inhibit oxidation. The antioxidant agent is incorporated in the packaging material, which scavenges the undesirable compounds from the headspace of the food package or by releasing antioxidant compounds to the food [27].

Recently the trend has shifted towards utilizing natural products due to the presence of biologically active compounds. Ascorbic acid (vitamin C), carotenoids, and tocopherols (vitamin E) are natural antioxidants obtained from fruits and vegetables. These compounds have immense potential to be used as an alternative to synthetic antioxidants for the polymer used for food packages. Phytochemicals and phenol derivatives obtained from plant sources also have a significant role in human physiology. However, the complex chemical nature of phenols and their method to get accurate structure limit the detailed study of their effect on the food product. Nevertheless, natural antioxidants such as tocopherol, Eugenol, thymol, green tea extract, and cinnamaldehyde have been successfully encapsulated into packaging films with the further possibility of enhanced activity in the food product [28].

PROBIOTICS

Nowadays, probiotics are gaining more attention because of their several health benefits to human health. Probiotics perform several functions like producing antimicrobial substances, modulation of the immune response, strengthening the intestinal barrier, and inhibiting pathogenic microorganisms either by producing antimicrobial peptides or competing for binding sites, nutrients, and growth factors (FAO, 2001). Probiotics were originally defined as a "mono- or mixed culture of live microorganisms which, when applied to man or animal, affects the host beneficially by improving the properties of indigenous microflora" (FAO, 2002) [34]. The most common bacteria used commercially for probiotic production are *Streptococcus thermophilus, Saccharomyces boulardii,* Lactobacillus, and Bifidobacterium. Dairy products such as yogurt, cheese, curd, and dairy desserts are popular food products added with probiotics culture. However, yogurt and curd are the most popular matrices for incorporating probiotic microorganisms [35].

Due to their antimicrobial activity and other health benefits, probiotics have the potential scope to be used in polymer matrices used in food packaging. In this way, probiotics have been encapsulated into packaging film or edible coating to develop bioactive food packaging materials as an alternative method for inhibiting the growth of pathogenic microorganisms.

Table 4. Polymeric matrix having probiotic culture

S.No.	Polymeric matrix.	Probiotic	Food product	Viability of culture	References
1	CMC-sodium caseinate	*Lactobacillus acidophilus, L. reuteri, L. casei, L. rhamnosus, and Bifidobacterium bifidum*	Raw trout fillets	10^9 -10^6 log CFU/g	M. Mozaffarzogh, et al. 2020
2	Agar	*Lactobacillus paracasei* L26 and *Bifidobacterium lactis* B94	hake fillets	10^6 log CFU/g	A.M. López de Lacey et al. (2014)
3	Sodium Caseinate	*Lactobacillus acidophilus* and *Lactobacillus casei*	-	>10^4 CFU/cm^2	E, Abdollahzadeh. et al. (2018)
4	Sodium carboxymethyl cellulose and hydroxyethyl cellulose	*Lactobacillus rhamnosus* GG	-	10 log CFU/ml	P, singh. et al. (2018)
5	cellulose nanofiber and inulin incorporated CMC	*Lactobacillus plantarum*	chicken fillet	6 log CFU/g	N, Zabihollahi et al. (2020)
6	Gelatin and low methoxyl pectin	*Lactobacillus plantarum, L. casei,* and *Saccharomyces boulardii*	-	10^6 CFU/g	D, Khodaei. et al. (2020)
7	Cassava starch and sodium-CMC	*Lactobacillus plantarum and Pedocococcus pentosaceu*	Banana	8 log CFU/g	S, Li. et al. (2020)

Besides protecting food from pathogenic attacks, probiotics have the potential to improve human gut health. Cell survival in the final product is a major issue while processing food that contains probiotic bacteria [35]. A viable cell count of 10^{8-9} per day is necessary to benefit the consumer. Aiming to improve existing technologies to deliver or develop probiotics, researchers investigate a new possible application by incorporating them into food packaging matrices or edible film/coatings [36].

Microencapsulation is a new technology used to encapsulate probiotics into the packaging film. It may be defined as the method to pack functional compounds into small capsules at a micro-level. These capsules are made up of an encapsulant material like protein, lipid, or carbohydrate. The microencapsulation technique is intended to protect the viable culture of probiotics from detrimental food conditions (e.g., low pH, high temp, oxygen availability, and light). As a result, it can decrease the contact between probiotic culture and the external environment and prevent its degradation, thus maintaining its final product activity [35].

FLAVOURS

Packaging films and edible coatings can encapsulate the active ingredients, thus helping in improving the food product's taste and nutritional value. Generally, edible films and coatings are used to enhance flavor, but fewer studies are reported to integrate nutraceuticals compounds into edible film or coating.

The US Department of 'Agriculture's (USDA's) Agricultural Research Service (Albany, California), in association with Origami Foods (Pleasanton, California), has developed vegetable and fruit edible films as an alternative to the seaweed sheets (nori), which has been originally used for sushi and other Asian foods. These edible wraps contain spices, seasoning colorants, flavors, and vitamins which enhance nutritional quality and give variety to the consumer. Tomato, apple, pear, mango broccoli, and various other fruits and vegetables are used to formulate soft and pliable sheets using infrared drying technology [44].

NUTRIENT

Micronutrient malnutrition among children and adults is of major concern in the entire world. It is of significant importance because it affects a large number of people and remains a risk for many other diseases. Some vitamins and minerals like vitamin E, vitamin C, calcium, and zinc can be incorporated into edible films and coatings. Thus, they enhance the nutritional quality of food products that may not be inherently present in food or degrade during processing [45]. Magnesium, calcium, and vitamin E are the most studied nutrients used in incorporating packaging material as they play essential roles in the human body to prevent certain diseases. Gluconal Cal (GC Glucona America Inc., Janesville, WI) is a mixture of calcium lactate, gluconate, and α-tocopheryl acetate is encapsulated in edible coatings made up of protein and carbohydrate. These edible coatings have shown increased nutritional value, good bioavailability of nutrients, and better barrier properties [46]. These edible films may be used for wrapping or coating to enhance the nutritional content of food products. (Mei et al. 2002) enhanced the calcium and vitamin E content of xanthan gum coating used to wrap food products. Edible coating of xanthan gum contains 5% Gluconal Cal, and 0.2% α-tocopherol acetate (vitamin E) by weight. After incorporation, calcium and vitamin E content increases from 2.6% to 6.6%, and from 0 to about 67% of the Dietary Reference Intakes (DRI) values, respectively [46, 47].

(Massilia et al. 2015) developed alginate-based film to enrich tropical fruits with calcium and vitamin D. Incorporation of vitamin-D and calcium into the alginate films contributes by 32-50% and 83.2% respectively of the daily requirement of calcium and vitamin-D prescribed for school-going children [48].

Encapsulation of Nutraceutical Compound in Food Packaging

Encapsulation of bioactive compounds into the packaging material can protect and deliver active compounds to the consumer. It is defined as a process for entrapping active compounds obtained from the plant (e.g., vitamins, phenols, antioxidants, and omega-3, omega -6 fatty acids, or viable microorganisms such as probiotics) within a polymer matrix (e.g., carbohydrate, protein, lipid). The main advantages of the encapsulation techniques include the protection of active compounds by encapsulating them in the polymer matrix, which helps them withstand external environmental conditions, thus making them more stable in packaging material. It also helps in masking unwanted odour and taste [50]. In addition, the encapsulation technique increases the nutraceutical compound's bioavailability besides its sustained release in the food product. Nutraceutical compounds can be incorporated into the melted polymer matrix by employing different encapsulation techniques and extruded in a thin film [28].

Different delivery systems or carriers are developed to incorporate active compounds or nutraceuticals into the bioactive packaging systems. This system includes nanoparticles, microcapsules, emulsions, cyclodextrins, and liposomes. However, the techniques should be incongruent with the packaging material and do not degrade the quality of packaging material in terms of its mechanical properties and barrier properties. Thus, selecting an appropriate delivery system for encapsulating nutraceutical compounds is of major concern to preserve the primary function of nutraceuticals and all functionalities of a package [50].

Liposomes are lipids compounds used extensively in cosmetics and pharmaceuticals products, but their utilisation in the food industry is limited. Liposomes are microscopic molecules with spherical-shaped vesicles composed of a wall of amphipathic lipid bilayers arranged in one or more concentric spheres with an aqueous phase inside and outside the lipid bilayers (Figure 1). The capability of liposome molecules to carry both hydrophilic or lipophilic compounds has allowed these microstructure compounds to be used as a potential delivery system for carrying biological active compounds [50]. Phospholipids, the main compound forming liposomes, can be made from lecithins produced from various foods: milk, egg, soy, sunflower, and canola) [51]. A chitosan-based coating with liposomes molecules containing laurel essential oil and nanosilver particles was formulated by (Z. Wu et al. 2018). The liposomes helped sustain the release of functional compounds and had shown good antioxidant and antimicrobial activity to preserve pork [52]. Nano-encapsulated cinnamaldehyde liposomes show significant antimicrobial activity against foodborne pathogens. Thus, it could be a potential alternative for antimicrobial packaging films. The major disadvantage of liposomes is that they can lead to the production of chromatic films due to lecithin, leading to instability in packaging material and high cost [50].

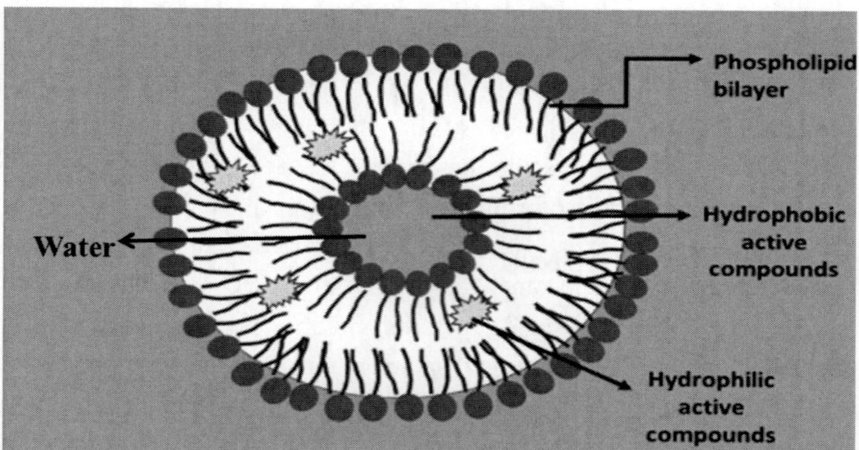

Figure 1. Liposomes loaded with nutraceuticals compounds.

Nanotechnology was first applied in the food sector to manufacture nanosized packaging materials with improved functionalities. Nanotechnology could develop foods with modifying colour, flavour, or nutritional properties with significantly different from their inherited characteristics. With advancements in scientific research, extrusion of food packaging material having nanostructured active molecules helps increase the shelf life of food products. Packaging films formed using nanotechnology have the ability to control air and moisture exchange with the external environment. Besides formulation of the biodegradable outer covering of food, edible nano-coatings (~5 nm thin coatings) can also be developed using nanotechnology [53]. These edible nano-coatings encapsulated with active compounds are applied to various meat, confectionery products, cheese, fruits, and vegetables, where they act as moisture and gas barrier. In addition, to prolong shelf life, they provide flavour, colour, enzymes, antioxidants, anti-browning compounds to the manufactured food products. The encapsulation of active nanosized particles into the polymer is used to fabricate packaging material which may help to increase barrier functionalities towards the external environment [54].

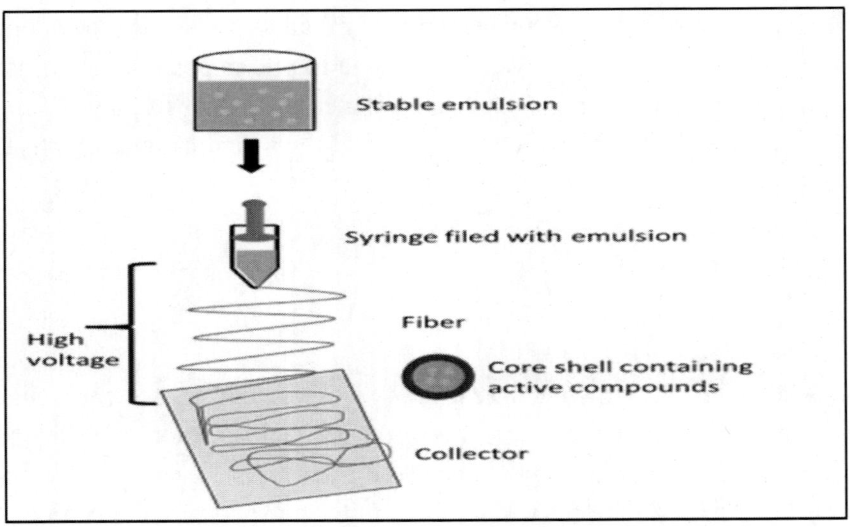

Figure 2. Electrospinning process producing core shell nano fibers containing active compounds.

Electrospinning is a novel technique for the development of bioactive packaging material containing nutraceutical compounds. It is a cost-effective, versatile technique used to develop a variety of continuous micro or nanoscale fibre from various biopolymers. The Figure 2 shows electrospinning process producing core shell fibers containing active compounds.

Advantages of using electrospun fibers over traditional polymers are its large area to volume ratio, submicron to the nanoscale diameter of molecules, high sensitivity to changes in external atmosphere, and capability to carry heat-sensitive active compounds [49]. This non-mechanical technique involves charging the polymer droplet using a strong electric field, which helps in the ejection of a liquid jet through the spinneret to form packaging fiber. Emulsion electrospinning is a novel technique to fabricate core-shell fibers to the encapsulation of different active compounds (flavonoids, essential oils) into the packaging material.

This technology leads to the development of nanofibers with core-shell morphology [55]. This core-shell structure helps to contain bioactive compounds and minimizes their volatilization, oxidation, and release ratio. Despite the advantages of this technique, minimal works have been done in the food industry. However, it shows an immense potential to be used in the developingtbioactive packaging materials, demonstrating the high effectiveness of the active compounds and their control release into the package.

CONCLUSION

There is a growing interest of food scientists to encapsulate nutraceuticals compounds into the food packaging material. Packaging material is a suitable carrier to deliver these active compounds to the consumer. In addition, while performing the essential function of packaging, a bioactive packaging system may also impart several health benefits to the consumer. Plant origin essential oils, probiotics, and nutrients are incorporated into polymeric matrices to enhance food quality, safety, and nutritional content. Microencapsulation, nanoencapsulation, and electrospinning are novel techniques used to incorporate active compounds in polymeric films. Despite all these technologies, researchers still need to investigate new methods to optimize bioactive packaging systems at the retail level and for different food products.

FUTURE TRENDS

The bioactive packaging system is at a nascent stage. More research needs to be carried out to optimize these systems at the commercial level and for different food products. In addition, compatibility between the packaging material and nutraceuticals compounds must be considered while fabricating bioactive packaging for the specific food product. Homogenous dispersion of active compounds into packaging material, their solubility in food products, and their sustained release to perform that particular functions are some areas that require more attention. Thus, more investigation needs to be done to fabricate a packaging material having all the basic functionality (barrier against the physical, chemical, and biological stimulus, carrying and protecting food protect during transportation and storage) along with active compound or nutraceutical compound that confer a health benefit to the consumer.

REFERENCES

[1] Kalra, E. K. (2003). Nutraceutical-definition and introduction. *AAPS PharmSci*, 5(3): 27–28. doi:10.1208/ps050325.

[2] Nasri, H., Baradaran, A., Shirzad, H. and Rafieian-Kopaei, M. (2014). New concepts in nutraceuticals as alternative for pharmaceuticals. *International Journal of Preventive Medicine*, 5(12): 1487–1499.

[3] Chacko, S. M., Thambi, P. T., Kuttan, R. and Nishigaki, I. (2010). Beneficial effects of green tea: a literature review. *Chinese Medicine*, 5:13. https://doi.org/10.1186/1749-8546-5-13.

[4] El-Guendouz, S., Aazza, S., Anahi Dandlen, S., Majdoub, N., Lyoussi, B., Raposo, S., Dulce Antunes, M., Gomes, V. and Graça Miguel, M. (2019). Antioxidant Activity of Thyme Waste Extract in O/W Emulsions. *Antioxidants* (Basel, Switzerland), 8(8): 243. https://doi.org/10.3390/antiox8080243.

[5] Topuz, Fuat and Uyar, Tamer. (2019). Antioxidant, Antibacterial and Antifungal Electrospun Nanofibers for Food Packaging Applications. *Food Research International*. 130. 108927. 10.1016/j.foodres.2019.108927.

[6] López-Rubio, Amparo & Gavara, Rafael & Lagaron, Jose Maria. (2006). Bioactive packaging: Turning foods into healthier foods through biomaterials. *Trends in Food Science & Technology*. 17: 567-575. 10.1016/j.tifs.2006.04.012.

[7] Gialamas, Haralampos & Zinoviadou, Kiki and Biliaderis, Costas. (2010). Development of a novel bioactive packaging based on the incorporation of *Lactobacillus sakei* into sodium-caseinate films for controlling *Listeria monocytogenes* in foods. *Food Research International*. 43: 2402-2408. 10.1016/j.foodres.2010.09.020.

[8] Otoni, C. G., Espitia, P. J. P., Avena-Bustillos, R. J. and McHugh, T. H. (2016). Trends in antimicrobial food packaging systems: Emitting sachets and absorbent pads. *Food Research International*. 83: 60.

[9] Ahvenainen, R. (2003). *Novel food packaging methods*. Boston, North America; Woodhead publishing limited.

[10] Appendini, P. and Hotchkiss, J. H. (2002). Review of antimicrobial food packaging. *Innovative Food Science & Emerging Technologies*. 3(2): 113–126.

[11] Barboza, J. N., da Silva Maia Bezerra Filho, C., Silva, R. O., Medeiros, J. V. R. and de Sousa, D. P. (2018). An Overview on the Anti-inflammatory Potential and Antioxidant Profile of Eugenol. *Oxidative Medicine and Cellular Longevity*, 1–9.

[12] El-Guendouz, S., Aazza, S., Anahi Dandlen, S., Majdoub, N., Lyoussi, B., Raposo, S., Dulce Antunes, M., Gomes, V. and Graça Miguel, M. (2019). Antioxidant Activity of Thyme Waste Extract in

O/W Emulsions. *Antioxidants* (Basel, Switzerland), 8(8): 243. https://doi.org/10.3390/antiox8080243

[13] Makwana, S., Choudhary, R., Dogra, N., Kohli, P. and Haddock, J. (2014). Nanoencapsulation and immobilization of cinnamaldehyde for developing antimicrobial food packaging material. *LWT - Food Science and Technology*. 57: 470–476. 10.1016/j.lwt.2014.01.043.

[14] Abdollahi, M., Rezaei, M. and Farzi, G. (2012). Improvement of active chitosan film properties with rosemary essential oil for food packaging. *International Journal of Food Science & Technology*, 47(4): 847–853. doi:10.1111/j.1365.

[15] Kashiri, M., Cerisuelo, J. P., Domínguez, I., López-Carballo, G., Muriel-Gallet, V., Gavara, R. and Hernández-Muñoz, P. (2017). Zein films and coatings as carriers and release systems of *Zataria multiflora* Boiss. Essential oil for antimicrobial food packaging. *Food Hydrocolloids,* 70: *260–268.* doi:10.1016/j.foodhyd. 2017.02.021.

[16] Ali, A., Noh, N. M. and Mustafa, M. A. (2015). Antimicrobial activity of chitosan enriched with lemongrass oil against anthracnose of bell pepper. *Food Packaging and Shelf Life*. 3: 56-61. doi:10.1016/j.fpsl.2014.10.003.

[17] Ahmad, M., Benjakul, S., Sumpavapol, P. and Nirmal, N. P. (2012). Quality changes of sea bass slices wrapped with gelatin film incorporated with lemongrass essential oil. International *Journal of Food Microbiology*, 155(3): 171–178.

[18] Mulla, M., Ahmed, J., Al-Attar, H., Castro-Aguirre, E., Arfat, Y. A., & Auras, R. (2017). Antimicrobial efficacy of clove essential oil infused into chemically modified LLDPE film for chicken meat packaging. *Food Control*, 73, 663–671. doi:10.1016/j.foodcont. 2016.09.018.

[19] Quesada, J., Sendra, E., Navarro, C., & Sayas-Barberá, E. (2016). Antimicrobial Active Packaging including Chitosan Films with *Thymus vulgaris* L. Essential Oil for Ready-to-Eat Meat. *Foods,* 5(4): 57. doi:10.3390/foods5030057.

[20] Fasihi, H., Noshirvani, N., Hashemi, M., Fazilati, M., Salavati, H. and Coma, V. (2019). Antioxidant and antimicrobial properties of carbohydrate-based films enriched with cinnamon essential oil by Pickering emulsion method. *Food Packaging and Shelf Life*, 19: 147–154. doi:10.1016/j.fpsl.2018.12.007.

[21] Llana-Ruiz-Cabello, M., Pichardo, S., Bermúdez, J. M., Baños, A., Núñez, C., Guillamón, E. and Cameán, A. M. (2016). Development of PLA films containing oregano essential oil (*Origanum vulgare* L.virens) intended for use in food packaging. *Food Additives & Contaminants: Part A*, 1–13.

[22] Passarinho, A. T. P., Dias, N. F., Camilloto, G. P., Cruz, R. S., Otoni, C. G., Moraes, A. R. F. and Soares, N. de F. F. (2014). Sliced Bread Preservation through Oregano Essential Oil-Containing Sachet. *Journal of Food Process Engineering*, 37(1), 53–62.

[23] Espitia, P. J. P., Soares, N. de F. F., Botti, L. C. M., Melo, N. R. de, Pereira, O. L., & Silva, W. A. da. (2012). Assessment of the efficiency of essential oils in the preservation of postharvest papaya in an antimicrobial packaging system. *Brazilian Journal of Food Technology*, 15(4), 333–342. doi:10.1590/s1981-.

[24] Oral, N., Vatansever, L., Sezer, C., Aydin, B., Guven, A., Gulmez, M. and Kurkcuoglu, M. (2009). Effect of absorbent pads containing oregano essential oil on the shelf life extension of overwrap packed chicken drumsticks stored at four degrees Celsius. *Poultry Science,* 88(7):1459–.

[25] Otoni, C. G., Soares, N. de F. F., da Silva, W. A., Medeiros, E. A. A. and Baffa Junior, J. C. (2014). Use of Allyl Isothiocyanate-containing Sachets to Reduce *Aspergillus flavus* Sporulation in Peanuts. *Packaging Technology and Science*, 27(7): 549–558.

[26] Han, J. H., Patel, D., Kim, J. E. and Min, S. C. (2014). Retardation of Listeria Monocytogenes Growth in Mozzarella Cheese Using Antimicrobial Sachets Containing Rosemary Oil and Thyme Oil. *Journal of Food Science*, 79 (11): E2272–.

[27] Gómez-Estaca, J., López-de-Dicastillo, C., Hernández-Muñoz, P., Catalá, R. and Gavara, R. (2014). Advances in antioxidant active

food packaging. *Trends in Food Science & Technology*, 35(1): 42–51. doi:10.1016/j.tifs.2013.10.008.

[28] Wrona, Magdalena & Nerín, Cristina & Alfonso, Maria & Caballero, Miguel. (2017). Antioxidant packaging with encapsulated green tea for fresh minced meat. *Innovative Food Science & Emerging Technologies*. 41. 10.1016/j.ifset.2017.04.001.

[29] Piñeros-Hernandez, D., Medina-Jaramillo, C., López-Córdoba, A. and Goyanes, S. (2017). Edible cassava starch films carrying rosemary antioxidant extracts for potential use as active food packaging. *Food Hydrocolloids*, 63: 488–495. doi:10.1016/j.foodhyd.2016.09.034.

[30] López de Dicastillo, C., Castro-López, M. del M., López-Vilariño, J. M. and González-Rodríguez, M. V. (2013). Immobilization of green tea extract on polypropylene films to control the antioxidant activity in food packaging. *Food Research International*, 53(1): 522–528. doi:10.1016/j.foodres.2013.05.022.

[31] Rambabu, K., Bharath, G., Banat, F., Show, P. L. and Cocoletzi, H. H. (2018). Mango leaf extract incorporated chitosan antioxidant film for active food packaging. *International Journal of Biological Macromolecules*. Vol- 126. doi:10.1016/j.ijbiomac.2018.12.196.

[32] De Moraes Crizel, T., de Oliveira Rios, A., D. Alves, V., Bandarra, N., Moldão-Martins, M. and Hickmann Flôres, S. (2018). Active food packaging prepared with chitosan and olive pomace. *Food Hydrocolloids*, 74: 139–150. doi:10.1016/j.foodhyd.2017.08.007.

[33] Gaikwad, K. K., Lee, J. Y. and Lee, Y. S. (2016). Development of polyvinyl alcohol and apple pomace bio-composite film with antioxidant properties for active food packaging application. *Journal of food science and technology*, 53(3): 1608–1619. https://doi.org/10.1007/s13197-015-2104-9.

[34] Veld, J. and Havenaar, Robert. (1991). Probiotics and health in man and animal. *European Journal of Clinical Nutrition*. 45: 29-31. 10.1002/jctb.280510419.

[35] Espitia, P., Batista, R., Azeredo, H. & Otoni, Caio. (2016). Probiotics and their potential applications in active edible films and coatings.

Food Research International. Vol-90: 42. doi:10.1016/j.foodres. 2016.10.026.

[36] Kechagia, M., Basoulis, D., Konstantopoulou, S., Dimitriadi, D., Gyftopoulou, K., Skarmoutsou, N. and Fakiri, E. M. (2013). Health benefits of probiotics: a review. *ISRN nutrition*. 2013(5). doi.org/10.5402/2013/481651.

[37] Mozaffarzogh, M., Misaghi, A., Shahbazi, Y. and Kamkar, A. (2020). Evaluation of probiotic carboxymethyl cellulose-sodium caseinate films and their application in extending shelf life quality of fresh trout fillets. *LWT*, 126, 109305. doi:10.1016/j.lwt. 2020.109305.

[38] López de Lacey, A. M., López-Caballero, M. E. and Montero, P. (2014). Agar films containing green tea extract and probiotic bacteria for extending fish shelf-life. *LWT - Food Science and Technology*, 55(2): 559–564. doi:10.1016/j.lwt.2013.09.028.

[39] Abdollahzadeh, E., Ojagh, S. M., Imani Fooladi, A. A., Shabanpour, B. and Gharahei, M. (2018). Effects of Probiotic Cells on the Mechanical and Antibacterial Properties of Sodium-Caseinate Films. *Applied Food Biotechnology*, 5(3): 155-162. https://doi.org/10. 22037/afb.v5i3.20360.

[40] Singh, P., Magalhães, S., Alves, L., Antunes, F., Miguel, M., Lindman, B. and Medronho, B. (2018). Cellulose-based edible films for probiotic entrapment. *Food Hydrocolloids*. 88. doi:10.1016/ j.foodhyd.2018.08.057.

[41] Zabihollahi, N., Alizadeh, A., Almasi, H., Hanifian, S. and Hamishekar, H. (2020). Development and characterization of carboxymethyl cellulose based probiotic nanocomposite film containing cellulose nanofiber and inulin for chicken fillet shelf life extension. *International Journal of Biological Macromolecules*. 160. doi:10.1016/j.ijbiomac.2020.05.066.

[42] Khodaei, D., Hamidi-Esfahani, Z. and Lacroix, M. (2020). Gelatin and low methoxyl pectin films containing probiotics: Film characterization and cell viability. *Food Bioscience*, 36(8):100660 doi:10.1016/j.fbio.2020.100660.

[43] Li, S., Ma, Y., Ji, T., Sameen, D., Ahmed, S., Qin, W., Dai, J., Li, Suqing and Liu, Y. (2020). Cassava starch/carboxymethylcellulose edible films embedded with lactic acid bacteria to extend the shelf life of banana. *Carbohydrate Polymers*. 248(1): 116805. doi:10.1016/j.carbpol.2020.116805.

[44] Quirós-Sauceda, A. E., Ayala-Zavala, J. F., Olivas, G. I. and González-Aguilar, G. A. (2014). Edible coatings as encapsulating matrices for bioactive compounds: a review. *Journal of food science and technology*, 51(9): 1674–1685. https://doi.org/10.1007/s1 3197-013-1246-x.

[45] Genevois, C., Flores, S. and de Escalada Pla, M. (2014). Effect of iron and ascorbic acid addition on dry infusion process and final color of pumpkin tissue. *LWT - Food Science and Technology*, 58(2): 563–570.

[46] Mei, Y. and Zhao, Y. (2003). Barrier and Mechanical Properties of Milk Protein-Based Edible Films Containing Nutraceuticals. *Journal of Agricultural and Food Chemistry*, 51(7): 1914–1918.

[47] Mei, Y., Zhao, Y., Yang, J. and Furr, Harold. (2006). Using edible coating to enhance nutritional and sensory qualities of baby carrots. *Journal of Food Science*, 67 (5), 1964-1968, doi:10.1111/j.1365-2621.2002.tb08753.x.

[48] Raybaudi-Massilia, Rosa & Vásquez, Florangel & Reyes, Ana and Troncone, Graciela & Tapia, Maria. (2015). Novel Edible Coating of Fresh-cut Fruits: Application to Prevent Calcium and Vitamin D Deficiencies in Children. *Journal of Scientific Research and Reports*. 6,142-156. doi:10.9734/JSRR/2015/15964.

[49] Bhushani, Anu and Chinnaswamy, Anandharamakrishnan. (2014). Electrospinning and electrospraying techniques: Potential food based applications. *Trends in Food Science & Technology*. 38: doi:10.1016/j.tifs.2014.03.004.

[50] Becerril, R., Nerín, C. and Silva, F. (2020). Encapsulation Systems for Antimicrobial Food Packaging Components: An Update. *Molecules,* 25(5): 1134. doi:10.3390/molecules25051134.

[51] Shade C. W. (2016). Liposomes as Advanced Delivery Systems for Nutraceuticals. *Integrative medicine* (Encinitas, Calif.), 15(1): 33–36.

[52] Wu, Z., Zhou, W., Pang, C., Deng, W., Xu, C. and Wang, X. (2019). Multifunctional chitosan-based coating with liposomes containing laurel essential oils and nanosilver for pork preservation. *Food Chemistry*. doi:10.1016/j.foodchem.2019.05.114.

[53] Bernela, M., Kaur, P., Ahuja, M. and Thakur, R. (2018). Nano-based Delivery System for Nutraceuticals: The Potential Future. *Advances in Animal Biotechnology and Its Applications*, 103–117. doi:10.1007/978.

[54] Nile, S. H., Baskar, V., Selvaraj, D., Nile, A., Xiao, J. and Kai, G. (2020). Nanotechnologies in Food Science: Applications, Recent Trends, and Future Perspectives. *Nano-Micro Letters*, 12(1): doi:10.1007/s40820-020-0383-9.

[55] Zhang, C., Feng, F. and Zhang, H. (2018). Emulsion electrospinning: Fundamentals, food applications and prospects. *Trends in Food Science & Technology*. 80: 10.1016/j.tifs.2018.08.005.

[56] Alfadul, S. & Elneshwy, A. A. (2010). Use of nanotechnology in food processing, packaging and safety review. *African Journal of Food, Agriculture, Nutrition and Development* (ISSN: 1684-5358) Vol 10 (6): 10. 10.4314/ajfand.v10i6.58068.

About the Editors

Anita Kumari is presently working as an Assistant Professor in the Department of Nutrition Biology, Central University of Haryana, Mahendergarh. She did her PhD in Food Science and Nutrition from CSK Himachal Pradesh Agricultural University, Palampur (H.P). She is having more than 10 years of teaching and research experience. She has guided several MSc dissertation students and PhD students are also working under her guidance. Her research is focused on the development and quality evaluation of functional food products from underutilized crops. She has more than 30 research papers, articles, and book chapters in national and international reputed journals. She has also been invited as a resource person by various national and international organizations to deliver talks on various aspects of nutrition. She is a member of various academic and professional bodies.

Gulab Singh, presently working as an Assistant Professor in Maharaja Agrasen University, Baddi, Himachal Pradesh. He is having more than 10 years of research and 2 years of teaching experience at postgraduate and graduate levels. Presently, one PhD student is working under his guidance. He has published more than 32 research papers, reviews, and book chapters in various journals and books of national and international repute. He is also editor of two international books "Microbiology and Biotechnology for the Sustainable Environment" (Nova Science Publishers) and "Aptamers – Biotechnological Application of Next-Generation Tool" (Springer Nature). He is a member of various academic and professional bodies. He has 1 national patent granted and 5 national patents filed in the field of food adulteration detection and environment contaminant detection.

INDEX

A

C

D

G

H

I

K

L

M

Q

R

S